T0179719

The Wiley 5G REF: Security

The Wiley 5G REF: Security

Editors-in-Chief

Rahim Tafazolli
University of Surrey, UK

Chin-Liang Wang
National Tsing Hua University, Taiwan, R.O.C.

Periklis Chatzimisios
International Hellenic University, Greece

Section Editor

Madhusanka Liyanage
University College Dublin, Ireland

The right of Rahim Tafazolli, Chin-Liang Wang, Periklis Chatzimisios, and Madhusanka Liyanage to be identified as the authors of the editorial material in this work has been asserted in accordance with law.

Registered Offices
John Wiley & Sons, Inc., 111 River Street, Hoboken, NJ 07030, USA
John Wiley & Sons Ltd, The Atrium, Southern Gate, Chichester, West Sussex, PO19 8SQ, UK

Editorial Office
The Atrium, Southern Gate, Chichester, West Sussex, PO19 8SQ, UK

For details of our global editorial offices, customer services, and more information about Wiley products visit us at www.wiley.com.

Wiley also publishes its books in a variety of electronic formats and by print-on-demand. Some content that appears in standard print versions of this book may not be available in other formats.

Library of Congress Cataloging-in-Publication Data Applied for:
ISBN: 9781119820314

Cover Design: Wiley
Cover Image: © Blackboard/Shutterstock

Set in 10/12pt WarnockPro by Straive, Chennai, India

SKY86CF8895-5B52-4AC7-A114-006F1D65E7E0_063021

Contents

Foreword

The Wiley 5G Ref: Security offers a stellar collection of articles selected from the online-only Work, The Wiley 5G Reference. It aims to provide a solid educational foundation for researchers and practitioners in the field of 5G Security and Privacy to expand their knowledge base by including the latest developments in these disciplines.

Security and Privacy have become the primary concern in the 5G and Beyond 5G (B5G) network as risks can have high consequences. The security and Privacy of 5G and B5G networks can be broadly categorized into three main components. First, most of the security threats and security requirements related to pre-5G mobile generations are still applicable in 5G and B5G networks. Second, 5G will have a new set of security challenges due to connectivity offered to new IoT-based vertical industries such as healthcare, smart grid, smart transportation, smart industries, etc. Each vertical will have a new set of security and privacy concerns due to increased users, heterogeneity of connected devices, new network services, and new stakeholders. Third, network softwarization and utilization of new technologies such as SDN (Software Defined Networking), NFV (Network Function Virtualization), MEC (Multi-access Edge Computing), and NS (Network Slicing) will introduce a brand-new set of security and privacy challenges.

However, most of the security models in pre-5G (i.e., 2G, 3G, and 4G) networks cannot be directly utilized in 5G due to new architecture and new service. Also, most of the security challenges in 5G are yet to be identified since 5G is under the development phase. Moreover, the 5G and B5G security solutions must address the new technical and performance requirements such as high scalability to connect billions of devices, high throughput to support Gbps connectivity, and ultra-low latency to support sub-millisecond range applications. Therefore, it is necessary to identify these threats and integrate the security and privacy solutions already at the design phase. Furthermore, the EU coordinated risk assessment of the cybersecurity of 5G networks report1 published in October 2019 highlights several necessary processes or configuration-related vulnerabilities that should be considered. For example, from the perspective of a mobile network operator, a poor network design and architecture and the lack of security requirements in the procurement process may lead to unwanted exposure or a disruption of service.

In addition, present-day mobile networks suffer from many security limitations such as securing only the perimeter, distributed and uncoordinated security mechanisms, security functions tightly coupled to physical resources, lack of adaptation and inter-operability, over-provisioned security mechanisms, scalability issues, vulnerability to various IP based attacks, lack of visibility and high network monitoring overhead, and unexpected privacy breaches. The introduction of centralized controllers with the separation of the control and data planes, programmability, NFV, the introduction of new network functions, and even introducing new stakeholders such as Local 5G Operators (L5GO) are expected to solve the security and privacy limitations of current telecommunication networks. On the other side, relying on a single controller creates a potential single point of failure. Moreover, integration of smart and agile monitoring and decision frameworks enabled via Artificial Intelligence (AI) are promising to tackle these security challenges. The security techniques and schemes devised and implemented for this massively connected and pervasive network structure will also have an overreaching impact on B5G systems since these traits are expected to be apparent in those systems. In that regard, AI-based resilience techniques, Machine Learning (ML) driven cognitive and self-X network management techniques, and privacy management for highly sensitive data running on B5G networks should also be considered.

Well-known experts in security and Privacy for the telecommunication network field wrote the book you are about to read. It provides the methodologies, guidance, and suggestions on how to deliver the possible ways of developing novel security and privacy solutions to protect the 5G telecommunication networks from strengthening critical infrastructures and novel research directions and open challenges that will encourage future research.

The book introduces the security landscape of 5G, and significant security and privacy risks associated with the 5G networks. Then, the security solutions for different segments of the 5G network, i.e., radio network, edge network, access network, and core network, are discussed. Since 5G is developed based on network softwarization, security threats associated with key network softwarization technologies such as SDN, NFV, NS, and MEC are also presented in detail. Then, the security issues related to the new 5G and IoT services are delivered. Finally, a detailed discussion on the privacy of 5G networks is presented by considering Datafied Society.

This book is another step in closing this educational gap. It is intended to provide additional learning opportunities for a wide range of readers, from graduate-level students to seasoned engineering professionals. We are confident that this book and the entire collection of selected articles will continue Wiley's tradition of excellence in technical publishing and provide a lasting and positive contribution to the teaching and practice of security and privacy of 5G and beyond networks.

Dr. Madhushanka Liyanage
Section Editor
University College Dublin
Ireland

List of Contributors

Iris Adam
Nokia Bell Labs Germany
Stuttgart
Germany

Ijaz Ahmad
VTT Technical Research Centre of
Finland
Espoo
Finland

Hirley Alves
6G Flagship
University of Oulu
Oulu
Finland

João André
National Laboratory for
Civil Engineering
Lisbon
Portugal

Shanay Behrad
Orange Labs Caen
Caen
France

Emmanuel Bertin
Orange Labs Caen
Caen
France

Carla Fabiana Chiasserini
Politecnico di Torino
CNR-IEIIT
Torino
Italy

Noel Crespi
Institut Mines-Telecom
Telecom SudParis
Paris
France

Aaron Yi Ding
Delft University of Technology (TU Delft)
Delft
The Netherlands

Silke Holtmanns
Nokia Bell Labs
Espoo
Finland

Jyrki Huusko
VTT Technical Research Centre of
Finland
Oulu
Finland

Anca D. Jurcut
School of Computer Science
University College Dublin
Dublin
Ireland

Aapo Kalliola
Nokia Bell Labs
Espoo
Finland

Vikramajeet Khatri
Nokia Bell Labs Finland
Helsinki
Finland

Gabriela Limonta
Nokia Bell Labs
Espoo
Finland

Madhusanka Liyanage
Centre for Wireless Communication
University of Oulu
Oulu
Finland

and

School of Computer Science
University College Dublin
Dublin
Ireland

Maode Ma
Nanyang Technological University
Singapore
Singapore

Francesco Malandrino
CNR-IEIIT
Torino
Italy

Mahesh K. Marina
The University of Edinburgh
Edinburgh
UK

Marja Matinmikko-Blue
University of Oulu
Oulu
Finland

Yoan Miche
Nokia Bell Labs
Espoo
Finland

Rupendra Nath Mitra
The University of Edinburgh
Edinburgh
UK

Mehrnoosh Monshizadeh
Nokia Bell Labs France
Nozay
France

and

Aalto University
Helsinki
Finland

Edgardo Montes de Oca
Montimage
Paris
France

Ian Oliver
Nokia Bell Labs
Espoo
Finland

Diana P.M. Osorio
Federal University of São Carlos (UFSCar)
Center for Exact Sciences and
Technology
São Carlos
SP
Brazil

and

6G Flagship
University of Oulu
Oulu
Finland

Alican Ozhelvaci
Nanyang Technological University
Singapore
Singapore

German Peinado
Nokia
Warszaw
Poland

Pawani Porambage
Centre for Wireless Communication
University of Oulu
Oulu
Finland

Pasika Ranaweera
School of Computer Science
University College Dublin
Dublin
Ireland

José D.V. Sánchez
Department of Electronics
Telecommunications and Information
Networks
Escuela Politécnica Nacional
Quito
Ecuador

Tiago de J. Souza
American Tower of Brazil
São Paulo
Brazil

Jani Suomalainen
VTT Technical Research Centre of
Finland
Espoo
Finland

Sebastiao Teatini
University of Oulu
Oulu
Finland

Rui Travanca
DIGAMA Engineering Consultants
Lisbon
Portugal

Lina Xu
School of Computer Science
University College Dublin
Dublin
Ireland

1

5G Mobile Networks Security Landscape and Major Risks

Rupendra Nath Mitra and Mahesh K. Marina

The University of Edinburgh, Edinburgh, UK

Introduction

Mobile communication networks since its emergence in the mid-1980s have been rapidly transforming our everyday life. Beginning with a first-generation analog voice communication system on the go, the fourth generation of the mobile networks currently guarantees high-speed Internet and ubiquitous connectivity among numerous smart devices. The present-day cellular networks cater to diversified services such as banking, health, governance, e-commerce, education, and mobile TV, to name a few (Liyanage et al. 2017). Yet today's mobile network infrastructure requires another quantum leap to its next-generation called the 5th generation (5G) to fulfill the growing demand of higher data rates in mobile Internet and the need for extremely low-latency communication links, enabling futuristic applications such as mission-critical communications, augmented reality, self-driving cars, virtual reality, and so on (Qayyum et al. 2019).

The 5G mobile networks are not just a technologically advanced version of currently deployed 4th generation (4G) systems but a completely new telecommunication paradigm powered by software-defined networking (SDN), network function virtualization (NFV), and cloud computing. The adoption of these new technologies in telecommunication system design brings unprecedented network agility and offers service providers the ability to implement 5G infrastructures incredibly rapidly, and that is too at a low cost. However, along with the unmatched network flexibility and advantages mentioned above, 5G networks bring in an increased concern for security and privacy. The unrestricted usage of open-source software and machine learning techniques, provision of virtualized services over shared physical infrastructure, connecting almost everything, and finally, the softwarization of the telecommunication network functions increase the cybersecurity threat to 5G networks (Belmonte Martin et al. 2016).

5G will play a pivotal role in building a networked society because critical services such as public transportation (i.e. railways and city transports), energy sectors, national security, governance, and healthcare, which currently operate on exclusive networks, will heavily rely on the pervasive 5G infrastructures. Therefore, any adversary may launch malicious attacks to the 5G networks and, if successful, can bring large-scale

The Wiley 5G REF: Security. Edited by Rahim Tafazolli, Chin-Liang Wang, Periklis Chatzimisios and Madhusanka Liyanage.
© 2021 John Wiley & Sons Ltd. Published 2021 by John Wiley & Sons Ltd.

catastrophe to the society (Kumar et al. 2018). So, the hasty adoption of open-source software, authentication protocols, large-scale machine learning (ML) models, and multi-tenant public cloud-based network operations need to be overseen and scrutinized formally before being integrated to the 5G infrastructure. A systematic study on privacy-preserving techniques and policies in 5G securing it from innumerable cyber threats is essential and timely (Hodges 2019).

In this article, we provide a fresh insight into the end-to-end (E2E) 5G security threat landscape, analyze the recommendations to secure the 5GS, identify evolved attack vectors, recognize the vulnerabilities, and categorize the major risks to highlight the crucial gaps currently prevailing in the way of 5G to become a trustworthy ecosystem. We also introduce the research groups working toward a safe and secure 5G network infrastructure and briefly review their lines of works.

4G to 5G: A Paradigm Shift

Today's cyber world is well connected through physical transoceanic cables, numerous satellite links, long haul optical or unguided media, wide-area connectivities, and local wireless and wired networks. Traditionally, end users connect to the worldwide cyber networks using user equipment, typically a general-purpose computer or a smartphone through the mobile network's radio access network or the locally available access network. With the advent of smart devices, exponential growth has been projected in the number of connected Internet of Things (IoT) in the coming years, as illustrated in Figure 1. With the growing demand for staying always connected everywhere and accessing high volume traffic on the go, cellular data service emerged as the predominant way to provide ubiquitous connectivity to the users. Thus, 5G network is designed in a very different way than its 4G counterpart so that 5G can cater to the massive amount of connectivities to numerous futuristic services and many smart IoT devices for billions of its subscribers, providing high-speed cellular broadband (University of Surrey 2017).

5G System Architecture

5G architecture is proposed to be fundamentally different than the network architecture of 4G because 5G necessitates backward compatibility and an agile, modular architecture for an expeditious low-cost deployment and smooth technical augmentation in

Figure 1 The projected boom in the number of connected gadgets (Statista Research Department 2016).

the future. The present-day 4G network inherits architectural features from its previous generations of traditional mobile networks. The 4G network has three segments, the user equipment (UE), the Radio Access Network (RAN) that connects to the UE over the air interface, and Evolved Packet Core (EPC) networks that connect to the RAN through transmission networks. However, 5G system (5GS) architecture has UE, multifaceted radio access networks called New Radio (NR), Multi-access Edge Cloud for reduced network latency, and the cloud-native 5G core networks (5G-CN) (Arfaoui et al. 2018; Belmonte Martin et al. 2016; Ericsson 2018). A piece of mobile equipment (ME) equipped with a universal subscriber identity module (USIM) constitutes the UE by which the users can access the 5G network over the air interface.

The important entity of NR is gNB, which hosts three main functional units called Radio Unit (RU), Distributed Unit (DU), and Centralized Unit (CU). Thus, NR becomes a flexible logical architecture, unlike 4G RAN, which has a monolithic architecture. 5G radio network (RN) can be deployed over the serving cell site and mobile edge server premise in different combinations of functional units as required by the application scenario. For example, only RU can be installed in the cell site with minimum installation cost and complexity, and DU-CU can be installed on the edge site (in a local exchange office) to build a latency tolerant access network. In contrast, a DU split from the CU can be installed in the cell site associated with RU for improved latency but at a higher installation cost and complexity. The latter is a lower layer split of RN functions, and the prior topology is an example of a higher layer functional split. DU-CU could also be integrated with the RU in the cell site for ULLC in the case of industrial automation.

5G is meant to coexist with legacy networks as well as stand-alone network architecture. Therefore, the 5G-CN has two different implementable provisions, stand-alone (SA) and non-stand-alone (NSA) modes. Besides, 5G-CN implements the network functions much like a 4G EPC does (3GPP 2019). Still, 5G-CN is defined as a Service-Based Architecture (SBA) framework for increased modularity and function reusability. 5G-CN hosts various network functions (NFs) interfaced among them instead of various core network entities of a traditional mobile network. SBA architecture applies only to interfaces among control plane (CP) functions within the 5GC. The control plane network functions of the 5G-CN interact using service-based interfaces (SBIs). This functional core architecture helps an implemented technology in the system to evolve and makes it seamlessly replaceable by its suitable future counterparts. 5G-CN brings more scalability, flexibility, and upgradability in the network architecture, enabling the operators to build and maintain the network at a low cost. Cloud-based core network functionalities and NFV become the significant enablers of 5G mobile network architecture because of the SBA and the network slicing.

Figure 2 looks into a 5G network from a functional perspective, where the cloud-native 5GC NFs interact among themselves over defined interfaces. The main NFs are shown in the diagram and discussed below.

The Network Functions Repository Function (*NRF*) registers and discovers NF services. An NF identifies other NFs through NRF.

The Unified Data Management (*UDM*) performs significant functionality of HSS from EPC. It is responsible for user identity management, subscription management, and generation of AKA credentials. The Authentication credential Repository and Processing Function (ARPF) retains the authentication credentials. Only the Subscription Identifier

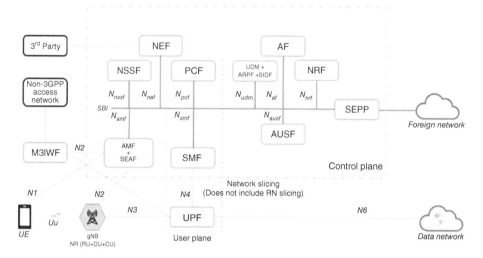

Figure 2 5G-CN SBA representation as specified by 3rd Generation Partnership Project (3GPP) (3GPP 2019).

De-concealing Function (SIDF) decrypts a Subscription Concealed Identifier (SUCI) to obtain its long-term identity, namely the Subscription Permanent Identifier (SUPI).

The Policy Control Function (*PCF*) is synonymous with PCRF of the EPC framework. It provides policy decisions to control plane functions.

The Network Exposure Function (*NEF*) is a new network function created for 5G-CN and was not present in 4G EPC. It ensures secure information exchange to and from an external application to 3GPP core and works like an exposure function.

The Application Function (*AF*) replicates the functionalities of AF from EPC architecture.

The Authentication Server Function (*AUSF*) is the authentication server that completes the Home Subscriber Server's (HSS) functionalities of an EPC along with UDM. The AUSF keeps a key that is derived after authentication for reuse in case of simultaneous registration of a UE in different access networks.

The Access and Mobility Function (*AMF*) is a subfunction of the Mobility Management Entity (MME) of 4G EPC. AMF tackles all the connection and session information from UEs or RAN and performs connection and mobility management tasks. The Session Management Function (SMF) takes care of the session management tasks. The AMF is collocated with the SEcurity Anchor Function (SEAF) which permits the secure introduction of future functions.

The *SMF* performs the interaction with the decoupled data frame and also acts as an Internet Protocol (IP) address manager and a Dynamic Host Configuration Protocol (DHCP) server.

The Network Slice Selection Function (*NSSF*) offers services to the AMF and NSSF in a different Public Land Mobile Network (PLMN) via the Nnssf service-based interface within 5G-CN. NSSF is responsible for network slice selection and management.

The Non-3GPP Interworking Function (*N3IWF*) is similar to the ePDG (for Evolved Packet Data Gateway) in the 4G EPC. N3IWF allows the UE to access 5G-CN through untrusted non-3GPP access networks. It is responsible for establishing secure communication between UE and 5G-CN and routing messages outside the 5G RAN. However, 3GPP does not categorize any non-3GPP technologies as untrusted. This decision is left on the operator.

The User Plane Function (*UPF*) is the functional point between the mobile network and the Data Network (DN). UPF facilitates Control and User Plane Separation (CUPS) for non-stand-alone (NSA) 5GS which allows scalable and efficient management of packet traffic in the mobile edge.

The Security Edge Protection Proxy (*SEPP*) is a new NF called Security Edge Protection Proxy introduced in 5GS by 3GPP SA3 to ensure secure signaling traffic between different operator networks. N32 interface between the SEPPs is designed to protect sensitive data flowing through it and to ensure secure authentication between SEPPs.

The NSA mode is an intermediate step toward a full 5G network from the legacy network. In this configuration, 5G gNB facilitates the access network by carrying user plane (UP) traffic along with 4G LTE eNB but connects to the 4G EPC (and not 5G-CN) over standard S1-U interface. The control plane traffic still routes through the 4G network. In other words, in NSA mode, 5G access network increases the capacity to the existing 4G services as a secondary serving cell being a slave to the 4G network which controls the 5G access network as the master. Since NSA mode does not demand the all-new 5G-CN but still capable of offering a 5G experience to the subscribers, telecom operators adopt NSA configuration as their first step to commercial 5G network.

On the other hand, SA mode demands 5G NR be connected to the 5GC; hence, it is the true 5G only network capable of offering low-latency mobile broadband service. 5G-CN is seen as the common core to 5G NR, 4G RAN, and any other type of access network in the future.

From the above description, we can understand that a majority of the 5G-CN NFs are created by refactoring the 4G EPC functions, MME, SPWG, HSS, and PCRF. At the same time, a set of brand new NFs such as NSSF, N3IWF, SEPP, or the SBA itself are introduced for enhanced network programmability, agility, and security. A full CP and UP separation in the 5G-CN is another unique trait where UPF processes UP data, and all other nodes tackle CP functionalities, enabling separate scale-up of CP and UP operations.

Salient Features of 5G Network

5GS is a step toward the converged network of IT and telecommunication infrastructure for the future. Thus, 5GS has several new features that the previous generations of telecom networks did not have. From a security perspective, 5G being a telecom network more similar to the IT infrastructure opens up a wider boundary of cyberattacks on the new unfortified sides of the 5G network (Shaik et al. 2015). The unique architectural amendments are briefly introduced as follows (5G Americas 2019; Hodges 2019):

- *New physical layer:* 5GS physical layer is significantly different from that of LTE. The use of a massive number of active antennae for efficient beamforming, high-frequency millimeter-wave signals for higher data rate over the air, and nonorthogonal multiple access for higher spectral efficiency makes 5G new radio conspicuous from other radio communication systems (3GPP 2018).

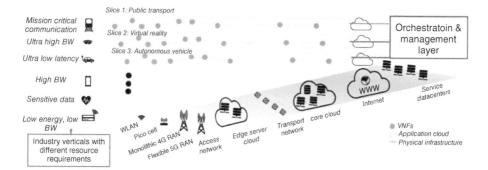

Figure 3 Schematic diagram of network slicing in an evolved network architecture of 5G coexisting with legacy networks.

- *Multiaccess Edge Computing:* 5G enables operators to host latency-critical services closer to the end user through the Multiaccess Edge Computing (MEC) capability. MEC helps to reduce the traffic load on the transport network and meet the Service Level Agreements (SLA) of an ultra-low-latency service by local hosting.
- *Network Slicing:* E2E network slicing is a significant feature of the 5G network. A network slice is an optimized set of network resources that enables a specialized service (Foukas et al. 2016). In principle, any given kind of slice of an operator can interconnect with the same kind of slice from another operator if the slice-sharing policy requirements are met. The NSSF enables the slice selection procedure. Figure 3 schematically describes the E2E network slicing over a 5G network infrastructure layer. Network slicing can be described as a horizontal network virtualization technique that offers the flexibility to mobile operators to logically separate its different users with different SLAs. Mobile virtual operators can borrow their virtual network infrastructure in a pay-as-you-go fashion where E2E network slicing is in operation.
- *Cloud-native architecture:* In principle, cloud-native architecture promotes software services that are broken down into a compact, more convenient pieces of constituent software, which is achieved using a microservice architecture. 5G-CN adopts a cloud-native architecture so that each piece of such microtechnologies can be individually scaled, reconfigured, and upgraded. The actual cloud-native core network is a unique feature of 5G.
- *Multi-RAT connectivity:* 5G supports a UE to stay connected to more than one 5G eNB, which also includes different Radio Access Technology (RAT) such as Wi-Fi. In such scenarios, the connected gNB will act as a master node and the other as a secondary node. This unique feature of 5G requires new security mechanisms and procedures to protect user traffic between UE and master and secondary nodes.
- *Support for non-3GPP access:* Apart from 3GPP access networks, 5G-CN opens its door to the non-3GPP access networks with enhanced security function defined by N3IWF. A UE can be served through Wi-Fi like access networks directly connected to the 5GC for reduced latency. The provision of such many non-3GPP accesses to the 5GC requires more robust authentication than that of 4G ePDG.

The Threat Landscape and Security Roadmap

5G network will entice an unprecedented number of cyberattacks due to its pervasive reachability into high-value targets such as governmental and financial organization, its capability to serve a wide variety of industry verticals, and its accessibility to a massive amount of user data. Previous generations of the mobile network have seen several malicious attacks (Rafiul Hussain et al. 2019), (Rupprecht et al. 2018), but 5G will face more aggravated threats. The attackers have continued to grow with time with inexpensive yet smarter AI-based attack tools, innovative side-channel attack techniques, and a plethora of malicious software capable of evading intrusion detection systems. The number of attacks that 5GS will experience is an order of magnitude higher than that of its previous generation counterparts (Belmonte Martin et al. 2016).

Before we further explore the different aspects of the security threats on 5G, a few closely related technical vocabularies need to be clarified for a better understanding of the reader. Table 1 defines some basic cybersecurity terms which have occurred frequently in this article after that.

5G Network Threat Surface

In the past few years, a broad set of attacks on GSM, UMTS, and LTE systems have been reported, which threatened user privacy, data secrecy, and business integrity. Rupprecht et al. analyzed the LTE layer two security vulnerabilities and designed attack vectors that exploited the fact that LTE user payload is encrypted in counter mode (AES-CTR) but not integrity protected to modify user data (Rupprecht et al. 2019). Hussain et al. have shown the security loophole in downlink paging occasion and exploited the weakness to leak victim's soft identities and coarse-grained location (Arapinis et al. 2012). So, the 5G defense mechanism needs further strengthening by underlying E2E network monitoring, reimplementing Artificial Intelligence (AI)-driven

Table 1 Definitions of some common cybersecurity terms.

Vulnerability	A weakness or lacuna which can be potentially exploited to damage the system or to have unauthorized access to the system
Threat	Anything that can cash in on the vulnerabilities to destroy or paralyze the system. A threat can result from unintentional accidents as well as from a planned attack
Attack	A planned activity that exploits security vulnerabilities and threatens the system by obtaining unauthorized access or causing disruptions in system operation
Risk	A composite factor which is a function of vulnerability, threat, and the amount of damage caused. If a threat breaks in exploiting an existing vulnerability, then how much potential damage it may cause to the system or in general to the business is called the risk associated with the vulnerability and threat. Likelihood of occurrence of the event is also another constituting factor of risk
Trust	A psychological confidence in our general life, but in the digital world, trust is often described as a binary model. Either one has complete trust or no trust in a digital system
Privacy	The confidentiality of the user identity and concealment of sensitive data and personal information that is often sent over a network to which a user is connected to

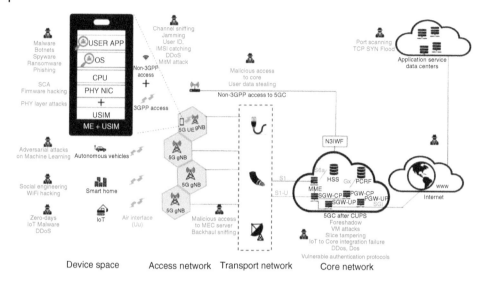

Figure 4 Evolved security threat landscape for 5G network.

intrusion detection systems, updating authentication protocols, and fixing security bugs in its previous versions (Amine Ferrag et al. 2018). In this section, we present a pragmatic view of the evolving 5G threat space by zooming in into the transfigured network architecture and interfaces.

Vulnerabilities of the 5G telecommunication ecosystem are spread all over the network (Ahmad et al. 2017). As elucidated in Figure 4, the UE, the air interface between the UE and the base stations, the physical and application interfaces, and the core networks all can be quietly scanned for exploitable vulnerabilities by hackers.

Security Threats to the UE: The UE has been a soft target by the hackers for quite a long time. Mobile operating systems, for instance, Android, have been reported with stealthy interapplication data leakage. In the 5G scenario, a plethora of malware along with countless phishing websites and nearby Bluetooth devices target end-user devices with vicious objectives (Antonioli et al. 2019). Wu et al. (2019) reported an extensive analysis of android open ports that revealed vulnerability patterns in open ports of popular apps, including Skype, Instagram, and Samsung Gear. Hardware chipsets and sophisticated sensor arrays can often be tampered to steal user data, and the high-speed 5G Internet connectivity to the devices expedites the unauthorized data transfer, leaving the users clueless about the harm to them.

The 5G architecture accommodates various use cases and business verticals with a wide variety of SLAs. The threat surface for the use cases which includes self-driving vehicles, low-energy low bandwidth IoTs, ultra-low-latency tactile Internet applications, and high bandwidth Virtual Reality (VR) applications, to name a few, is unique in many ways. For instance, the machine learning models that control the behavior of an autonomous car are vulnerable to the plentiful of adversarial attacks (Qayyum et al. 2019; Cao et al. 2019; Usama et al. 2019; Usama et al. 2019). Billions of IoT apparatus are often compromised to create a large-scale network of infected machines called botnet, which can then perform baleful activities such as Distributed Denial of Service (DDoS) attack. Shaik et al. demonstrated DoS attacks against UEs by exploiting the fact that certain tracking area update messages sent from the LTE network are accepted by UEs without any integrity protection (Shaik et al. 2015).

Security Threats to the Air Interface: The air interface being open to everyone undergoes a massive number of maleficent attacks day in and day out. Hence, securing communication over the air is the most challenging. The Man in the Middle (MitM) attack is one of the commonly found attacks that happen in the air interface by intercepting or decrypting the victim's data surreptitiously. Another common threat is intentionally jamming the physical channel with a transmission of a high-power signal of the same frequency as that of the cellphones, which outmuscles the phone's signal causing a communication lapse. International Mobile Subscriber Identity (IMSI) catching has been a severe privacy threat to the subscribers of the previous generation telecom networks where the downlink paging messages are sniffed for side-channel data leakage and identify or track a connected device (Rafiul Hussain et al. 2019). Jover reported a DoS attack against the LTE EPC using a compromised UE or another eNodeB that sends enormous number of attach request messages to the EPC to eventually choke the service (Jover 2013).

However, the new 5G standard enhanced the privacy safeguards against such attacks by frequently refreshing and encrypting the user identifier called Subscription Concealed Identifier (SUCI) and dynamically adjusting the paging timings. Although fundamental security mechanisms of the entire telecommunication system depend on cryptographic algorithms and security protocols in the bit level in the digital domain, the need for another layer of secrecy of the propagation channel is often required from the information-theoretic perspective, which is called the security of physical layer.

The RAN and the mobile edge servers for 5G are crucial for ensuring ultra-low-latency services. A rouge base station attack, where a false base station convinces the end users to get attached to it instead of a bona fide gNB, can severely downgrade the service experienced by the users. On the other hand, the transmission networks that are responsible for bringing the aggregated mobile traffic to the 5GC can be passively sniffed for possible UP or CP meta-data.

Security Threats to the Core Networks: 5G packet core, which deploys and manages large-scale virtual machines for network functions virtualizations and slice management, is susceptible to information leakage through covert channels between various instances of virtual machines and hence vulnerable to attacks trying to gain access to the system-level information of the whole infrastructure (Michalopoulos et al. 2017). Since the network slicing does not require strict hardware separations, shared hardware resources such as cache memory can be exploited to gain side-channel access.

5G-enabled end-user devices can connect to the 5G-CN through the non-3GPP access network via a network function called N3IWF. The securely encrypted tunnel between the UE and the N3IWF is vulnerable to cyber threats if the lightweight encryption fails to encounter brute-force attacks often faced by Virtual Private Network (VPN) or Secure Shell (SSH) tunnels over insecure Internet.

It has been well known that Signaling System 7 (SS7) and Diameter protocol, which have been widely used in wireless and wired telecommunication networks for signaling in call processing, value-added services, traffic routing, and information exchange, have exploitable vulnerabilities (Rao et al. 2015). The major weakness of SS7, which has been in use until 3G network systems, had a lack of encryption. Current 4G systems use an IP-based improved signaling protocol called Diameter, where the encryption is compulsory in principle but often not practiced by the service providers within the network. 5GS specifications proposed enhancement in the procedure of accepting the resource reservation requests using an HTTP/XML-based interface from other services. However, the PCF maintains traditional Diameter protocol to exchange information with AF, keeping the vulnerability alive in the 5GS.

Security Threats to the Application Servers: The conventional Internet and the application servers, external to the 5GS, remain vulnerable to the full range of cyber threats ranging from Domain Name Server (DNS) tunneling to DDoS attacks such as Transmission Control Protocol (TCP) Synchronize (SYN) flood attack (Chen et al. 2016). 5G subscribers will experience poor application responsiveness or no service at all if the application server is under such an attack, no matter how ultra-high-speed 5G connection they are subscribed to Wu et al. (2018) and Vairam et al. (2019).

5G network architecture has a new set of threat vectors, which is exceptionally challenging to be tackled by the 5G system designers. Network softwarization has merged the telecommunication networks with Information Technology (IT) infrastructures and thus become a great contributor to the broad range of 5G security threats that we are facing today. To mitigate the threats to 5GS, 3GPP has published recommendations for security and privacy procedures that are open to all for further inspection. In the following section, we investigate the 3GPP security recommendations to secure 5G networks.

Inspecting 3GPP Security Proposals

The 3GPP that comprises seven Standards Development Organizations (SDOs) from across the world is responsible for developing the technical specifications for mobile communication systems starting from 3G onward. The relevant partner SDOs then convert the specifications into proper deliverables, which becomes the standards. According to 3GPP announcements, 5G will be spelled out in two or more phases starting from the Rel-15, which is the first phase of technical documentation of 5G (3GPP 2019). Technological Specification Group-Service and System Aspects 3 (TSG-SA3), a technological specification working group within 3GPP that specifies and standardizes the security architecture, privacy requirements, and relevant protocols for 5G, has recently published specifications where they mentioned the security procedures for the 5G core and the 5G NR and in general 5G security architecture (ETSI 2019). 3GPP Release 15 (R-15) depicts the 5G security as a domain-based architecture that serves both SA and NSA modes of 5G operations.

- *Network access security* domain includes the set of security requirements that ensure a UE to authenticate and access the 5G network through both 3GPP and non-3GPP access networks. Access security measures are responsible for securing the most vulnerable part, the RAN, where a UE exchanges all forms of data with the network over the air interface.
- *Network domain security* is the set of security features that help the network nodes to communicate the user and control plane data securely among themselves.
- *User domain security* ensures a user to have secure access to the mobile equipment.
- *Application domain security* offers reliability so that user applications and providers' applications interact securely.

Table 2 Key 5G security recommendations by 3GPP.

Requirements of UE

- UE should be capable of exchanging encrypted messages with gNB using the ciphering algorithms NEA0, 128-NEA1, and 128-NEA2, as presented in R-15
- The UE must support 5G-GUTI, an 80-bit-long unique identifier allocated by the AMF to the UE during network registration to keep the subscriber's IMSI safe

Requirements of gNB

- gNB must support the encryption and ciphering algorithms a UE is required to have
- Commissioning gNB should be done through secured O & M systems so that no attacker can stealthily modify the gNB configuration
- Clear text key storing is strongly discouraged. A secured and trusted environment should be used to store sensitive user data and to execute privacy-preserving algorithms

Requirements of core

- Network operators are required to divide the network into several trust zones so that the subnetworks of different operators lie in the different trust zones
- Message exchanges among SBA NFs should be confidential and done only after successful authentication
- E2E core network interconnection security solution should be activated

---

- *SBA domain Security* is the set of security procedures that enable the NFs of the SBA architecture to exchange messages.
- *Visibility and configurability of security* is vital to let the user know if a security feature is turned on and in use or not.

3GPP documented the security requirements on the UE, gNB, and the core networks toward a more transparent and trustworthy ecosystem (Surridge et al. 2018). Table 2 highlights a few key security requirements as prescribed by the 3GPP. Security requirements of the 5G physical layer are not specified in the R-15 security proposals, and security of narrowband IoT devices, vehicle-to-everything services, codec, and streaming techniques are to be specified in release 16 due in June 2020.

Security Research Groups and Visions

5G security enablers are proposed, developed, delivered, and maintained by a handful of technical working groups. Within the 3GPP, Technical Specification Group Service and System Aspects (TSG-SA) take care of the overall system architecture, and service specifications based on the 3GPP recommendations and the SA3 working group are exclusively responsible for security and privacy of 5G. Similarly, the Technical Committee (TC) Cyber is the technical group within ETSI responsible for designing reliable cybersecurity solutions and is an active contributor to 5G security and privacy deliverables. Another such group is 5G-Ensure that belongs to the 5G Infrastructure Public–Private Partnership (5GPPP), which is a joint initiative between the European Commission (EC) and the European Information and Communications Technology (ICT) industry to deliver 5G solutions. Table 3 briefs the security goals of the above-mentioned working groups along with other prominent technical working groups working in 5G security and privacy area and their primary responsibilities to enable a safe and trustworthy 5G network.

Table 3 A synopsis of 5G Security Research Groups and deliverables.

3GPP-SA3	SA WG3 is responsible for specifying security and privacy in 3GPP system's architectures and protocols and defining cryptographic algorithms for specifications
ETSI TC Cyber	TC Cyber works on quantum-safe cryptography and cybersecurity for national safety and ensures privacy for enterprise and individual data
5G-ENSURE Security	5G-ENSURE sec is responsible for realizing the 5G security specification, development, and documentation of the challenges and needs for 5G security (Morant et al. 2016)
NGMN 5G SCT	SCT focuses on security issues raised by NGMN, revision, and updation of deliverables by security work stream
ETSI ISG MEC	Industry Specification Group on Multi-access Edge Computing creates standardized, open environment that allows flawless integration of various applications from third parties across different vendors' Multi-RAT Edge Computing (MEC) platforms
STIX Community	STIX Community develops a structured language called STIX for formal description of new cyber threats for efficient information sharing and analysis in a standard way
GSMA FAS group	The FAS group analyzes the global fraud and security threat landscape and predicts risks for network operators and subscribers
Trusted Computing Group	TCG is a nonprofit organization which develops, defines, and promotes vendor-independent, open industry specifications for interoperable trusted computing platforms

Why Securing 5G is Challenging?

Identifying potential cybersecurity threats and providing prudent solutions to resolve them is not a whole new area of research in computer networking. However, with the advent of 5GS and its pervasive revolutionary nature to connect anything anywhere, the scope of possible risk has gotten more extensive. The evolved threat space brings a fresh and unique set of challenges for cybersecurity experts in securing the 5GS and establishing it as a trustworthy ecosystem (Siddiqi et al. 2017). This section summarizes the vital challenges of securing a hyperscale pervasive network such as 5G.

Billions of Hackable IoT Devices

New vulnerabilities are contributed significantly by billions of insecure smart devices, which are tiny computers equipped with high-end sensors. These devices give rise to the massive 5G IoT traffic (Salman et al. 2017). However, these small devices are often hackable by side-channel attacks, exploiting software security weaknesses, and doing reverse engineering. Once an attacker manages to get access to the device, he can virtually do many malicious activities to capture or modify sensitive user data stealthily and can even install unwanted software for continuous unsolicited user activity monitoring. Thus, IoTs accessing 5G networks open up more point of entries to hack into subscribers' privacy domain.

Many attacks on IoT devices are known as *zero-day attacks*. Zero-day attacks received the name from the number of days the device developers have known about the vulnerabilities (Palani et al. 2016). Several zero-day attacks have been reported with speculated fix time as high as 4 months, which is enough to bring disastrous loss to the user or the network by an attacker. Not just attacks, interactions between coexisting IoT applications may also lead to unsafe system behavior. For instance, an application that turns the garage lights on detecting the owner's car may interact with another application that turns off lights when it detects higher power consumption. Another popular way to attack a network is called "botnets," which is essentially hacking a weak IoT and then using them to paralyze the network or steal user data. Through large-scale measurement studies, Herwig et al. (2019) showed how IoT botnets compromise other IoT devices, and what countries are susceptible to botnet attacks. They detailed the analysis of Hajime, a new peer-to-peer IoT botnet that targets different processor architectures such as ARM (Advanced RISC Machine) and MIPS (Million Instructions Per Second)-based devices. When attackers take control of several IoT devices connected to 5GS, they can generate a massive amount of service requests than the network is designed for and thus initiate a DDoS attack. Due to the high download speed of 5G, stealing user data and network information by an attacker having illegal access to it becomes much faster than ever.

The Convergence of IT and Telecom Infrastructure

5GS inspired the telecom network to merge with IT infrastructure by virtualizing high-level network functions traditionally carried out by dedicated physical entities. Network functions are pushed to the cloud of connected servers throughout the network. SDN, NFV, and cloud computing are the key enablers of the 5G network toward a programmable, scalable, all IP infrastructure for voice and broadband (Blanco et al. 2017). Schneider and Horn (2015) pointed out that due to a flat IP-based network architecture, cyberattacks such as IP spoofing and port scanning can harm the 5G network if not appropriately secured from the very beginning. Moreover, in cloud-native computing architecture, user information is stored, processed, and shared by many colocated services following techniques such as replication, distributed file synchronization, and controlled data flow, to name a few. However, a few of these techniques only have intrinsic security features (Akhunzada et al. 2015; He et al. 2016). A recent contribution in network systems research explicates that the public cloud for implementing 5GC functionalities is feasible (Nguyen et al. 2018), although another set of security research reveals that the data leaks among the virtual machines are quite possible due to the lack of hardware isolation.

Open-source Softwarization

5GS standardization is largely implemented by open-source software alliances such as Openair Interface (Open Air Interface 2017) along with many proprietary vendor-specific implementations. Small-scale implementations often opt for open-source software implementation to keep costs low. The advantage of using an open-source software suite is that anybody from the community can review the codes and flag potential bugs. Many reviewers can report those bugs so that the vulnerabilities can be detected early for patching.

On the other hand, the National Vulnerability Database (National Vulnerability Database 2019) publicly lists recent exploits, which could be potential targets by the attackers. Another risk of using open-source software modules at enterprise scale is sluggish development practices (such as copying code from unreliable sources) and slow process of security patching. Hence, it is indeed a challenge to secure 5GS, which mainly adopts several open-source software (Liyanage et al. 2016).

AI vs AI

5GS adopts a massive number of third-party services, network management and orchestration system, and intrusion detection systems that unleash the power of machine learning, which itself is not secured enough (Marino et al. 2018; Barreno et al. 2006). Adversarial machine learning has recently gained popularity, and it is well recognized now that machine learning models are vulnerable to adversarial attacks. In the 5G context, AI and its subfields such as machine learning and deep learning have been adopted to accomplish a variety of tasks, including resource management, carrier sensing, cross-channel learning, user requirement profiling, and anomalous traffic detection for cyber defense.

On the other hand, machine learning techniques are increasingly getting popular also among hackers to initiate smarter and faster yet low-cost attacks. Hackers can use publicly available AI tools to generate fake images or videos to fool users. For instance, Deepfake can produce a video that contains words on a celebrity's mouth that she never told (Wikipedia 2019). GANs can guess passwords better than a brute force password-guessing tool. Modern spammers use AI tools to generate unsolicited emails with spam-contents that can fool the spam filters in operation and can make their way to a user's inbox. Fuzzing, which is a popular means to discover exploitable vulnerabilities of a system by monitoring the system behavior after putting a massive set of random inputs to it, has been more potent after being driven by AI. Hence, new age cyber criminals are empowered and ready to fight the AI-based cyber defense techniques with another set of AI-based cyberattack tools. The cybersecurity game is on, and it is AI versus AI.

Real-time E2E Security Monitoring at Scale

Securing the 5G network requires intrinsic security features to be developed as an integral part of the 5GS development process. Every module should have secure communication protocols and interfaces over which they transfer user and control data. We cannot envisage 5G security as a separate layer of shield that can save the whole 5G system after its deployment (Liyanage et al. 2015). This stand-alone security idea is not suitable mainly because of the highly modular and agile network architecture of 5GS, providing non-3GPP access to the 5G core network, and hosting highly sensitive public health and national security data services (Kumari et al. 2018). But there is no ultimate security by design guaranteed for a complicated, evolving, and a multivendor system such as 5G (Liyanage et al. 2017). Hence, continuous monitoring for incoming attacks is more appropriate than just securing the system during its development.

Physical Layer Security

Securing the real-life noisy physical channel in the presence of active eavesdropper is challenging because the snooper can contaminate the pilot signaling and hence can ruin channel measurement, which is crucial for massive MIMO and beamforming techniques to work. Moreover, the nonorthogonal multiple access techniques are hard to implement securely, since users need to share unprotected broadcast messages among them (Wu et al. 2018).

Can 3GPP Recommendations Solely Secure the 5GS?

5G security architecture, as proposed by 3GPP, throws light only on a few selected aspects of network security procedures and privacy-preserving protocols (Blanc et al. 2018). However, cybersecurity research continues to identify weaknesses in existing defense mechanisms and new protocols (Gross 2018). In this section, we highlight a few research needs and corner cases that 3GPP security proposals left out of scope.

ML security: ML models are going to be an integral part of future networks. Various applications, such as autonomous cars, are controlled by ML algorithms, although many other network operations use ML to enhance user experience. On the other hand, network monitoring and security software adopt ML systems for intrusion detection, malicious activity recognition, and understanding user behavior for optimized utilization of network resources (Usama et al. 2019). Nevertheless, well-designed adversaries can fool the ML algorithms (Biggio and Roli 2018). If the ML models of a 5G network security software undergo an adversarial attack, the services that rely on the 5G connectivity immediately become insecure. 3GPP does not propose any minimum safety requirements for ML, especially deep learning-based algorithms for 5G network monitoring.

A machine learning model should qualify specific performance tests and be proved as highly reliable, dependable, and understandable before they are allowed to serve in a critical network system such as 5G. Also, the ML models need to be categorized based on their resiliency to adversarial attacks, reliability under edge cases, availability during limited network coverage, and capacity. A true *carrier-grade AI* standard needs to be agreed soon to keep 5GS free from poor and vulnerable ML software because deep learning and other machine learning models are being adopted in 5GS increasingly. Directing resilient machine learning research does not come under the 3GPP purview; still, the safety of ML models has critical implications on the 5G network performance and security to the society.

Government requirements: Lawful interception of user data is often a requirement in scenarios where law enforcement authority wants to track an individual for criminal offenses committed. Hence, it is mandatory not to altogether abolish support for null encryption or unencrypted mode of communications from the 3GPP specifications. This requirement from a government leaves a vulnerable corner case that may potentially lead to an infrequent occasion of security threat to the networks (Ahmad et al. 2018).

Coexistence with vulnerable LTE network: Various security research works have recently disclosed several loopholes in LTE's privacy and authentication procedures (Shaik et al. 2015). 5G network will initially coexist with the LTE counterpart, where a set of known vulnerabilities, including RNTI tracking and DNS traffic hijacking in layer 2, exist without being adequately addressed by current safety proposals. Thus, the existing loopholes of LTE may become a potential weakness to the new 5G networks, and there is no countermeasure recommended by 3GPP.

Vendor-specific hardware security: 3GPP security proposal assumes that the user-specific sensitive data such as credentials and long-term keys and authentication algorithms are stored and executed in secure hardware. However, the security auditing scheme to ensure the tamper-resistant secure hardware component falls outside the scope of the specifications (Jover and Marojevic 2019).

Thus, 3GPP security recommendations have limitations, presumptions, and a handful of protocol corner cases that can leave 5G networks vulnerable to future cyberattacks (Dehnel-Wild and Cremers 2018). To make 5G a trusted mobile network ecosystem, cross-domain research endeavors are essential along with the efforts from the telecom standardization bodies represented by 3GPP.

Current State of the Arts and Open Research Areas

The widespread security threat landscape and shortcomings in the existing systems offer exciting opportunities for the cybersecurity and communication system experts to indulge in innovative research that resolves the current security flaws and establishes 5G as a ubiquitous mode of reliable communication. In this section, we highlight a few decisive aspects of 5G security and state-of-the-art systems currently available to secure the network. We also throw light on possible future research endeavors aiming towards an improved and safer 5G networks.

Security of IoT: The security of the increasing number of smart IoT devices connected to the network is a challenging research topic. IoT security specifications demand robust and verified protocols yet to be declared. Hence, 3GPP pushed IoT security as a topic for their release 16. The current state-of-the-art IoT security solutions such as IoTGUARD apply dynamic security strategy on all interacting commodity devices in a home or a small network because treating individual IoT device is no longer sufficient in a multiapp environment (Berkay Celik et al. 2019). IoTFuzzer is another successful app-based technique to uncover the IoT device's memory corruption vulnerability (Chen et al. 2018).

Formal Verification of 5G Protocols: The various protocols through which the 5G network functions exchange various authentication key and control messages need to be formally analyzed and proven to sufficiently meet a set of security objectives. Basin et al. provided a formal analysis of 5G Authentication and Key Exchange (AKA) protocol and showed that active privacy attacks are still possible (Basin et al. 2018). TAMARIN Prover is a state-of-the-art protocol verification tool that helps verify the reliability of the stateful security protocols such as 5G-AKA. However, extensions of verification tools and exhaustive formal analysis of other security protocols specified for 5GS need further attention from the researchers.

Adversarial Attacks on ML Systems: The considerable excitement for adopting AI in the mobile networking domain gives rise to the need for a "telecom-grade" AI, which is reliable, explainable, and secure enough to serve the customers. The ML algorithms are vulnerable to adversarial attacks, and the latest attacks have proved that not only the image analysis ML algorithms but industrial automation algorithms are also vulnerable (Zizzo et al. 2019). Deep learning algorithms are the nervous systems of connected autonomous cars, but carefully engineered adversarial inputs can fool these systems (Sitawarin et al. 2018). Thus, it is evident that 5GS needs well-developed scanning tools for AI safety to ensure users' security. Cleverly designed attacks to expose the weaknesses of popular ML algorithms are necessary, and, on the other hand, research toward building attack resilient AI systems needs to grow.

Security of SDN: Recent attacks, such as SSH over robust cache covert channels in the cloud, divulge the inherent weaknesses of cloud computing on which a majority of the SDN solutions depends (Maurice et al. 2017). The current state of the art of SDN security tool is ForenGuard, which is capable of recording the runtime dependencies and activities involving both the SDN's control and data plane and detect any control plane attack on the SDN (Wang et al. 2018). Dixit et al. developed AIM-SDN that divulges the vulnerabilities of widely used SDN datastores and addressed the weaknesses (Dixit et al. 2018). Veriflow is another popular SDN-based network DDoS defense mechanism (Khurshid et al. 2013). However, a scalable security monitoring tool ensuring the safety of the microservices running on a public cloud is needed to facilitate the secure and rapid deployment of the 5G.

Apart from the digital domain security measures, the physical analog propagation channel also needs to be secured. Although 3GPP assumes that no particular physical layer security is necessary for above 28 GHz 5G radio, the possibility of an eavesdropper spoiling pilot signaling needs to be examined further.

Conclusion

E2E security and privacy standards are vital necessities for a hyperscale, pervasive, and high-performance telecommunication network such as 5G so that it can withstand the future threat vectors and emerge as a trusted network architecture. 3GPP formally approaches 5G security and privacy; however, many aspects of it remain unattended by the security standardization bodies. This article argues that this is the right time to consider 5G security and privacy as a design obligation and not to see it as an overhead to a faster system standardization because it is often not adequate and economical to annex security and privacy features afterward.

5G infrastructures revamp the traditional cellular network architecture radically to cater to the exponentially growing demand for high-speed, low-latency mobile Internet spurring exciting new business opportunities. Hence, the current 5G standardization should include security and privacy concerns to thwart early attacks on the 5GS. Standardized security procedures should test the reliability of ML systems before they are adopted in the network operations. Cross-discipline research areas including cryptographic security, formal device testing, systematic and exhaustive testing of security procedures of 5G, and cyberattacks designing for 5G network components are necessary to bridge the security gaps that currently exist in present-day telecommunication systems.

The beneficiaries of a disruptive technology such as 5G are the people from all the cross sections of the society. At the same time, any security defect can impact a significant section of the society that depends on the technology. Therefore, this article highlights the emerging threat landscape and major security risks associated with the advent of the 5G technology, anticipate impacts of unresolved security weaknesses, and spotlights new research directions essential to establish 5G as the most trusted network architecture for the years to come.

Acronyms

4G	4th Generation	ng-eNB	5G-eNodeB
5G	5th Generation	NOMA	Nonorthogonal Multiple Access
5G-CN	5G-Core Network	NR	New Radio
5G-GUTI	5G-Global Unique Temporary Identifier	NRF	NF Repository Function
5GS	5G System	NSA	Non-stand-alone
AF	Application Function	NSSF	Network Slice Selection Function
AI	Artificial Intelligence	NVF	Network Function Virtualization
AMF	Access and Mobility Management Function	O&M	Operations and Monitoring
AUSF	Authentication Server Function	PCF	Policy and Charging Function
CU	Centralized Unit	PCRF	Policy and Charging Rule Function
CUPS	Control and User Plane Separation	PLMN	Public Land Mobile Network
DDoS	Distributed Denial of Service	R-15	3GPP Release-15
DHCP	Dynamic Host Configuration Protocol	RN	Radio Network
DoS	Denial of Service	RNTI	Radio Network Temporary Identifier
DU	Distributed Unit	RU	Radio Unit
E2E	End-to-end	SA3	Service and System Aspects 3
EPC	Evolved Packet Core	SBA	Service-based Architecture
GAN	Generative Adversarial Network	SBI	Service-based Interface
gNB	next-generation NodeB	SDN	Software-defined Networking
GSM	Global System for Mobile Communication	SDO	Standards Development Organization
HSS	Home Subscriber Server	SEPP	SEcurity Protection Proxy
IoT	Internet of Things	SLA	Service-level agreement
IT	Information Technology	SSH	Secure Socket Shell
LTE	Long-term Evolution	SUCI	SUbscription Concealed Identifier
ME	Mobile Equipment	TSG	Technological Specification Group
MIMO	Multiple Input Multiple Output	UDM	Unified Data Management
ML	Machine Learning	UE	Users Equipment
MME	Mobility Management Entity	ULLC	Ultra-low-latency Communication
N31WF	Non-3GPP Internetworking Function	UMTS	Universal Mobile Telecommunication System
NEF	Network Exposure Function	USIM	Universal Subscriber Identity Module
NF	Network Function	VPN	Virtual Private Network

Related Articles

5G-Core Network Security
Physical-Layer Security for 5G and Beyond

References

3GPP (2018). 5G; NR; Physical layer; General description. Technical specification (ts), 3rd Generation Partnership Project (3GPP), 09 2018.

3GPP (2019). 3rd Generation Partnership Project; Technical Specification Group Services and System Aspects; Release 16 Description; Summary of Rel-16 Work Items (Release 16). Technical report (tr), 3rd Generation Partnership Project (3GPP), 09 2019.

3GPP (2019). 3rd Generation Partnership Project; Technical Specification Group Services and System Aspects; Release 15 Description; Summary of Rel-15 Work Items (Release 15). Technical specification (ts), 3rd Generation Partnership Project (3GPP), 09 2019.

5G Americas (2019). The Evolution of Security in 5G. White paper (wp), 5G Americas, 08 2019.

Ahmad, I., Kumar, T. and Liyanage, M. et al. (2018). Overview of 5G security challenges and solutions. *IEEE Communications Standards Magazine* 2 (1), 36–43.

Ahmad, I., Kumar, T., Liyanage, M. et al. (2017). 5G security: analysis of threats and solutions. *2017 IEEE Conference on Standards for Communications and Networking (CSCN)*, 193–199. IEEE.

Akhunzada, A., Ahmed, E. and Gani, A. et al. (2015). Securing software defined networks: taxonomy, requirements, and open issues. *IEEE Communications Magazine* 53 (4), 36–44.

Amine Ferrag, M., Maglaras, L. and Argyriou, A. et al. (2018). Security for 4G and 5G cellular networks: a survey of existing authentication and privacy-preserving schemes. *Journal of Network and Computer Applications* 101, 55–82.

Antonioli, D., Tippenhauer, N.O. and Rasmussen, K. (2019). Nearby Threats: Reversing, Analyzing, and Attacking Google's' Nearby Connections' on Android,

Arapinis, M., Mancini, L., Ritter, E. et al. (2012). New privacy issues in mobile telephony: fix and verification. *Proceedings of the 2012 ACM Conference on Computer and Communications Security*, CCS '12, 205–216, New York, NY, USA. ACM.

Arfaoui, G., Bisson, P. and Blom, R. et al. (2018). A security architecture for 5G networks. *IEEE Access* 6, 22466–22479.

Barreno, M., Nelson, B., Sears, R. et al. (2006). Can machine learning be secure? *Proceedings of the 2006 ACM Symposium on Information, Computer and Communications Security*, 16–25. ACM.

Basin, D., Dreier, J., Hirschi, L. et al. (2018). A formal analysis of 5G authentication. *Proceedings of the 2018 ACM SIGSAC Conference on Computer and Communications Security*, 1383–1396. ACM.

Belmonte Martin, A., Marinos, L. and Rekleitis, E. et al. (2016). Threat Landscape and Good Practice Guide for Software Defined Networks/5G, Technical report (tr), European Union Agency for Network and Information Security (ENISA).

Belmonte Martin, A., Marinos, L., Rekleitis, E. et al. (2016). *Threat Landscape and Good Practice Guide for Software Defined Networks/5G*. White paper (wp), Huawei Technologies.

Berkay Celik, Z., Tan, G. and McDaniel, P.D. (2019). Iotguard: dynamic enforcement of security and safety policy in commodity iot, *NDSS*.

Biggio, B. and Roli, F. (2018). Wild patterns: ten years after the rise of adversarial machine learning. *Pattern Recognition* 84, 317–331.

Blanc, G., Kheir, N., Ayed, D. et al. (2018). Towards a 5G security architecture: articulating software-defined security and security as a service. *Proceedings of the 13th International Conference on Availability, Reliability and Security*, page 47. ACM.

Blanco, B., Fajardo, J.O. and Giannoulakis, I. et al. (2017). Technology pillars in the architecture of future 5G mobile networks: NFV, MEC and SDN. *Computer Standards & Interfaces* 54, 216–228.

Cao, Y., Xiao, C. and Cyr, B. et al. (2019). Adversarial sensor attack on lidar-based perception in autonomous driving, *arXiv preprint arXiv:1907.06826*.

Chen, J., Diao, W. and Zhao, Q. et al. (2018). Iotfuzzer: discovering memory corruptions in iot through app-based fuzzing, *NDSS*.

Chen, M., Qian, Y. and Mao, S. et al. (2016). Software-defined mobile networks security. *Mobile Networks and Applications* 21 (5), 729–743.

Dehnel-Wild, M. and Cremers, C. (2018). Security Vulnerability in 5G-aka Draft, Department of Computer Science, University of Oxford, Tech. Rep.

Statista Research Department (2016). Iot-number-of-connected-devices-worldwide.

Dixit, V.H., Doupé, A., Shoshitaishvili, Y. et al. (2018). Aim-sdn: attacking information mismanagement in sdn-datastores. *Proceedings of the 2018 ACM SIGSAC Conference on Computer and Communications Security*, 664–676. ACM.

Ericsson (2018). 5G Security - Enabling a Trustworthy 5G System. White paper (wp), Ericsson.

ETSI (2019). System Architecture for the 5G System; (Release 15). Technical specification (ts), ETSI 3rd Generation Partnership Project (3GPP), 06 2019.

Foukas, X., Nikaein, N., Kassem, M.M. et al. (2016). Flexran: a flexible and programmable platform for software-defined radio access networks. *Proceedings of the 12th International on Conference on Emerging Networking EXperiments and Technologies*, CoNEXT '16, 427–441, New York, NY, USA. ACM.

Gross, M. (2018). Eth Researchers Uncover Security Gaps in the 5G Mobile Communication Standard,

He, D., Chan, S. and Guizani, M. (2016). Securing software defined wireless networks. *IEEE Communications Magazine* 54 (1), 20–25.

Herwig, S., Harvey, K. and Hughey, G. et al. (2019). Measurement and analysis of hajime, a peer-to-peer iot botnet, *NDSS*.

Hodges, J. (2019). Heavy Reading's 2019 5G Security Survey, Research report (rr), F5 Networks, Fortinet, NetNumber, and Palo Alto Networks, 02 2019.

Jover, R.P. (2013). Security attacks against the availability of lte mobility networks: overview and research directions. *2013 16th International Symposium on Wireless Personal Multimedia Communications (WPMC)*, 1–9. IEEE.

Jover, R.P. and Marojevic, V. (2019). Security and protocol exploit analysis of the 5G specifications. *IEEE Access* 7, 24956–24963.

Khurshid, A., Zou, X., Zhou, W. et al. (2013). Veriflow: verifying network-wide invariants in real time. *Presented as part of the 10th {USENIX} Symposium on Networked Systems Design and Implementation ({NSDI} 13)*, 15–27.

Kumar, A.D., Chebrolu, K.N.R. and Soman, K.P. (2018). A brief survey on autonomous vehicle possible attacks, exploits and vulnerabilities, *arXiv preprint arXiv:1810.04144*.

Kumar, T., Liyanage, M., Ahmad, I. et al. (2018). User Privacy, Identity and Trust in 5G. *A Comprehensive Guide to 5G Security*, page 267.

Kumari, A., Gowri, S., Radhika, E.G., et al. (2018). An approach for end-to-end (e2e) security of 5G applications. *2018 IEEE 4th International Conference on Big Data Security on Cloud (BigDataSecurity), IEEE International Conference on High Performance and Smart Computing (HPSC) and IEEE International Conference on Intelligent Data and Security (IDS)*, 133–138. IEEE.

Liyanage, M., Ahmed, I. and Okwuibe, J. et al. (2017). Enhancing security of software defined mobile networks. *IEEE Access* 5, 9422–9438.

Liyanage, M., Ahmed, I., Ylianttila, M. (2015). Security for future software defined mobile networks. *2015 9th International Conference on Next Generation Mobile Applications, Services and Technologies*, 256–264. IEEE.

Liyanage, M., Bux Abro, A., Ylianttila, M. and Gurtov, A. (2016). Opportunities and challenges of software-defined mobile networks in network security. *IEEE Security & Privacy* 14 (4), 34–44.

Marino, D.L., Wickramasinghe, C.S., and Manic, M. (2018). An adversarial approach for explainable ai in intrusion detection systems. *IECON 2018-44th Annual Conference of the IEEE Industrial Electronics Society*, 3237–3243. IEEE.

Maurice, C., Weber, M., Schwarz, M. et al. (2017). Hello from the other side: SSH over robust cache covert channels in the cloud. *NDSS*, vol. 17, 8–11.

Michalopoulos, D.S., Gajic, B., Crespo, B.G.-N. et al. (2017). Network resilience in virtualized architectures. In *Interactive Mobile Communication, Technologies and Learning* (eds M.E. Auer and T. Tsiatsos), 824–839. Springer.

Morant, S., Wary, J.-P. and Piri, E. (2016). 5G Enablers for Network and System Security and Resilience, Deliverable, EG-ENSURE, 06 2016.

National Vulnerability Database (2019). Online. nvd.nist.gov (accessed 26 November 2019).

Nguyen, B., Zhang, T., Radunovic, B. et al. (2018). Echo: a reliable distributed cellular core network for hyper-scale public clouds. *Proceedings of the 24th Annual International Conference on Mobile Computing and Networking*, 163–178. ACM.

ONF (2015). Principles and Practices for Securing Software-Defined Networks. Technical report (tr), Open Networking Foundation (ONF).

Open Air Interface (2017). 5G Software Alliance for Democratising Wireless Innovation.

Palani, K., Holt, E., and Smith, S. (2016). Invisible and forgotten: zero-day blooms in the iot. *2016 IEEE International Conference on Pervasive Computing and Communication Workshops (PerCom Workshops)*, 1–6. IEEE.

Qayyum, A., Usama, M., Qadir, J. and Al-Fuqaha, A.I. (2019). Securing Connected & Autonomous Vehicles: Challenges Posed by Adversarial Machine Learning and the Way Forward, *CoRR*, abs/1905.12762.

Qayyum, A., Usama, M., Qadir, J. and Al-Fuqaha, A. (2019). Securing connected & autonomous vehicles: challenges posed by adversarial machine learning and the way forward, *arXiv preprint arXiv:1905.12762*.

Rafiul Hussain, S., Echeverria, M. and Chowdhury, O. et al. (2019). Privacy attacks to the 4G and 5G cellular paging protocols using side channel information, *NDSS*, 2019.

Rao, S.P., Holtmanns, S., Oliver, I., and Aura, T. (2015). Unblocking stolen mobile devices using ss7-map vulnerabilities: exploiting the relationship between imei and imsi for eir access. *2015 IEEE Trustcom/BigDataSE/ISPA*, vol. 1, 1171–1176. IEEE.

Rupprecht, D., Dabrowski, A. and Holz, T. et al. (2018). On security research towards future mobile network generations. *IEEE Communications Surveys & Tutorials* 20 (3), 2518–2542.

Rupprecht, D., Kohls, K., Holz, T. and Pöpper, C. (2019). Breaking lte on layer two, *IEEE Symposium on Security & Privacy (SP)*.

Salman, O., Kayssi, A., Chehab, A., and Elhajj, I. (2017). Multi-level security for the 5G/iot ubiquitous network. *2017 Second International Conference on Fog and Mobile Edge Computing (FMEC)*, 188–193. IEEE.

Schneider, P. and Horn, G. (2015). Towards 5G security. *2015 IEEE Trustcom/BigDataSE/ISPA*, vol. 1, 1165–1170. IEEE.

Scott-Hayward, S., Natarajan, S. and Sezer, S. (2015). A survey of security in software defined networks. *IEEE Communications Surveys & Tutorials* 18 (1), 623–654.

Sezer, S., Scott-Hayward, S. and Chouhan, P.K. et al. (2013). Are we ready for sdn? Implementation challenges for software-defined networks. *IEEE Communications Magazine* 51 (7), 36–43.

Shaik, A., Borgaonkar, R. and Asokan, N. et al. (2015). Practical Attacks Against Privacy and Availability in 4G/lte Mobile Communication Systems.

Shaik, A., Borgaonkar, R. and Asokan, N. et al. (2015). Practical attacks against privacy and availability in 4G/lte mobile communication systems, *arXiv preprint arXiv:1510.07563*.

Siddiqi, M.A., Khoso, M. and Aziz, A. (2017). Security Issues in 5G Network.

Sitawarin, C., Nitin Bhagoji, A. and Mosenia, A. et al. (2018). Darts: Deceiving autonomous cars with toxic signs, *arXiv preprint arXiv:1802.06430*.

Surridge, M., Correndo, G., Meacham, K. et al. (2018). Trust modelling in 5G mobile networks. *Proceedings of the 2018 Workshop on Security in Softwarized Networks: Prospects and Challenges*, SecSoN '18, 14–19, New York, NY, USA. ACM.

University of Surrey (2017). 5G Whitepaper: 5G Security Overview. White paper (wp), University of Surrey.

Usama, M., Asim, M., Latif, S. et al. (2019). Generative adversarial networks for launching and thwarting adversarial attacks on network intrusion detection systems. *2019 15th International Wireless Communications & Mobile Computing Conference (IWCMC)*, 78–83. IEEE.

Usama, M., Qadir, J., Al-Fuqaha, A.I. and Hamdi, M. (2019). The Adversarial Machine Learning Conundrum: can the Insecurity of ML become the Achilles' Heel of Cognitive Networks?, *CoRR*, abs/1906.00679,

Usama, M., Qayyum, A., Qadir, J., and Al-Fuqaha, A. (2019). Black-box adversarial machine learning attack on network traffic classification. *2019 15th International Wireless Communications & Mobile Computing Conference (IWCMC)*, 84–89. IEEE.

Vairam, P.K., Mitra, G., Manoharan, V. et al. (2019). Towards measuring quality of service in untrusted multi-vendor service function chains: balancing security and resource consumption. *IEEE INFOCOM 2019-IEEE Conference on Computer Communications*, 163–171. IEEE.

Wang, H., Yang, G., Chinprutthiwong, P. et al. (2018). Towards fine-grained network security forensics and diagnosis in the sdn era. *Proceedings of the 2018 ACM SIGSAC Conference on Computer and Communications Security*, 3–16. ACM.

Wikipedia (2019). Online. https://en.wikipedia.org/wiki/Deepfake (accessed 26 November 2019).

Wu, D., Gao, D. and Chang, R.K.C. et al. (2019). Understanding Open Ports in Android Applications: Discovery, Diagnosis, and Security Assessment.

Wu, Y., Khisti, A. and Xiao, C. et al. (2018). A survey of physical layer security techniques for 5G wireless networks and challenges ahead. *IEEE Journal on Selected Areas in Communications* 36 (4), 679–695.

Zizzo, G., Hankin, C., Maffeis, S., and Jones, K. (2019). Adversarial machine learning beyond the image domain. *Proceedings of the 56th Annual Design Automation Conference 2019*, page 176. ACM.

Further Reading

Dutta, A. and Hammad, E. (2020). 5G Security Challenges and Opportunities: A System Approach. *2020 IEEE 3rd 5G World Forum (5GWF)*. IEEE.

Ericsson.com. 5G standardization - Ericsson [online]. https://www.ericsson.com/en/tech-innovation/standardization/5G-standardization.

Fonyi, S. (2020). Overview of 5G security and vulnerabilities. *The Cyber Defense Review* 5 (1), 117–134.

Suomalainen, J., Juhola, A., Shahabuddin, S. et al. (2020). Machine learning threatens 5G security. *IEEE Access* 8, 190822–190842.

Usama, M., Mitra, R.N., Ilahi, I. et al. (2020). Examining machine learning for 5G and beyond through an adversarial lens. *arXiv preprint arXiv:2009.02473*.

2

SDMN Security

Edgardo Montes de Oca

Montimage, Paris, France

Software-defined Mobile Network (SDMN) Security

Introduction

Software-Defined Mobile Networks (SDMNs) is a rapidly emerging paradigm that involves designing and implementing mobile networks and functions as software allowing the introduction of generic, commodity, and off-the-shelf hardware and software in the core and radio access parts. The key concepts involved are (i) Software-Defined Networking (SDN), allowing dynamic programmable network management, and separating the data and control planes; and (ii) Network Function Virtualization (NFV), introducing an architecture based on the virtualization of network functions that can be connected and chained together to create communication services. The introduction of these concepts has an important impact on how the security of mobile networks is managed. They introduce new capabilities with respect to manageability, scalability, and dynamicity, but also new vulnerabilities that will be discussed in the next section, followed by the presentation of different solutions. In particular, distributed and highly flexible and programmable security monitoring architectures are important for providing the awareness of the state of the network at all levels (i.e. from the application to physical layers, at the data, and control planes) to enable the prevention and reaction to security breaches involving anomalies, sensitive data loss, and cyber-attacks.

SDN Concepts

To satisfy new requirements related to flexibility, dynamicity, and scalability, future mobile networks are adopting SDN and NFV-based architectures that allow introducing Cloud Computing, virtualization, Mobile Edge Computing (i.e. Fog Computing) technologies, as well as anything-as-a-service models, for improved resource use and management. SDN decouples the control and data planes of the network, placing the control functions outside of the network elements (e.g. software or hardware switches, routers) and enabling external control through a logical software entity called the network SDN Controller. SDN in mobile networks brings the ability to provide centralized control, hardware device abstraction, network virtualization, common device standards, and automation. These features are useful for designing scalable, dynamic, and

The Wiley 5G REF: Security. Edited by Rahim Tafazolli, Chin-Liang Wang, Periklis Chatzimisios and Madhusanka Liyanage.

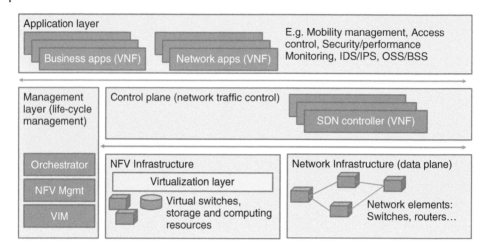

Figure 1 SDN layered architecture.

optimized security mechanisms (Liyanage et al. 2015a). With respect to security and privacy, this introduces new opportunities and concerns. Basic connectivity services need to be trustworthy in the context of new functions for industry (e.g. manufacturing, intelligent transport, smart grids, and e-health); involving multi-domain, multi-tenant, multi-provider business models; and implying the support for massive mobile internet devices. In this context, trust models need to be made dynamic; new devices need to be managed and misbehaving entities and devices need to be detected and neutralized; service delivery models need to be revised; the evolving threat landscape needs to be understood and managed; security assurance and compliance needs to be measured and assessed with respect to the risk levels (Baker et al. 2019); end-to-end security and security-by-design need to be considered; and privacy needs to be guaranteed.

SDN allows separating the SDMN into three layers (Pattaranantakul 2019). To have a more complete picture we added the Management layer and the NFV Infrastructure as can be seen in Figure 1:

- *Application layer* is where SDN, NFV, and business applications are located. These include networking management, analytics, or business applications that need to interact with the SDN controllers through Application Programming Interfaces. They inform the controllers on behavior and resources needed and obtain an abstract view of the network by collecting information from the controller for decision-making purposes. For instance, security analysis applications can be deployed here to detect abnormal behavior in the network, applications, and systems. The applications can communicate with the SDN controller through the northbound interface, e.g. RESTCONF (Bierman et al. 2013), Frenetic (Foster et al. 2011), Pyretic (Reich et al. 2013), or also with the NFV Orchestrators to manage the virtualized services or applications.
- *Management layer* is dedicated to the life-cycle management of the Virtualized Network Functions (VNFs). The SDN applications also need to interact with the NFV Orchestrators to manage (e.g. deploy, start, stop, and remotely configure) new or already deployed security functions.

- *Control plane* consists of one or more SDN controllers. It maintains a global view of network, provides a hardware abstraction to the SDN applications, and performs control tasks (e.g. routing, flow forwarding, and packet dropping) to manage the network devices in the physical and virtual infrastructure layers via the southbound interfaces (e.g. OpenFlow (Open Networking Foundation 2012; OpenFlow.org 2015)) according to the requests from the applications. The controller extracts information from the virtual/physical network elements and communicates them to the SDN applications, including statistics and network events. OpenFlow is the most widely used protocol that allows communicating instructions between the control and data planes.
- *Network Infrastructure layer (i.e. data plane)* is composed of network elements such as switches and routers that forward traffic flows based on rules deployed by the controllers. Both virtual and physical switches coexist in this layer. Note that software switches can be deployed on servers, dedicated hardware, or as VNFs managed by the NFV virtualization layer. SDN enabled OpenFlow switch implementations are, for instance, OpenvSwtich (Linux Foundation 2019) and Pica8 (Pica8, Inc. 2009).
- *NFV Infrastructure layer* provides the virtualization layer for the VNFs that includes virtual switches, storage, and computation resources.

SDN Vulnerabilities

In Ahmad et al. (2018), the authors provide an overview of the security and privacy challenges in 5G networks brought by SDN, NFV, and Cloud Computing. In this article, the authors first present the main security challenges from the 5G PPP Phase 1 Security Landscape whitepaper (5GPPP Security WG 2017): security isolation of network slices and of inter-slice communications; the need for strong authentication and authorization to avoid misuse of the network resources exposed through the control plane; and the vulnerabilities introduced by configuration errors of VNFs (Virtual Network Functions). They also identify other vulnerabilities inherited from software, virtualization, and the Internet. A detailed analysis is provided in Ahmad et al. (2018) and also in Scott-Hayward et al. (2016). In Ferrag et al. (2018), the authors provide a complete survey of 5G Security and specifically describe SDMN Security issues. Rawat and Lenkala (2017) summarize security attacks and countermeasures in SDN. In Table 1, we provide a summary of the type of vulnerabilities and threats related to SDMNs and the possible causes.

These attacks can be categorized as depicted in the following Table 2 where possible solutions are given.

As already indicated in Liyanage et al. (2015b), the main threats in SDMN are due to the Controllers but also to the open multi-stakeholder nature of the system. The location and access to the Controllers need to be monitored, and the communications between the controllers, the network elements, and the applications need to be secured through encryption techniques (e.g. TSL) and the keys need to be managed securely. But these techniques are not sufficient to assure high availability (counter DoS attacks), counter insider attacks, or involuntary vulnerable user activity (e.g. remote access from vulnerable sites, introduction of non-verified equipment). Therefore, every change and access needs to be monitored and audited for troubleshooting and forensics. In virtual and

Table 1 Attacks in SDMN, targets, and possible causes.

Attacks	Target	Example causes
Forged or faked traffic flows; Eavesdropping; Flow modification; Man-in-the-Middle attacks; Replay attacks; Side; Channel attacks; Asymmetric Routing attacks	Data plane	Compromised software/hardware network elements Compromised virtual machines and hypervisors Lack of TLS or other strong authentication technique
Cloning or deviating traffic; Forged requests to overload the controller or neighboring switches; Overloading of Flow Tables; Fake controller-based attacks; Manipulating switch software; Buffer Overflow attacks	Switches	Compromised software/hardware switches Lack of TLS or other strong authentication technique Open APIs
TCP and other protocol-specific attacks (e.g. ARP spoofing); TLS/SSL attacks; SDN scanner attacks; Message modification attacks; Controller switch communication flooding	Control plane	Unauthorized access Lack of TLS or other strong authentication technique No physical systems and boundaries exist that facilitate security control Open APIs
Denial of Service attacks; Fake switches; Software vulnerabilities; Backdoor entries; Attacks of East-West Channels; Attacks via compromised apps	Controller	Centralized controllers become single points of failure Hardware/software overloading
Admin Stations Eavesdropping; DoS attacks Replay attacks; Back door entrance; Message modification attacks; Fraudulent rule insertion; Controller hijacking	Application plane	Lack of mechanisms to ensure trust between the controller and management apps Unauthorized access via apps Vulnerabilities in software and Operating Systems Insecure storage of apps
Lack of trusted resources for forensics and remediation; Monitoring issues; Virtualization related issues	Network management	Visibility in virtualized environments is reduced. Virtualization creates boundaries that could be breached by exploiting vulnerabilities and bugs in the virtualization code (e.g. bugs in the hypervisor). Virtualization can help isolate systems but can also be abused to make malicious systems more difficult to detect. The system becomes a collection of files that can be stolen more easily. Virtualization facilitates changes making it necessary for security applications to keep up with this new dynamicity

Table 2 Threats and possible solutions.

Security/privacy threats	Possible solutions					
	Protocol security and encryption	Isolation and efficient transport of protected data	Malware scanning and elimination	Intrusion detection and prevention	Behavior analysis and Business activity monitoring	Full International Mobile Subscriber Identity (IMSI) protection
Attacks on authentication, e.g. Password reuse, Brute force, intrusions	x			x	x	x
Attacks on integrity, e.g. Tampering, Message insertion and modification, Compromised elements	x	x	x	x	x	x
Attacks on privacy, e.g. Man-in-the-middle, Eavesdropping, Replay, Tracing, Spoofing, Scanning attacks, impersonation, and social engineering attacks	x	x		x	x	x
Attacks on availability, e.g. Redirection, Distributed Denial of Service				x	x	
Insider attacks, e.g. Unintended, Malicious			x	x	x	

dynamic environments, where visibility is often reduced, this has to be done using very flexible and adaptable distributed monitoring solutions.

To deal with these vulnerabilities many solutions are proposed in the literature and commercial products. In the following sections, we provide an overview of some of the most interesting and recent ones.

Intrusion detection and prevention, and behavior analysis are security enablers that can help secure against all types of attacks. On the other hand, they imply high costs if used systematically and are less effective when encrypted traffic is used, even though they are able to analyze some encrypted flows to detect for instance encrypted tunneling or act as Man-in-the-Middle to decrypt traffic at the endpoints. The other techniques also imply sacrifices in costs (e.g. latency, bandwidth, processing, energy). Thus, it is important to be able to dynamically configure the security functions and mechanisms depending on changing needs and risks, i.e. find the right balance between security level, risks, and costs.

SDN Security

SDN transforms mobile networks into programmable, software, and centralized control networks that allow network operators and administrators to adjust and automate the management of network resources dynamically to address the changing needs of new business-driven network applications. The combination of SDN and NFV brings new opportunities for improving the security. Security network functions (e.g. firewalls, load balancers, Deep Packet Inspection (DPI), network monitoring, Identity and Access Management, Intrusion Detection and Prevention, network isolation, data protection, VPNs, Secure DNSs) can be decoupled from the hardware or software network elements (e.g. routers, switches), constructed by chaining basic functionalities, and migrated to NFV to achieve better distribution in virtualized environments, improved adaptability to network slicing, dynamic configuration, and automated deployment to satisfy particular security requirements.

In Liyanage et al. (2015c, 2017), the authors focus on the security challenges of SDMN based on SDN and NFV. They propose a multi-tier component-based security architecture to address the vulnerabilities that include (i) securing the control and data plane communications, (ii) security policy-based mitigation, (iii) DPI and SDN-Based traffic monitoring techniques to improve security threat detection and manage the security and monitoring functions. The authors validated the solution and demonstrate the ability of the proposed architecture to prevent attacks and automate the countermeasures.

Research work dealing with security in SDN mobile networks is not yet extensive and is confined basically to three main trends (i) implementation of security functions (e.g. firewalls and Intrusion Detection Systems (IDSs)/Intrusion Prevention Systems (IPSs)) in SDN environments, (ii) algorithms for improved detection (e.g. based on Machine Learning), and (iii) data plane extensions.

Security Function Implementations

Examples of IDS/IPS implementations are SDNIPS (Xing et al. 2014) and TIPS (Joldzic et al. 2014). More recently, the authors Manso et al. (2019) analyze the use of Snort IDS (Cisco Snort 2019) interacting with the Ryu SDN Controller (https://osrg.github .io/ryu/) and identify scalability problem such as the generation by Snort of too many DDoS alerts that overload the controller. In Scott-Hayward and Arumugam (2018), the authors propose a stateful security data plane solution, OFMTL-SEC, designed to provide protection against SDN configuration-based attacks and ARP spoofing attacks without controller intervention. This can be seen as a solution that avoids bottlenecks due to centralized controllers.

Advanced Detection Algorithms

Anomaly detection algorithms are proposed in Mehdi et al. (2011) for home and Small Office Home Office (SOHO) networks. In Meng et al. (2019), the authors propose a deep learning approach based on a pooling scheme that uses the Term Frequency-Inverse Document Frequency (TF-IDF) to weight the characteristics of network traffic. In Albahar (2019), the authors propose a recurrent neural network (RNN) model based on a new regularization technique (RNN-SDR) for intrusion detection within SDNs. The algorithm is embedded in the SDN controller reducing slightly the throughput and latency depending on the number of switches, implying that there is a trade-off between the performance and security. In Tang et al. (2018), the authors propose a Gated Recurrent Unit Recurrent Neural Network (GRU-RNN) for anomaly detection. In Sultana et al. (2019), the authors present a survey of the use of Machine Learning approaches for SDN-based IDSs. Besides scalability problems, the authors identify the need for further research on how to secure SDN itself that can be subjected to DDoS, forged traffic flows, vulnerabilities in switches, and attacks on the control plane.

A major challenge that is not addressed by any of these works is the relevance of the data sets used to evaluate the solutions that are always synthetic or not SDN-specific. Another challenge is that none of the works actually consider the following key challenges: the distributed nature of SDN/NFV-based environments; the optimization of the security functions with respect to the cost and the level of risk; the vulnerabilities of the APIs used; the evasion and adversarial machine learning techniques that can be used to reduce the efficiency of the detection techniques; and how different security solutions focusing on different aspects can be integrated into one comprehensive framework that is easy to manage.

Data Plane Extensions
In Park et al. (2019), the authors argue that security applications acting within the control plane allow applying network security measures on a large scale, with reduced costs, and simplified management. On the other hand, only simple security measures are possible, there is an overhead for the controllers and the performance is thus impacted. Another alternative studied by the authors are data plane middle-boxes (e.g. using NFV) that they believe can obtain better performance, enable richer functions (e.g. payload inspection), and do not imply controller overheads. Still, they identify other problems that include network overhead caused by traffic redirecting, and the need of flow steering and additional control channels for VNFs. In conclusion, the authors advocate that security services should be part of the packet processing logic, e.g. as a set of OpenFlow actions, where the processing of the packets is done without redirecting it. They evaluate their solution, called DPX (Data-Plane eXtensions for SDN Security Service Instantiation), and show that it can perform threshold-based DoS and rule-based DPI analysis. This solution embeds security functions directly into a switch as a set of actions. The limitations indicated by the authors are that the solution cannot directly enable distributed solutions that span the whole network, such as network-wide DoS detection. Other drawbacks are that the per-flow granularity of the security analysis and actions could imply important scalability problems that are not analyzed by the authors. Furthermore, the embedded functions seem to be hardcoded and proprietary and thus lack the dynamicity needed in SDN/NFV environments. Nevertheless, the idea of a programmable data plane is an emerging topic that shows much potential. In this sense, the P4 programming language, discussed below, offers a new improved alternative to OpenFlow.

SDN controllers and orchestrators, deployed as VNFs, will provide the network control functionality (i.e. traffic steering and forwarding, routing, path provisioning between VNFs), construct service chains, and direct the traffic in a flexible and auto-mated way. For the security functions, it is possible to programmatically filter, clone, and redirect traffic that needs to be analyzed to detect possible attacks or better protect it from malicious intent. Several open-source SDN controllers are available. Some are centralized, such as Floodlight (Big Switch Networks 2016), OpenDaylight (The Linux Foundation Projects 2013), NOX (Gude et al. 2008), POX (POX 2015), RYU (NTT Labs 2016); and others are distributed, such as Open Network Operating System (ONOS) (The Linux Foundation 2014). Distributed controllers need to share information when a global view of the state of the network is needed, but they reduce the single-point of failure risk of unique centralized controllers and are more scalable by allowing the sharing of communication and computation loads (Zerkane 2018).

OpenFlow is a platform-agnostic interface to the switches or routers. It allows match/action abstraction that consists in the specification of tables comprising {rule, action} pairs. If the rule is matched by the incoming packet, the action associated to the rule is executed. Thus it is only possible to select which action should be associated to the outcome of a limited set of actions (e.g. drop, output to port, push/pop Virtual Local Area Network/Multi-Protocol Label Switching (VLAN/MPLS) tag) pre-implemented by the device vendor. From the security point of view, it is possible to filter packets based on the header field.

As indicated in Scott-Hayward et al. (2016), the OpenFlow switch specification (OpenFlow Switch Specification Version 1.4 2013) proposes the use of Transport Layer Security (TLS) for the authentication between the controllers and switches. Unfortunately, the feature is optional, TLS is not completely specified and widely adopted, and it does not prevent DDoS, fraudulent rule insertion, and modification (Benton et al. 2013).

A problem with OpenFlow is that it does not support new header definitions, which are necessary for applying new packet encapsulations. In 2014, a new high-level lan-guage has been proposed called Programming Protocol-independent Packet Processors (P4) (Bosshart et al. 2014) that supports a fully programmable parser, facilitating the definition and deployment of new headers. P4 introduces a standardized, universal way for data plane programming. Currently, it is supported by commercial and open-source switches, such as BMv2 software switch (https://github.com/p4lang/behavioral-model), Netronome SmartNIC (https://www.netronome.com/products/smartnic/overview), TOFINO ASIC Switch (https://www.barefootnetworks.com/technology), and NetF-PGA (Zilberman et al. 2014). Figure 2 represents P4 in the SDN architecture.

Many research results are appearing since 2019 using the P4 language but only a few deal with security and concretely show how P4 can be used to secure networks and functions. Following are some examples:

P4Guard, presented in Datta et al. (2018), is a protocol-independent and platform-agnostic software-based firewall designed for NFV and SDN environments. It uses the P4 language to perform packet filtering. The authors show improvements in performance and flexibility of a P4-based firewall over an OpenFlow-based one called VNGuard (Deng et al. 2015).

Another firewall implementation using P4 is presented in Vörös and Kiss (2016). Here, the authors describe how to design a layer 3 firewall, with protocol, and port filtering,

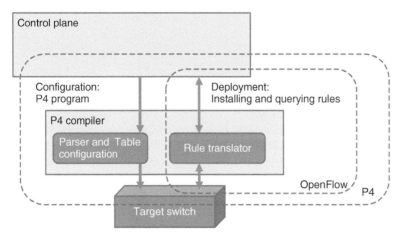

Figure 2 P4 in SDN.

flood attack detection, and the ability to make decisions about Ethernet, IPv4, IPv6, TCP, UDP header fields.

In Hill et al. (2018), the authors show how P4 can be used for tracking the full path a flow takes in a network. This is very useful for detecting and blocking malicious flows. Here, Bloom filters are implemented in P4 and the approach is evaluated for the detection of SYN attacks. The Bloom filter is implemented with an array of bits and a predetermined number of hash functions used to identify a flow and determine if it has not involved the control plane. As more items are added to the set, there is an increasing chance that an item that is not actually a member of the set will coincidently have the same bits set, causing a false positive. So Boom filters should only be used if false positives are not a crucial issue.

In Scholz et al. (2019), the authors propose an extension of the P4 Portable Switch Architecture with cryptographic hash functions for improving authentication and resilience of data flows. Currently, protocols and applications requiring cryptographic hashes for authentication or integrity cannot be described using P4. The implementation of cryptographic hash functions would increase the applicability of P4 to wide range of use cases.

In Pontarelli et al. (2017), the authors describe the implementation of a Packet Manipulation Processor (PMP) consisting of an array of small Reduced Instruction Set Computer (RISC) processors prototyped using FPGA with a specifically devised instruction set able to provide complex packet modification at line rate, such as manipulation/insertion of header fields, forwarding packets, and forging of new packets. The processor needs to be programmed in assembly language and a compiler for a high-level programming language is left as future work. Nevertheless, the authors show that it is possible to map P4 primitive actions to PMP instructions and that switch packet generation primitives are possible with PMP but are not supported by P4.

Security Monitoring and Policy Enforcement

A distributed monitoring solution composed of monitoring agents (or probes) and a centralized management application can help solve many security issues that are not solved by encryption of control and data traffic. These include assuring high availability of the network functions, countering insider attacks or disruptions due to compromised/vulnerable software/hardware elements, monitoring and auditing the changes in the configuration and software updates for troubleshooting and forensics. To do this effectively in virtual environments where visibility is often reduced, distributed, and flexible monitoring architectures need to be deployed that can provide the information necessary to detect, prevent and mitigate security breaches and denials of services. For this, effective traffic analysis rules and algorithms are needed that detect not only known attacks (i.e. signature-based DPI) but also analyze behavior and statistics to detect unknown or unforeseen attacks.

Attack signatures can be extremely difficult to formulate based on available malware information. The analysis rules are needed to detect traffic flows that are potentially caused by malware software components. The rule effectiveness is required also from the point of view of performance, because the analysis and information extraction must be done at the flow and packet levels which require a lot of processing.

Behavior analysis rules can be easier to formulate in systems where the set of actions are well categorized. They are more complicated to formulate in systems where actions are less predictable. The study of using Machine Learning to detect anomalies or differentiate normal from abnormal traffic in SDN-based systems is at a very early stage. For instance, in Abubakar and Pranggono (2017), the authors present flow-based anomaly detection based on Machine Learning to overcome the limitation of signature-based IDS like Snort. They obtain over 97% accuracy for the detection of attacks in SDN (i.e. DoS: Denial of Service attacks, R2L: Root to Local attacks, U2R: User to Root attack, and Probe: Probing attacks) using neural networks for pattern recognition using the NSL-KDD dataset (Ever et al. 2019) to implement the training and evaluation of the proposed model.

A monitoring system for SDN/NFV virtualized mobile networks has been implemented and validated in the CelticPlus SIGMONA (https://www.celticnext.eu/project-sigmona) (Liyanage et al. 2017) and SENDATE (www.sendate.eu) (Blanc et al. 2018) projects and is represented in Figure 3. It is composed of two main functions:

- *Probe*: a lightweight software application that captures events from the network, applications, and operating system (i.e. metadata from packets and flows, traces from applications, system logs) and processes them using Complex Event Processing, Deep Packet and Flow Inspection, Machine Learning, and behavior and trend analysis techniques. These probes can be deployed as independent VNF to analyze control and data flows (acting as an IDS or IPS) or can be embedded in network elements (e.g. VNF, software/hardware elements). They can be dimensioned and configured remotely by a centralized monitoring application (part of the MANO) to efficiently perform specific analysis as required. They can provide statistics, notifications, and alarms to the centralized monitoring application.
- *Centralized monitoring application*: receives the captured events and notifications from the probes through a secure channel, and performs security analysis (using the same techniques as the probe) to detect operational and security incidents. It allows the visualization of traffic statistics and alarms by operators so that they can trigger or configure automated countermeasures. For this, it will interact with the controllers (to redirect and filter network traffic) and the orchestrators (to (re)deploy

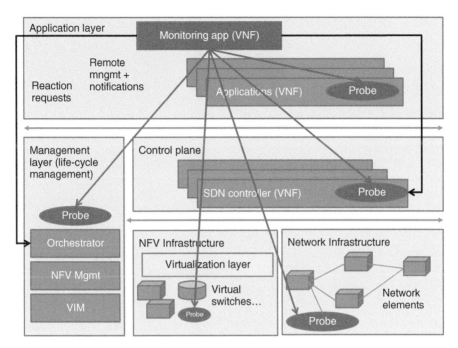

Figure 3 Monitoring architecture for SDN/NFV virtualized environments.

and (re)configure security functions, or carry out other actions to assure the reliability of the network services).

Other modules can be integrated to perform risk, liability, and root cause analysis (RCA) to improve the effectiveness and precision of the security countermeasures.

In a multi-domain, multi-domain and multi-provider business model enabled by SDN, security rules and properties need to be made comprehensive to the operators and administrators. IDS/IPSs usually require knowing proprietary and complex formalisms, e.g. Snort rules, that allow expressing only known threat signatures. The adoption of Security Service Level Agreements (Security SLAs) allows stating formally the security guarantees that are required (Casola et al. 2018). These formalisms have been used in the CelticPlus SENDATE for 4G/5G mobile network systems (www.sendate.eu) and H2020-MUSA projects (musa-project.eu) (Velasco et al. 2019) for federated cloud systems.

In order to guarantee the security of services and verify the respect of security-related Security SLAs it is necessary to capture pertinent metrics, correlate data, verify the defined SLAs for a given application or service, and notify or even react to any failure to fulfill the established SLAs. The Security Service Level Agreements (SSLAs) can specify threats that need to be detected and neutralized but also the security properties that need to be satisfied.

The defined SSLAs allow measuring the Security related KPI of a system. So the SSLA monitoring can be used to measure the KPIs of the other security functions during the operation of the system. Examples of SSLA are:

Performance SLA (but detected to be caused by DDoS attacks):

- Data and service availability (e.g. response time is not degraded more than a certain % of time);
- Attack resilience (e.g. if an attack is detected the service continues with a certain quality defined by QoS KPIs).

Security SLAs:

- geo-localization of data and services;
- service disruption: deviation from the normal use regarding different metrics (response time, number of requests, number of connected users, etc.) at different periods in the day or week, and for different network functions, interfaces, services, and applications; and
- frequency of security analysis (e.g. vulnerability scanning).

Real-time SSLA Monitoring has a cost and thus has its own KPI to measure its efficiency and effectiveness. These include the following:

- resource use in terms of CPU and memory (RAM and disk) of the probes;
- time it takes to deploy a new SSLA rule or revoke a deployed rule;
- % of false positives detected; and
- number of packets/sessions analyzed per second, with respect to the bandwidth.

Some of these metrics can be determined using network, system log, and application log monitoring and analysis techniques (e.g. attack detection, access controls, security protocols), some need the development of special modules (e.g. data and service availability, data location, vulnerability scans), some need the execution of active tests (stimulate the system to measure end-to-end metrics), and others need the correlation of data from different sources.

Figure 4 presents a high-level architecture of what is called a Security Service Level Agreement Assurance Function (SSAF) proposed and evaluated in the SENDATE project (www.sendate.eu) with (i) deployed probes that capture and analyze network, system and application events; and (ii) a centralized application for managing the probes, aggregating information and generating reports for the security management functions.

The SSAF provides the following main services:

- Visibility of the security measures and techniques enabled within the slice.
- Definition of the SSLA rules that can represent what should happen (i.e. security properties) or what should not happen (e.g. attacks, anomalies, vulnerabilities).
- Definition of reaction strategies (to be triggered automatically or by the operator), and the interactions with the Orchestrators (e.g. to manage security controls), Controllers (e.g. to manage network traffic), or directly with other security functions (e.g. Firewalls). Countermeasures can manifest themselves as a specific Security VNF deployed in a slice as a default service or during the operation of the slice (Liyanage et al. 2018a).

Conclusion and Future Trends

An updated overview of the vulnerabilities and opportunities introduced by SDN architectures in future Mobile Networks show that there is still much progress to be made

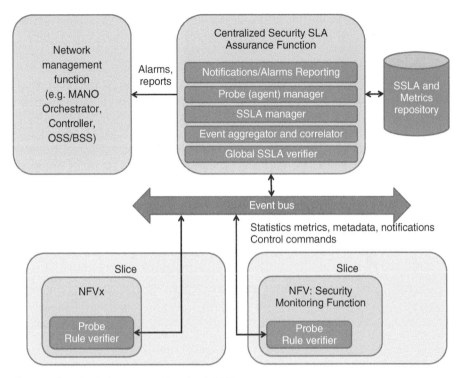

Figure 4 Security SLA Assurance Function (SSAF).

to obtain a comprehensive solution that integrates the many different security requirements and solutions.

Several topics that are not yet well covered by research and commercial efforts include

- *Balance between security, cost, and risks*: Security of mobile networks has a cost and thus involves finding the right balance between the required security (defined, for instance by security policies), the risks involved, the cost in resources (e.g. CPU, memory, Capital and Operation Expenditures), and the impact on performance (e.g. latency, bandwidth) (Liyanage et al. 2018b). New protocols, encryption techniques (e.g. modified Authentication and Key Agreement (AKA; Braeken et al. 2019)) improve traffic and security management adapted to SDMN needs.

- *New business models for security management*: The management framework for traffic control and service orchestration is being revised and new capabilities are being introduced, such as Software-Defined Security, Security as a Service, Security and SLAs, network isolation based on slicing, and service chaining. Other techniques involve security certification of slices and services, and security-by-design modeling (e.g. TOSCA 2016). The goal is to improve the self-protection, self-resilience, and self-management of future 5G and beyond mobile networks. To improve the management, cognitive Artificial Intelligence, and Machine Learning (AI/ML) techniques are seen as enablers, but adversarial ML needs also to be considered and avoided. Several surveys describe de different ML techniques used in SDN (Zhao et al. 2019), for SDN security (Herrera and Camargo 2019), and one provides more information on adversarial ML (Nguyen 2018).

- *Establishing trust domains*: Virtualized software-based NFV functions require Trusted Execution Environments (TEEs) for the secure provisioning and of sensitive operations (Lefebvre et al. 2018). This allows the definition and implementation of trust domains definition with real-time identification of compromises and security breaches at the software execution level. Leveraging distributed (Jung et al. 2019) technologies could also be used for establishing trust between the different stakeholders involved. Yazdinejad et al. (2020) use a combination of blockchains and P4 for detecting attacks by analyzing packet behavior.
- *RCA and establishing liabilities*: In the multi-tenant, multi-domain, and multi-provider environment that is even more stimulated by the network "softwareization" and virtualization, being able to determine the causes of malfunctions and security breaches becomes an important requirement for operators. Little work has been published on RCA for mobile SDN-based or 5G networks. In Novotný et al. (2018), the authors use Bayesian Networks for detecting faults in mobile ad hoc networks (MANETs). Mfula and Nurminen (2017) present a solution also based on Bayesian networks to perform automated evidence-based RCA for 5G networks. Gómez-Andrades et al. (2016) use unsupervised techniques for Self-Healing and Self-Organizing Long-Term Evolution (LTE) networks. Other works apply RCA to cloud-based systems (Wang et al. 2019; Garraghan et al. 2019). Thus, RCA techniques and determining the responsibilities and liabilities in SDN/NFV needs to be studied further.

Many of these topics (e.g. RCA, TEE, ML) are being considered for beyond 5G SDN/NFV-based SDMN by the recently started H2020 Inspire5G-plus project (www.inspire-5gplus.eu).

Related Articles

5G Security – Complex Challenges
Security Monitoring and Management in 5G

References

5GPPP Security WG (2017). 5G PPP Phase1 security landscape, Whitepaper, 5GPPP Papers.

Abubakar, A. and Pranggono, B. (2017). Machine learning based intrusion detection system for software defined networks. EST, 138–143.

Ahmad, I., Kumar, T., Liyanage, M. et al. (2018). Overview of 5G security challenges and solutions. *IEEE Communications Standards Magazine* 2 (1): 36–43.

Albahar, M.A. (2019). Recurrent neural network model based on a new regularization technique for real-time intrusion detection in SDN environments. *Security and Communication Networks* 2019: 8939041:1–8939041:9.

Baker, D., Liu, H., and Roberts, C. (2019). (NASA): Ensuring flexibility and security in SDN-based spacecraft communication networks through risk assessment. 2019 IEEE International Symposium on Technologies for Homeland Security.

Benton, K., Camp, L.J., and Small, C. (2013). OpenFlow vulnerability assessment. Proceedings of the second ACM SIGCOMM workshop on Hot topics in software defined networking. ACM, 151–152.

Bierman, A., Bjorklund, M., Watsen, K., and Fernando, R. (2013). RESTCONF Protocol. https://www.ietf.org/archive/id/draft-bierman-netconf-restconf-04.txt (accessed 4 November 2020).

Big Switch Networks (2016). Project floodlight. www.projectfloodlight.org (accessed 4 November 2020).

Blanc, G., Kheir, N., Ayed, D. et al. (2018). Towards a 5G security architecture: articulating software-defined security and security as a service. ARES, 47:1–47:8.

Bosshart, P., Daly, D., Gibb, G. et al. (2014). P4: programming protocol-independent packet processors. *ACM SIGCOMM Computer Communication Review* 44 (3): 87–95.

Braeken, A., Liyanage, M., Kumar, P., and Murphy, J. (2019). Novel 5G Authentication protocol to improve the resistance against active attacks and malicious serving networks. *IEEE Access* 7: 64040–64052.

Casola, V., De Benedictis, A., Modic, J. et al. (2018). Per-service security SLAs for cloud security management: model and implementation. *IJGUC* 9 (2): 128–138.

Cisco Snort (2019). Network intrusion detection and prevention system. www.snort.org (accessed 6 March 2019).

Datta, R., Choi, S., Chowdhary, A., and Park, Y. (2018). P4Guard: Designing P4 based firewall. MILCOM, 1–6.

Deng, J., Hu, H., Li, H. et al. (2015). VNGuard: An NFV/SDN combination framework for provisioning and managing virtual firewalls. Proceedings of IEEE Conference on Network Function Virtualization and Software Defined Network (NFV-SDN), San Francisco, CA.

Ever, Y.K., Sekeroglu, B., and Dimililer, K. (2019). *Classification Analysis of Intrusion Detection on NSL-KDD Using Machine Learning Algorithms*, vol. 2018, 111–122. MobiWIS.

Ferrag, M.A., Maglaras, L.A., Argyriou, A. et al. (2018). Security for 4G and 5G cellular networks: a survey of existing authentication and privacy-preserving schemes. *Journal of Network and Computer Applications* 101: 55–82.

Foster, N., Harrison, R., Freedman, M.J. et al. (2011). Frenetic: a network programming language. ICFP, 279–291.

Garraghan, P., Xue, O., Yang, R. et al. (2019). Straggler root-cause and impact analysis for massive-scale virtualized cloud datacenters. *IEEE Transactions on Services Computing* 12 (1): 91–104.

Gómez-Andrades, A., Luengo, P.M., Serrano, I., and Barco, R. (2016). Automatic root cause analysis for LTE networks based on unsupervised techniques. *IEEE Transactions on Vehicular Technology* 65 (4): 2369–2386.

Gude, N., Koponen, T., Pettit, J. et al. (2008). NOX: Towards an operating system for networks. *SIGCOMM Computer and Communication Review* 38 (3): 105–110.

Herrera, J.A. and Camargo, J.E. (2019). A survey on machine learning applications for software defined network security. International Conference on Applied Cryptography and Network Security (ACNS).

Hill, J., Aloserij, M., and Grosso, P. (2018). Tracking network flows with P4. INDIS@SC, 23–32.

Joldzic, O., Djuric, Z., and Vukovic, D. (2014). Building a transparent intrusion detection and prevention system on SDN. *Norsk informasjonssikkerhetskonferanse (NISK)* 7 (1).

Jung, Y., Peradilla, M., and Agulto, R. (2019). Packet key-based end-to-end security management on a Blockchain control plane. *Sensors* 19 (10): 2310.

Lefebvre, V., Santinelli, G., Müller, T., and Götzfried, J. (2018). Universal trusted execution environments for securing SDN/NFV operations. ARES, 44:1–44:9.

Linux Foundation (2019). OVS: Open vSwitch. www.openvswitch.org (accessed 4 November 2020).

Liyanage, M., Ahmad, I., Ylianttila, M. et al. (2015a). Leveraging LTE security with SDN and NFV. 2015 IEEE 10th International Conference on Industrial and Information Systems (ICIIS). IEEE, 220–225.

Liyanage, M., Gurtov, A., and Yliantilla, M. (ed.) (2015b). Chapter 18: security aspects of SDMN. In: *Software Defined Mobile Networks (SDMN): Beyond LTE Network Architecture*. Wiley Book.

Liyanage, M., Ahmed, I., Ylianttila, M. et al. (2015c). Security for future software defined mobile networks. 2015 9th International Conference on Next Generation Mobile Applications, Services and Technologies. IEEE. 256–264.

Liyanage, M., Ahmed, I., Ylianttila, M. et al. (2017). Enhancing security of software defined mobile networks. *IEEE Access* 5: 9422–9438.

Liyanage, M., Ahmed, I., Okwuibe, J. et al. (2018a). Software defined security monitoring in 5G networks. In: *A Comprehensive Guide to 5G Security*. Wiley.

Liyanage, M., Ahmed, I., Abro, A.B. et al. (2018b). *A comprehensive Guide to 5G Security*. Chapter 10. Wiley.

Manso, P.C., Moura, J., and Serrão, C. (2019). SDN-Based intrusion detection system for early detection and mitigation of DDoS attacks. *Information* 10 (3): 106.

Mehdi, S.A., Khalid, J., and Khayam, S.A. (2011). Revisiting traffic anomaly detection using software defined networking. Recent Advances in Intrusion Detection, 161–180.

Meng, Q., Zheng, S., and Cai, Y. (2019). Deep learning SDN intrusion detection scheme based on TW-pooling. *JACIII* 23 (3): 396–401.

Mfula, H. and Nurminen, J.K. (2017). Adaptive root cause analysis for self-healing in 5G networks. HPCS, 136–143.

Nguyen, T.N. (2018). The challenges in SDN/ML based network security: a survey. *arXiv preprint arXiv* 1804.03539.

Novotný, P., Ko, B.-J., and Wolf, A.L. (2018). Locating faults in MANET-Hosted software systems. *IEEE Transactions on Dependable and Secure Computing* 15 (3): 452–465.

NTT Labs (2016). RYU SDN Framework. osrg.github.io/ryu (accessed 4 November 2020).

Open Networking Foundation (2012). OpenFlow switch specification. https://www .opennetworking.org/wp-content/uploads/2014/10/openflow-spec-v1.3.0.pdf (accessed 4 November 2020).

OpenFlow Switch Specification Version 1.4 (2013). Open networking foundation www .opennetworking.org (accessed 4 November 2020).

OpenFlow.org (2015). OpenFlow definition. http://archive.openflow.org/wp/learnmore (accessed 4 November 2020).

Park, T., Kim, Y., Yegneswaran, V., et al. (2019). DPX: Data-plane eXtensions for SDN security service instantiation. DIMVA, 415–437.

Pattaranantakul, M. (2019). Moving towards software-defined security in the era of NFV and SDN. PhD thesis. University of Paris-Saclay, France. https://www.theses.fr/2019SACLL009.pdf (accessed 27 December 2019).

Pica8, Inc. (2009). OpenFlow-enabled Ethernet Switches. docs.pica8.com (accessed 4 November 2020).

Pontarelli, S., Bonola, M., and Bianchi, G. (2017). Smashing SDN "built-in" actions: programmable data plane packet manipulation in hardware. NetSoft, 1–9.

POX (2015). POX Wiki. https://openflow.stanford.edu/display/ONL/POX+Wiki (accessed 3 October 2017).

Rawat, D.B. and Lenkala, S.R. (2017). Software defined networking architecture, security and energy efficiency: a survey. *IEEE Communications Surveys and Tutorials* 19 (1): 325–346.

Reich, J., Monsanto, C., Foster, N. et al. (2013). Modular SDN programming with pyretic. *Login* 38 (5): 128–134.

Scholz, D., Oeldemann, A., Geyer, F. et al. (2019). Cryptographic hashing in P4 data planes. ANCS, 1–6.

Scott-Hayward, S. and Arumugam, T. (2018). OFMTL-SEC: State-based Security for Software Defined Networks. NFV-SDN, 1–7.

Scott-Hayward, S., Natarajan, S., and Sezer, S. (2016). A survey of security in software defined networks. *IEEE Communications Surveys and Tutorials* 18 (1): 623–654.

Sultana, N., Chilamkurti, N., Peng, W., and Alhadad, R. (2019). Survey on SDN based network intrusion detection system using machine learning approaches. *Peer-to-Peer Networking and Applications* 12 (2): 493–501.

Tang, T.A., Mhamdi, L., McLernon, D.C., Ali Raza Zaidi, S., Ghogho, M. (2018). Deep recurrent neural network for intrusion detection in SDN-based networks. NetSoft, 202–206.

The Linux Foundation (2014). ONOS (Open Network Operating System). https://wiki.onosproject.org/display/ONOS/Wiki+Home (accessed 4 November 2020).

The Linux Foundation Projects (2013). OpenDayLight project. www.opendaylight.org (accessed 4 November 2020).

TOSCA (2016). Simple profile for network functions virtualization (NFV version 1.0). http://docs.oasis-open.org/tosca/tosca-nfv/v1.0/csd03/tosca-nfv-v1.0-csd03.pdf (accessed 20 January 2018).

Velasco, E.R., Iturbe, E., Larrucea, X. et al. (2019). Service Level Agreement-based GDPR Compliance and Security assurance in (multi)Cloud-based systems. In the IET (Institution of Engineering and Technology) Software Journal. Special issue on Security and Privacy in Cloud-based systems. Editor-in-Chief: Hana Chockler, King's College London, UK. DOI: 10.1049/iet-sen.2018.5293.

Vörös, P. and Kiss, A. (2016). Security middleware programming using P4. HCI (20), 277–287.

Wang, H., Nguyen, P., Li, J. et al. (2019). GRANO: Interactive graph-based root cause analysis for cloud-native distributed data platform. *PVLDB* 12 (12): 1942–1945.

Xing, T., Xiong, Z., Huang, D., and Medhi, D. (2014). SDNIPS: Enabling Software-Defined Networking based intrusion prevention system in clouds. 2014 10th International Conference on Network and Service Management (CNSM), 308–311.

Yazdinejad, A., Parizi, R.M., Dehghantanha, A., and Choo, K.-K.R. (2020). P4-to-blockchain: a secure blockchain-enabled packet parser for software defined networking. *Computers & Security* 88: 101629.

Zerkane, S. (2018). *Security Analysis and Access Control Enforcement through Software Defined Networks*. (Analyse de sécurité et renforcement de control d'accès à travers les réseaux programmables). Brest, France: University of Western Brittany.

Zhao, Y., Li, Y., Zhang, X. et al. (2019). A survey of networking applications applying the software defined networking concept based on machine learning. *IEEE Access* 7: 95397–95417.

Zilberman, N., Audzevich, Y., Covington, G.A., and Moore, A.W. (2014). NetFPGA SUME: Toward 100 Gbps as research commodity. *IEEE Micro* 34 (5): 32–41.

Further Reading

Chowdhary, A., Huang, D., and Pisharody, S. (2018). *Software-Defined Networking and Security: From Theory to Practice*. Boca Raton: CRC Press. doi: 10.1201/9781351210768.

Pujolle, G. (2015). *Software networks: Virtualization, SDN, 5G and security*. Wiley. doi: 10.1002/9781119005100.

3

5G Security – Complex Challenges

Silke Holtmanns[1], Ian Oliver[1], Yoan Miche[1], Aapo Kalliola[1], Gabriela Limonta[1], and German Peinado[2]

[1] *Nokia Bell Labs, Espoo, Finland*
[2] *Nokia, Warszaw, Poland*

Introduction 5G Security Building Blocks

The new fifth-generation network (5G) offers flexibility and bandwidth for a large range of services, users, and applications. Mobile networks have been technologically evolving and each generation offers more features for society and businesses. 5G is a platform for use cases like the following:

- Augmented reality
- Entertainment
- Connected mobility (cars, public transport, drones)
- Smart industry
- Critical infrastructure, e.g. electricity grids, water
- Connected health
- Smart cities

This wide range of use cases is achieved by using Internet protocols, application programming interfaces (APIs), and standardized interfaces. Some of those use cases can be considered critical, while others could be thought of as nice to have. Therefore, the security need and the attack risk going along with those services differ widely. Recently, there were concerns voiced about the security of 5G networks. With evolution of each technology the industry improved the security of the communication systems. But the security and reliability of a network depends on many different items, some of them are not even of technical level. We now discuss some of them, where approaches and solutions exist, and then move on to the items, where technology and business are still in the evolutionary process.

Old Challenges in New Context for 5G

Managing risks and challenges for companies is not a new field. There exist a range of specifications outlining approaches how to evaluate and tackle risks. The International

The Wiley 5G REF: Security. Edited by Rahim Tafazolli, Chin-Liang Wang, Periklis Chatzimisios and Madhusanka Liyanage.
© 2021 John Wiley & Sons Ltd. Published 2021 by John Wiley & Sons Ltd.

Organization for Standardization ISO 27 specification (International Organization for Standardization 2019) family offers good guidance for risks related to

- Management of personnel
- Role and resource assignment
- Incident management
- Access control
- Asset management
- Physical security
- Cryptographic key management
- Contractual risks

Those risks can be tackled with known procedures, e.g. human resource-related procedures for revoking access rights for a leaving employee, building access control, role-based access systems, cryptographic key management lifecycle, etc. For those items the risks do not depend on the special network type and it is not of important if it is a 2G or 5G network. The security of a 5G network and users therein is not just about to solve the pure technical issues, but rather to address all surroundings like definition of security service level agreements (SLAs) or financial liabilities in trust chains among others.

In general, 5G systems are going to be service oriented, so it is quite natural to offer also differentiated security levels to different services to accommodate their specific security needs. Moreover, when slicing is introduced, the 5G network operator may inherit security requirements from consumers of the slices not only due to SLAs but also due to relevant regulations. For example, the European Payment Service Directive (PSD2) (European Commission 2019a) in the banking sector may pose specific security measures to a set of APIs required to manage a slice in a 5G network operator that is used for specific financial services. Today, already operators have contracts with each other for roaming and those contain elements regarding to security and fraud.

On the technical side, each network generation is an evolution of its predecessor and the learnings are taken into account for the next generation and also existing networks are continuously improved. For this reason, 5G has a substantially improved air interface security, where many research results went into the improvement (Shair et al. 2019). The user privacy on the air interface is now much better protected, e.g. from location-tracking attacks (3rd Generation Partnership Project (3GPP) 2019a). For a network to achieve the full flexibility and speed the 5G network architecture is based on the usage of Information Technology (IT) protocols for control information. The 5G architecture is a service-based architecture (SBA) TS 23.501 (3rd Generation Partnership Project (3GPP) 2019b), where the telecommunication nodes (called network functions) are connected to a service bus (depicted in green in Figure 1), which utilizes the representational state transfer (REST) API (Standards.REST n.d.) and HTTP2 (Belshe et al. 2015). Each node can communicate via the green service bus to another, e.g. below the unified data management (UDM) node communications with the session management function (SMF).

REST API and HTTP2 are well-known tools and protocols used, for example commonly in online web stores. This brings in some operational challenges how to handle the access authorization and authentication for partners, e.g. roaming partners or service providers. While the REST API is a classical web protocol and testing tools are

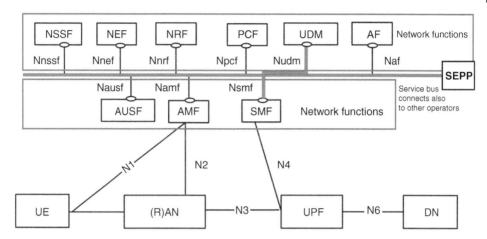

Figure 1 Service-based architecture in 5G.

known, the usage scenario is telecommunication specific. This means that some parts of the security testing and configuration verification can be automated, but other parts are currently not automatically testable. With each generation of mobile networks, the networks opened up more. This opening up need to be configured in a secure manner, for that the right expertise and validation in form of penetration tests and tools are needed. New and old roaming partners, service providers, and carriers use those APIs as the service bus also extends across organizations and toward other partners. There lies one of the risks, that in a formerly closed environment the understanding of potential risks of being victim to attacks from malware, ransomware, nation states, or normal hackers is not sufficient. Operational security of mobile networks needs to expand to cover the step from formerly closed systems to open systems. This includes

- monitoring of control traffic;
- monitoring virtual machine (VM) management traffic; and
- security zoning and deployment of specific firewalls, e.g. IP level, for control or user plane traffic.

For these kind of challenges solutions exist or are in active development, e.g. by the GSM Association (GSMA). Now we turn into the area of active research.

New Security Research Challenges for 5G

Recently, we have seen a large discussion around the concept of trust in mobile networks. We now delve first into the concept how to ensure trust in a mobile network and its elements, then we continue with the usage of machine learning for security, and close with a thought opener how to approach cyber security risk assessment for 5G.

One research challenge is the trust that we have in the infrastructure where all those virtualized network nodes or physical network nodes are placed. How to detect malicious modifications and system compromises is a large challenge.

While trusted computing concepts have existed for a long while and indeed do already play a part – in some minor role – in existing equipment, i.e. secure boot, secure key storage, etc., expanding this out to the larger ecosystem of 5G systems is critical. The reasoning for this is the increased scope of 5G especially into critical domains, such as industry 4.0, medicine, and transport, as well as into less or non-secure domains, such as Internet of Things (IoT), where devices may be little more than microcontrollers with 5G connectivity. Going out of the device space, we have increased reliance on virtualized workload, microservices, cloud and fog computing, as well as the necessary update (firmware) infrastructure around that.

We present here a model of the "trusted computing landscape" for at least attestation-related areas.

Trust

Dimension 1: Trust Stack

This is the traditional trust stack, which starts with a single machine and single root-of-trust, typically a piece of hardware, such as Trusted Platform Module (TPM), and a suitable measurement mechanism, e.g. Trusted Computing Group's (TCG) x86 BIOS/UEFI boot standard (Trusted Computing Group (TCG) 2017).

The idea is that, based on a root of trust (RoT), each component in the boot sequence is measured and added to the trust chain, as depicted in Figure 2. We start with the RoT, which will measure the firmware before it is executed. This extends upwards, building a chain of trust, through measurements of the operating system and hypervisor layers, e.g. using Intel Intel Trusted Execution Technology (TXT); and then into some aspects of run-time integrity, e.g. using Linux IMA+SELinux tags for monitoring the integrity of static parts of the file system. After this, there are relationships to remote attestation that are described later.

In situations where the virtualized nodes are responsible for critical tasks, it is desirable for that virtualized workload to run on trusted infrastructure. In cloud management environments, such as OpenStack, there is a process called workload placement, by which VMs can be instantiated, suspended, and migrated only on a trusted environment. While this is possible, it is not widely used due to the lack of management environments, issues relating to the trust of the whole VM/virtual network function (VNF) supply chain, and the lack of any reasonable mechanism for identifying VMs globally.

Beyond this is the idea that data can be "annotated" or "notarized" by coupling it with the underlying chain of trust. If a source node sends data that are notarized by

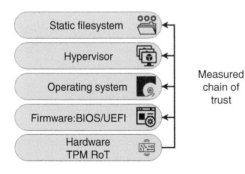

Figure 2 Creating a measured chain of trust.

including the chain-of-trust information, the following things can be established and proven:

- The provenance of the data
- The identity of the source of the data
- The integrity of the source of the data
- The integrity of the data itself

This ensures that the trust in the source and data can be verified and no malicious modification took place.

Dimension 2: E2E Trust

Single-machine trust is fine, except we need to move into larger environments, such as server rooms. The TCG TPM 2.0 standard (Trusted Computing Group (TCG) 2016) supports a mechanism for communicating attestation data with remote attestation mechanisms, which allow validation of the state from e.g. a central control panel.

An attestation server has the responsibility of deciding whether a given element is trusted or not, according to a given set of trust criteria, which define what are acceptable deviations.

This can be expanded in two directions: one down to the attestation of the code running and data executed in a secure environment, such as that provided by enclaving technologies (e.g. Intel Software Guard Extensions (SGX) and ARM TrustZone). It should be emphasized that such Central Processing Unit (CPU) enclaving technologies do not provide trust but can be utilized in the establishment of a trusted environment.

As we move further from the core toward Edge and IoT, we start to get into the idea of establishing trust over a distributed set of machines. In this respect, we have a similar concept to that of network slicing. We utilize the term security attributed slicing, and in this specific case a particular specialization of this called trusted slicing.

Such slices are orthogonal to network slicing, but they have interactions which are system and use case dependent. Within a slice, the attestation server is now responsible for enforcing a specific trust policy over a set of devices, as well as maintaining the possible trust relationships between them. How to secure these slices and ensure their trustworthiness remains an active research area.

Masses of IoT devices are being deployed, but the integrity of those devices is often questionable. While for some IoT devices the investment in security is not justified, for other critical IoT devices it is of utmost importance. There is ongoing research on how to ensure the integrity and trustworthiness of IoT devices. Currently, the best case is a Raspberry Pi with TPM, but still this does not ensure that the device is fully trustworthy. There is surprisingly little work here, but it will take a combination of everything in all the dimensions here to solve.

Dimension 3: Supply Chain Integrity

Supply chains in a global economy are very complex. In particular, if the supply chain contains software and hardware. The software may contain components, such as open source and dynamic libraries, and tools, such as debuggers or compilers, which are used to form the software product. On the hardware side, tiny elements, chip design software, cables, and boards are sourced all over the world. The supply chain is a like a tree with many different branches. The farther the final product is from the root, the lower the risk is.

How to ensure the trustworthiness of the whole chain is a very complex question. Solutions like blockchain might play a part, but that is name-dropping a technology rather than providing a solution. Partial certification, reputation, SLAs, and reviews may help to control "branches" of that tree. There are so many challenges here that make any reasonable discussion exceptionally difficult. How to secure such a complex environment remains an open challenge.

Dimension 4: Dynamicity

Trust so far assumes static systems. However, for some seemingly well-regulated systems, such as rail and automotive, this is not the case. Many systems in critical domains incorporate the use of devices which are dynamic. An example of such system can be found in the medical domain, where devices are constantly plugged in and out or even moved in or out of the hospital. The dynamicity of these devices cause havoc with static trusted concepts, including the identity, integrity, and update management infrastructure of such devices. Due to the criticality of these industries, where failures may cause (human) harm, it is important to extend and adapt our notions of trust and trustworthiness to include these scenarios. The work here is in its early stage, but some potential ideas exist (Figure 3).

Malware in Mobile Core Networks

Malware is a critical threat for any entity deploying communication services and devices. In the context of telecommunication core networks (CN), there is a wide

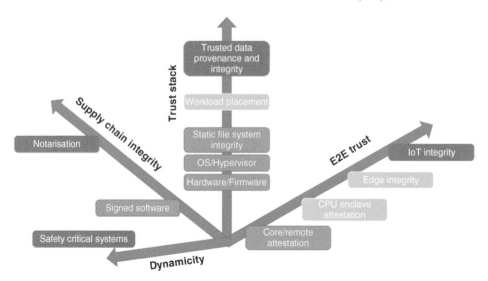

Figure 3 Trust dimensions.

range of potential attack vectors from external connectivity, inside network access, mobile customers and network tenants. Unfettered access to a telecommunication CN by a malicious actor can have extremely damaging consequences ranging from exposed user data to life-threatening disruption of network-dependant cyberphysical systems. While commonly propagated malware in the form of ransomware, worms, and automated attacks that form the background threat level of any internet-connected system remains a threat to telecommunication systems; highly targeted and heavily resourced attacks are of significant concern for the telecommunication network environment (Cybereason 2019). A challenge in the threat mitigation is that while telecommunication systems are of extremely high value, the implementations are somewhat closed and there is limited real-time threat sharing globally.

Signature-based methods of malware detection are of increasingly limited value especially against targeted attacks, as tailored executable obfuscation techniques, polymorphic malware, and even zero-day exploits are utilized by an attacker. To defend against this ever-changing threat, various machine learning methods ranging from clustering, K-nearest neighbors algorithm to deep neural networks can be used. In an ideal case, a detection method is capable of accurately identifying malicious system activity or code execution while giving as few false positives as possible.

In addition to detection and prevention of malware execution on a live environment, it is essential to deploy honeypots for proactive threat intelligence. As telecommunication CNs are highly specialized systems with largely dedicated protocols and network elements, a honeypot needs to be created specifically for this environment and further tailored to fit the surrounding deployment specifics. A telecommunication CN ideally pretends to be a specific network element with the capability to communicate utilizing the telecommunication-specific protocols, e.g. GPRS Tunneling Protocol Control plane (GTP-C). This honeypot is effectively isolated from the production environment and exists solely to attract attacks and consequently enable the attacker activity analysis. One part of the analysis is the gathering of malware samples, which are typically infiltrated into the system prior to execution. These samples can then be shared for wider threat intelligence availability and incorporated into any malware sample sets for improving the capabilities of malware detection systems.

As honeypots are preferably separate from real systems but should remain plausible from attacker point of view, they need to implement relevant parts of the telecommunication CN and protocols internally. Any directly attacker-accessible interfaces must be implemented by the honeypot, and any telecommunication network internals effecting attacker-visible responses should be simulated with sufficient fidelity to maintain the deception that the attacker is accessing a real system. This honeypot architecture is presented in Figure 4.

The data collected from such a mobile CN honeypot could give insights into new undiscovered attack patterns which are not visible today in normal internet traffic.

Measuring Security

Security measurement is a latent challenge for any Security Office organization. Mobile communication networks connect toward each other through the Interconnection Network (called IPX). As those networks are independent of each other and act in different

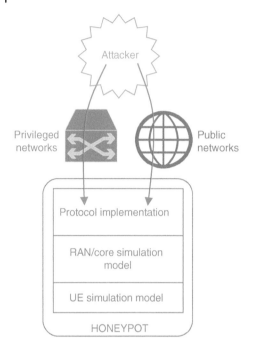

Figure 4 Honeypot architecture.

Privileged networks

Public networks

Protocol implementation

RAN/core simulation model

UE simulation model

HONEYPOT

jurisdictions, there is no central authority to secure the overall system. In the Interconnection environment, the problem is to achieve a reliable technical measure of the security in the SLAs between partners (i.e. network operators, IPX service providers, service providers, etc.). There may be different levels of security depending on the grade of interconnection partnership an operator may have. Typically, in the telecommunication industry the interconnections projects are implemented case by case, and commonly without a consistent methodology across different operators, vendors, or regions. Such methodology would be required to support in the near future the previously mentioned interconnection agents in order to make practical decisions on the security mechanisms to be implemented and enforced via *security SLAs*. For example, one of key slices supported by 5G infrastructure is dedicated to providing service to a critical infrastructure like railways, and there may be a need to interconnect this slice with other railway network in a neighbor country. Reasonably, the level of security needs to be higher (e.g. high availability and redundancy), and even the information elements to be secured in IPX would very likely be different from other type of interconnection, for instance consumers on roaming. Higher availability and fault tolerance can be achieved in many ways, e.g. duplication of data, higher testing requirements or otherwise. But a Security SLA would turn security into a product with financial incentive, like bandwidth.

To enforce different levels of security, we need to define security measurements to make sound decisions on those levels. Now the question is how to monitor and measure the security levels when there are not industry-accepted standards yet for how monitoring in telecom networks should be implemented. Operators still need to work with their vendors to establish an approach valid for today and flexible enough to align later on with industry standards. The US Federal Communication Commission subcommittee working on the subject (US Federal Communication Commission (FCC) 2017; US Federal

Communication Commission (FCC) 2018) recommends monitoring and analytics to be implemented, and indeed strong analytics with machine learning and artificial intelligence may be necessary to detect anomalies between network functions. This is key to maintain the lifecycle of a risk management approach.

The existing security risk evaluation framework based on the ISO 27 series (International Organization for Standardization 2019) allows us to evaluate part of the risks related to 5G networks. It builds a good starting point, while still not giving a full picture of the risks e.g. related to IPX.

Need for a Comprehensive and Regular Risk Management Approach

Currently security of networks relies on technical standards (e.g. 3rd Generation Partnership Project (3GPP), European Telecommunications Standards Institute (ETSI), GSMA), multiple security assurance schemas (e.g. security by design processes) and product security features provided by vendors, and best practices adopted by the Information and Communication Technology (ICT) industry during last years, such as network perimeter protection, zoning, traffic separation, and secure network topologies as well as secure operations and maintenance (e.g. security operation centers).

The standards-specified network and security functions are created to build a robust security design, but its deployment as a usage is not always enabled, and indeed in the specifications we find a significant number of security requirements cataloged as "mandatory to support, optional to use." This means in technical terms, that vendors are obliged to implement, while the deploying operator of the equipment is not mandated to use it or switch on the security function. With respect to security assurance and product security, even if industry initiatives like GSMA Network Equipment Security Assurance Scheme (NESAS) (GSM Association (GSMA) 2019) and 3GPP Security Assurance Specification (SCAS) (3rd Generation Partnership Project (3GPP) 2019c) appear in the near horizon as common accepted schemas, the reality today is mostly based on vendor proprietary approaches. Last but not least, data centers and infrastructure need to be secured by operators and their operational teams, yet here again, these additional measures are optional and not always implemented. Thus, the mere existence of proper security considerations in standards, products, and architectures may not be sufficient to achieve the security level required by the criticality of 5G assets (i.e. network, services, and applications). The need of comprehensive and regular risk management approaches is easily inferred to support the decision on the security measures to be deployed based on the asset's criticality.

Status Regulatory Approaches for Risk Management

In 2018, the European Union established two pillars in the cybersecurity environment, first in May with the reform of EU data protection rules through the entry into application of the General Data Protection Regulation (GDPR) (European Commission 2016), and second in December with the agreed EU Cyber Security Act (CSA). CSA has just entered into force on Thursday, 27 June 2019 (European Commission 2019b). It means CSA becomes a fundamental piece of the cyber security legislation in Europe together with GDPR more focused on data privacy aspects. GDPR broadens the relevance of risk, as it is explicitly based on the notion of a risk-based approach. It effectively incorporates a risk-based approach to data protection, requiring organizations (both controllers and processors) to assess the "likelihood and severity of risk" of their personal data processing operations to the fundamental rights and freedoms of individuals. In short, any organization that is required to comply with GDPR needs to conduct regular risk assessments.

The official journal of the EU on 7 June 2019 published the new regulation (EU) 2019/881 of the European Parliament and of the Council, known as EU CSA (European Commission 2019b). In the clause (49), part of the initial assumptions, we found specifically a clear reference to the need of establishing risk management–based approaches and measuring the security:

> Efficient cybersecurity policies should be based on **well-developed risk assessment methods**, in both the public and private sectors. Risk assessment methods are used at different levels, with no common practice regarding how to apply them efficiently. Promoting and developing best practices for risk assessment and for interoperable risk management solutions in public-sector and private-sector organisations will increase the level of cybersecurity in the Union. To that end, ENISA should support cooperation between stakeholders at Union level and facilitate their efforts relating to the **establishment and take-up of European and international standards for risk management and for the measurable security of electronic products, systems, networks and services which, together with software, comprise the network and information systems**.

How to Proceed Today in 5G Risk Management Approach

As the use cases expand that 5G networks are going to serve and more and more critical applications will utilize mobile networks for their communication, there is a need for a broad, holistic approach to risk management. Regulation and SLA most likely will even require it. Applications will have different security needs and different resources available that they can invest in security, therefore a leveled approach to take into account practical business reality is required. Now, in the context of 5G, the different stakeholders, and specially the network operators, shall respond the imminent questions, where to start? How to proceed?

Considering that the technical specifications for hardening individual CN nodes took years to develop by a group of experts, a comprehensive risk assessment itself may be a little bit premature and ambitious. 5G is in the early stages of its evolution, we will most likely see first 5G radio network connected to a 4G CN, nevertheless the holistic approach to the security is required, and there are certainly starting points to consider. We encourage to start with the existing specifications developed for node security from 3GPP (GSM Association (GSMA) 2019) and GSMA (3rd Generation Partnership Project (3GPP) 2019c) and then extend the scope with deeper consideration of the following items:

- *Risk assessment on common infrastructure*, such as transport networks and cloud infrastructure (i.e. software-defined networks, network function virtualization). A sound risk assessment on the common infrastructure supporting the 5G New Radio (NR) and CN will bring not only the awareness and knowledge on new risks and corresponding mitigation mechanisms, but also the security advantages that this common infrastructure will add for 5G networks, i.e. flexibility, programmability, or automation among others.

- *Risk assessment in supply chain governance.* 5G opens the network to new businesses, so the number of stakeholders is vastly multiplied. On the other side, the very extended and continuously increasing supply chain produces a lack of visibility of the risks across the complex ecosystem, entailing suppliers, partners, regulators, and customers, often with movements around the globe, i.e. subject to different regulations. Assessing the risk toward each external business partner and supplier could indeed become a heavy burden for many organizations, if a formal risk assessment methodology (e.g. Supply Chain Risk Management (SCRM), ISO/IEC 27036 (International Organization for Standardization 2014), SCRM as defined by NIST SP 800-161 (National Standards Institute (NIST) 2015), see next figure) focused on mitigating the risk is not applied as part of the overall risk management activities (Figure 5).

- *Intrinsic Risks of new 5G infrastructure and network.* Technical aspects of the new 5G technology in first commercial deployments, e.g. in NR – Dual Connectivity in the Non-Standalone Architecture (NSA) architectures, are to be currently considered,

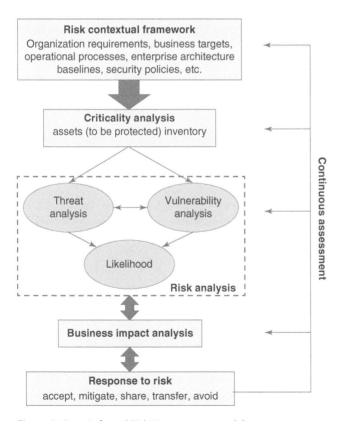

Figure 5 Generic formal Risk Management model.

and stepwise upcoming aspects in 5G CN, devices and/or slicing will be as well part of the overall technical risk assessment. Please note that the coexistence of 5G with 4G/3G/2G as well as with other non 3GPP infrastructures and fixed networks, is also a risk factor to take into account in the overall risk management framework. In 2017, 5G ENSURE program (EU-funded project for the development of 5G under 5G-PPP umbrella) (5G ENSURE 2017) through one of the deliverables referenced as 5G-ENSURE D2.6 Risk Assessment, Mitigation and Requirements (5G ENSURE 2017) took the first steps toward the definition of a risk assessment and mitigation methodology, having proposed an approach for the full set of 5G-ENSURE security use cases. The risk management context is then defined, looking first at the 5G assets and actors, which is followed by the identification of threats. The risk evaluation methodology for use case analysis is also introduced with some possible approaches to risk likelihood and impact estimation. Although the study is focused on the 5G-ENSURE-specific use cases, at the end of the document a "holistic" approach to define additional risks of 5G system infrastructure is proposed.

The methodology which will be used within the 5G-ENSURE project for the risk assessment is based on the Risk Management Process (ISO 27005) (International Organization for Standardization 2019) and, especially, on its simplification represented by NIST SP-800-30 (National Standards Institute (NIST) 2012).

Latest Proposal for 5G Risk Management

In May 2019, more than 150 cybersecurity experts from EU member states, Australia, Canada, the United States, Israel, Japan, and other countries met in Prague to discuss 5G security. During the Prague 5G Security Conference, the Czech conference chairman published a voluntary series of recommendations called the Prague Proposals (Government of the Czech Republic 2019). One of the key perspectives considered in the discussions with the aim of decreasing the security risks associated with developing, deploying, operating, and maintaining complex communication infrastructures such as 5G networks was the need of a proper risk assessment, i.e. "Systematic and diligent risk assessment, covering both technical and non-technical aspects of cyber security, is essential to create and maintain a truly resilient infrastructure. A risk-based security framework should be developed and deployed, taking into account state-of-art policies and means to mitigate the security risks."

Summary

5G is a new technology evolution that will serve a wide range of use cases, which is achieved through a new network architecture and NR capabilities. Those new features and the opening up of networks require to change the approach of today for security and risks. We outline that for many potential risk areas there exist solutions. For the currently unchartered areas, like trusted cloud systems, advanced anomaly detection in CN, and risk management active research is ongoing. The research areas outlined are just a snapshot, other active areas are highly secure slicing and network management, usage of blockchains, etc.

Related Articles

5G Mobile Networks Security Landscape and Major Risks
5G-Core Network Security

References

3rd Generation Partnership Project (3GPP), Technical Specification TS 33.501 v15.5.0
(Release 15) (2019a). Security architecture and procedures for 5G System. https://www
.3gpp.org/DynaReport/33501.htm (accessed 15 September 2019).

3rd Generation Partnership Project (3GPP), Technical Specification TS 23.501 v16.1.0
(Release 16) (2019b). System architecture for the 5G System (5GS). https://www.3gpp
.org/DynaReport/23501.htm (accessed 15 September 2019).

3rd Generation Partnership Project (3GPP) (2019c). Security assurance specification
(SCAS) technical specifications TS 33.511 – TS 33.519. https://www.3gpp.org/
DynaReport/33-series.htm (accessed 15 September 2019).

5G ENSURE (2017). 5G Enablers for Network and System Security and Resilience Project
(ENSURE). http://www.5gensure.eu/ (accessed 15 September 2019).

Belshe, M., Peon, R., and Internet Engineering Task Force (IETF) (2015), Request For
Comments RFC 7450. Hypertext Transfer Protocol Version 2 (HTTP/2). https://tools
.ietf.org/html/rfc7540 (accessed 15 September 2019).

Cybereason (2019). Operation softCell: A worldwide campaign against telecommunications
providers. https://www.cybereason.com/blog/operation-soft-cell-a-worldwide-
campaign-against-telecommunications-providers (accessed 15 September 2019).

European Commission (2016). General data protection regulation (GDPR). https://ec
.europa.eu/info/law/law-topic/data-protection_en (accessed 15 September 2019).

European Commission (2019a). Payment services (PSD 2) – Directive (EU) 2015/2366.
https://ec.europa.eu/info/law/payment-services-psd-2-directive-eu-2015-2366_en
(accessed 15 September 2019).

European Commission (2019b). EU cyber security act (CSA). https://ec.europa.eu/digital-
single-market/en/eu-cybersecurity-act (accessed 15 September 2019).

Government of the Czech Republic (2019). Prague 5G Security Conference announced
series of recommendations: The Prague Proposals. https://www.vlada.cz/en/media-
centrum/aktualne/prague-5g-security-conference-announced-series-of-
recommendations-the-prague-proposals-173422/ (accessed 15 September 2019).

GSM Association (GSMA) (2019). Network equipment security assurance Scheme
(NESAS). https://www.gsma.com/security/network-equipment-security-assurance-
scheme/ (accessed 15 September 2019).

International Organization for Standardization, ISO/IEC 27036 (2014). Information
technology – Security techniques – Information security for supplier
relationships – Part 1: Overview and concepts. https://www.iso.org/standard/59648
.html (accessed 15 September 2019).

International Organization for Standardization (2019). IEC/ISO 27000
family – Information security management systems. https://www.iso.org/isoiec-27001-
information-security.html (accessed 15 September 2019).

National Standards Institute (NIST), Ronald S. Ross (2012). Guide for conducting risk assessments. https://www.nist.gov/publications/guide-conducting-risk-assessments (accessed 15 September 2019).

National Standards Institute (NIST), Jon M. Boyens, Celia Paulsen, Rama Moorthy, Nadya Bartol SP 800-161 (2015). Supply chain risk management practices for federal information systems and organizations. https://www.nist.gov/publications/supply-chain-risk-management-practices-federal-information-systems-and-organizations (15 September 2019).

Shair, A., Borgaonkar, R., Part, S. et al. (2019). New vulnerabilities in 4G and 5G cellular access network protocols: exposing device capabilities. ACM WiSec '19 Proceedings of the 12th Conference on Security and Privacy in Wireless and Mobile Networks, 221–231, Miami, Florida (USA), 15–17 May 2019. https://dl.acm.org/citation.cfm?id=3319728&qualifier=LU1041682.

Standards.REST (2019). Representational State Transfer (REST) Standards Overview. https://standards.rest (accessed 15 September 2019).

Trusted Computing Group (TCG) (2016). Trusted platform module library, part 1: Architecture (Trusted platform module library specification, Family 2.0 No. Level 00, Revision 01.38).

Trusted Computing Group (TCG) (2017). TCG PC client platform firmware profile (TCG PC client platform firmware profile specification, Family 2.0 No. Level 00 Revision 1.03 v51).

US Federal Communication Commission (FCC) (2017). Communications security, reliability and interoperability council V, working group (WG) 10. https://www.fcc.gov/about-fcc/advisory-committees/communications-security-reliability-and-interoperability (accessed 15 September 2019).

US Federal Communication Commission (FCC) (2018). Communications security, reliability and interoperability council VI, working group (WG) 3. https://www.fcc.gov/about-fcc/advisory-committees/communications-security-reliability-and-interoperability-council (accessed 15 September 2019).

Further Reading

Ghafir, I., Hammoudeh, M., Prenosil, V. et al. (2018). Detection of advanced persistent threat using machine-learning correlation analysis. *Future Generation Computer Systems* 89: 349–359.

Limonta, G. and Oliver, I. (2019). Analyzing trust failures in safety critical systems. Proceedings of the 29th European Safety and Reliability Conference, ESREL.

Nawrocki, M., Wählisch, M., Schmidt, T.C., Keil, C., and Schönfelder, J. (2016). A survey on honeypot software and data analysis. eprint arXiv:1608.06249, 1–38.

Oliver, I., Holtmanns, S., Miche, Y. et al. (2017). Experiences in trusted cloud computing. In: *Network and System Security. NSS 2017. Lecture Notes in Computer Science*, vol. 10394 (ed. Z. Yan, R. Molva, W. Mazurczyk and R. Kantola). Cham: Springer.

Oliver, I., Kalliola, A., Holtmanns, S. et al. (2018). A testbed for trusted telecommunications systems in a safety critical environment. In: *Computer Safety, Reliability, and Security.*

SAFECOMP 2018. Lecture Notes in Computer Science, vol. 11094 (ed. B. Gallina, A. Skavhaug, E. Schoitsch and F. Bitsch). Cham: Springer.

Yamaguchi, F., Lindner, F., and Rieck, K. (2011). Vulnerability extrapolation: assisted discovery of vulnerabilities using machine learning. In Proceedings of the 5th USENIX conference on Offensive technologies (WOOT'11). USENIX Association, Berkeley, CA, USA, 13.

4

Physical-Layer Security for 5G and Beyond

Diana P.M. Osorio[1,3], José D.V. Sánchez[2], and Hirley Alves[3]

[1] *Federal University of São Carlos (UFSCar), Center for Exact Sciences and Technology, São Carlos, SP, Brazil*
[2] *Department of Electronics, Telecommunications and Information, Networks, Escuela Politécnica Nacional, Quito, Ecuador*
[3] *6G Flagship, University of Oulu, Oulu, Finland*

Introduction

The International Telecommunication Union (ITU) has classified 5G network services into three categories: enhanced mobile broadband (eMBB), ultra-reliable and low-latency communications (URLLC), and massive machine-type communications (mMTC). These services are supposed to coexist in the same network architecture by allocating network resources in such a way that the isolation among different inner logical networks (slices) is ensured through network slicing (3GPP 2017). The three broad categories account for a myriad of highly diverse applications from different industry sectors. For instance, eMBB focuses on very high peak data rates in dense areas such as stadiums or urban centers, as well as moderate rates for cell-edge users, broadband everywhere, and high-speed mobility such as connected trains. mMTC comprehends scenarios with a large number of low-complexity and low-power Internet of things (IoT) devices, which are sporadically active and send small data payloads in order to allow for battery life savings. Then, mMTC focuses on high-density applications, such as smart wearables, smart agriculture, sensor networks, and connected city/home. On the other hand, URLLC refers to applications with stringent requirements on availability, low latency, and reliability, such as E-health services, augmented reality, vehicle-to-everything (V2X) networks, Tactile Internet, and Industry Automation.

Alongside the opportunities 5G bring due to such heterogeneous applications, major challenges regarding information security emerge raising more concern about privacy than ever before. In many use cases, 5G connects critical infrastructure with highly sensitive and confidential information being transmitted, posing a threat not only for the information conveyed but to the industry and society. In this sense, lightweight, efficient, and service-based security solutions to attend the diverse restrictions of 5G-and-beyond applications are required. Traditionally, network security is provided by bit-level cryptography-based techniques, carried out at upper layers. However, those methods are limited to satisfy the requirements of 5G-and-beyond applications due to the following reasons: (i) cryptographic methods based on public keys are extremely

The Wiley 5G REF: Security. Edited by Rahim Tafazolli, Chin-Liang Wang, Periklis Chatzimisios and Madhusanka Liyanage.
© 2021 John Wiley & Sons Ltd. Published 2021 by John Wiley & Sons Ltd.

challenging in large-scale and decentralized networks; (ii) secure links required for the exchange of private keys cannot be guaranteed in some scenarios; (iii) so far, public-key encryption has been unbreakable by using very long key pairs; however, the advance on computational capabilities, such as advanced quantum computers, could crack key pairs in just few hours; thus, eavesdropping and active attacks are a high risk in future networks; and (iv) demand for extra delay and complexity to provide strong security are undesirable for some 5G applications, especially those related to URLLC services.

A new paradigm for providing enhanced security in wireless networks is referred to as physical-layer security (PLS), which can potentially offer secure transmissions by efficiently exploiting the properties of wireless medium and high randomness of wireless channels (Poor and Schaefer 2017), thus being particularly attractive in resource-limited application scenarios. The basic idea behind PLS techniques is to degrade the channel for eavesdroppers, thus preventing them from gaining information about the confidential messages from the received signal. In this way, PLS techniques can offer an additional level of security, which, integrated with traditional cryptography techniques, can safeguard the highly sensitive data expected to be transmitted over future networks.

Notion on Physical-Layer Security

Secure communication notion dates to 1949, when, from the theoretic information theory, Shannon introduced the concepts on secrecy transmissions with his so-called noiseless cypher system in Shannon (1949). In that work, Shannon defined the concept of *perfect secrecy* as the condition when the eavesdropper (Eve) completely ignores the information transmitted from a legitimate transmitter (Alice) to a legitimate receiver (Bob), then the best Eve can do is just randomly guessing the original information bit by bit. For that purpose, Alice and Bob share a secret key K that Alice uses to encrypt a message M into a code word X to be transmitted to Bob, which is overheard by Eve. Then, the transmission is considered secure if there is a statistical independence between X and M, i.e. the mutual information is equal to $I(M; X) = 0$.

Later, in 1975, Wyner introduced the discrete memoryless wiretap channel as depicted in Figure 1 (Wyner 1975). In that model, noisy channels are considered, then the criterion for *perfect secrecy* is hard to attain, still a perfect secure communication can be

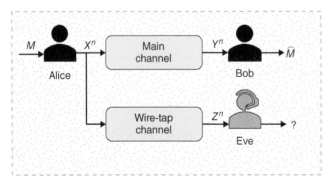

Figure 1 The wire-tap channel.

guaranteed if the eavesdropper channel is a degraded version of the main channel, without the need of a shared key. Therefore, Alice must encode M into a n-length code word X^n, and the outputs at Bob's and Eve's are Y^n and Z^n, respectively. Hence, the concept of *weak secrecy* was formulated, which establishes that the information leakage rate vanishes asymptotically with the block length n, i.e. when $\lim_{n \to \infty} (1/n)I(M; Z^n) = 0$, that is the mutual information leakage is not forced to zero on each channel use, but on average. Wyner's work was generalized by Csiszár and Körner (1973) for the broadcast channel. Maurer and Wolf (2000) determined that the criterion of *weak secrecy* is too weak to guarantee confidentiality of the secret information, as even if the leakage rate goes to zero as the block length approaches infinity, critical information bits can be leaked to an illegitimate receiver. Thus, they defined the *strong secrecy* criterion when the asymptotic mutual information goes to zero as the code word length n approaches infinity, thus disregarding the term $1/n$ from the *weak secrecy*. Therefore, the *strong secrecy* criterion guarantees that the decoding error probability at Eve approaches one exponentially fast. The *strong secrecy* criterion was further strengthened to derive into *semantic security*, through which Eve is not able to obtain any information about the secret message, then it is equivalent to accomplish *strong secrecy* for all message distributions, which can be mathematically expressed as $\lim_{n \to \infty} \max_{p_M} I(M; Z^n) = 0$, where p_M is the probability distribution of the secrete message M.

Secrecy Performance Metrics

Some of the main secrecy performance metrics most used in the literature are highlighted in this section.

Secrecy Capacity

This is the most widely used metric to evaluate the secrecy performance of wireless networks. In Wyner (1975), the secrecy capacity was defined as the maximal achievable rate at which a secret message is recovered reliably at Bob while remaining useless for Eve; thus, the trade-off between the information rate to the destination and the level of ignorance at the eavesdropper (measured by its equivocation) is characterized. It was shown that a nonzero secrecy capacity can be attained if Bob's channel is less noisy than Eve's channel, thus only working for discrete memoryless channels. In Leung-Yan-Cheong and Hellman (1978), the secrecy capacity for the Gaussian wiretap channel was studied, wherein it was established that the secrecy capacity is the difference between the capacities of the legitimate and eavesdropper channels; therefore, a secure communication is possible if and only if the signal-to-noise ratio (SNR) of the legitimate channel is larger than that of the eavesdropper channel. Later, Barros and Rodrigues (2006) considered the impact of fading on the secrecy capacity. Mathematically, the secrecy capacity for a channel realization of a quasi-static fading scenario is given by

$$C_S = [C_M - C_W]^+ = \max\{\log_2(1 + \gamma_M) - \log_2(1 + \gamma_W), 0\} \tag{1}$$

where C_M and C_W are the capacities for the main and the wire-tap links, respectively, and γ_M and γ_W are the corresponding received SNRs at Bob and Eve.

For ergodic fading channels, it is assumed that the channel rapidly transits through all fading states, so that Alice, Bob, and Eve might experience a different fading state for each channel use. Thus, by considering that all nodes have perfect knowledge of the instantaneous channel state information (CSI) about the current fading state, the ergodic secrecy capacity is given as (Poor and Schaefer 2017)

$$C_{ES} = \max_{\mathbb{E}_A[P(h)] \le P} \mathbb{E}_A[C_S] \tag{2}$$

where $A = \{(\gamma_M, \gamma_W) : \gamma_M > \gamma_W\}$, and $\mathbb{E}[\cdot]$ denotes expectation. Then, the expectation is taken over all fading realizations in which Bob experiences a better channel than Eve, and, correspondingly, the power $P(h)$ is allocated only to those fading realizations h in which Bob experiences a better channel than Eve, thus $P(h)$. It is worthwhile to mention that, for fading channels it is sufficient that $\Pr\{\gamma_M > \gamma_W\} > 0$ to have a positive secrecy capacity; thus, fading can be considered beneficial for as even if the wire-tap channel is better than the main channel on average, the ergodic secrecy capacity is positive, because whenever Bob experiences a better channel than Eve instantaneously, this fading realization can be exploited for secure communication.

Secrecy Outage Probability

The secrecy outage probability (SOP) is defined as the probability that the instantaneous secrecy capacity falls below a target secrecy rate R_S (Barros and Rodrigues 2006). It can be formulated as

$$SOP = \Pr[C_S < R_S] \tag{3}$$

where $\Pr[\cdot]$ indicates probability. The SOP in (3) indicates that whenever $R_S < C_S$, the wire-tap channel will be worse than the main channel, and then the wire-tap codes used by Alice will ensure perfect secrecy.

Alternative Secrecy Outage Formulation

According to the classical secrecy outage probability defined above, an outage event occurs whenever a transmission cannot be decoded by Bob or when there is some information leakage to Eve, that is, that metric does not distinguish between reliability and security. For instance, when Bob's channel cannot support R_S, i.e. $C_M < R_S$, this implies that $C_S < R_S$, thus accounting for an outage event. However, Alice can suspend the transmission in that case; thus, this does not represent a failure in achieving perfect secrecy. Considering this, in Zhou et al. (2011), an alternative metric for the secrecy outage probability was proposed to effectively measure the probability that the secret message fails to achieve perfect secrecy by conditioning the outage event upon a message actually being transmitted. Therefore, when the CSI of the main channel is available, Alice can decide whether or not to transmit and at which rate, thus attaining a considerable reduce on the secrecy outage probability. This alternative metric can be mathematically expressed as

$$SOP_A = \Pr[C_W < R_M - R_S | \text{Message Transmission}] \tag{4}$$

where R_M is the code word transmission rate.

Fractional Equivocation-Based Metrics

Classical SOP metric presents the following drawbacks: (i) it does not allow to quantify the amount of information leaking to the eavesdroppers when an outage event occurs; (ii) it cannot provide insights on the eavesdropper's ability to decode successfully confidential messages; (iii) it cannot be directly linked to the quality of service (QoS) requirements for different applications and services. Considering this, the authors in He et al. (2016) proposed novel metrics that provide a more comprehensive understanding of physical-layer security and how secrecy is measured. Those metrics focus on quasi-static fading channels, and they are based on the so-called partial secrecy regime, whereby a system is evaluated by means of the fractional equivocation, with regard to the level at which the eavesdropper is confused. The fractional equivocation for a given fading realization of the wireless channel is given by

$$\Delta = \begin{cases} 1, & \text{if } C_W \leq C_M - R_S \\ (C_M - C_W)/R_S, & \text{if } C_M - R_S < C_W < C_M \\ 0, & \text{if } C_M \leq C_W \end{cases} \tag{5}$$

Considering (5), the authors in He et al. (2016) defined the following metrics: generalized secrecy outage probability (GSOP), average fractional equivocation (AFE), and average information leakage rate (AILR). The GSOP characterizes that the information leakage ratio, $1 - \Delta$, is larger than a certain value, $1 - \theta$. It can be mathematically expressed as

$$\text{GSOP} = \Pr[\Delta < \theta] \tag{6}$$

This metric allows to specify different levels of secrecy requirements according to Eve's ability to decode the confidential messages, by changing the value of θ. Then, the classical SOP is a special case when $\theta = 1$.

On the other hand, the AFE gives a lower bound on Eve's decoding error probability, and it is given by

$$\overline{\Delta} = \mathbb{E}[\Delta] \tag{7}$$

Finally, the AILR provides a notion of how the information is leaked by Eve; then, by assuming a fixed-rate transmission, it can be defined as

$$R_L = \mathbb{E}\left[\frac{I(M; Z^n)}{n}\right] = (1 - \overline{\Delta})R_S \tag{8}$$

Intercept Probability

An intercept event occurs when the secrecy capacity C_S is negative, which means that the wiretap channel has a better SNR than the main channel; it can be expressed as

$$P_{INT} = \Pr[C_S(\gamma_M, \gamma_W) < 0] \tag{9}$$

Probability of Strictly Positive Secrecy Capacity

The probability of strictly positive secrecy capacity (SPSC) is the probability that the secrecy capacity C_S remains higher than 0, which means that security in communication has been attained. Mathematically, it can be written as

$$P_{\text{SPSC}} = \Pr[C_S(\gamma_M, \gamma_W) > 0] \tag{10}$$

Physical-Layer Security Techniques

Artificial Noise Generation

Goel and Negi (2008) proposed a technique where the wire-tap channel is made artificially degraded by injecting an artificial noise (AN). For this purpose, a trustworthy node, which can be Alice, Bob, or a third one, is in charge of sending an interfering signal (jamming) to intentionally degrade the wire-tap channel and thus hampering Eve's chances on gaining any information from the secret message, while the legitimate channel remains unaffected. Thus, by selectively degrading the eavesdropper's channel, secret communication can be guaranteed. In that work, two scenarios are considered, the first one considers that Alice has multiple transmit antennas as illustrated in Figure 2(a), and the second considers relays that simulate the effect of multiple antennas. It was demonstrated that the number of transmit antennas at Alice must be higher than that of the Eve for ensuring that the legitimate channel will not be affected. Then, a nonzero secrecy capacity can be guaranteed by using artificial noise, even if Eve is closer to Alice than Bob. Moreover, Goeckel et al. (2011) employed cooperative relays for generating artificial noise, and proposed a secret wireless communications protocol, where a relay was used for assisting the legitimate transmissions, while a group of relays were employed for jamming the eavesdroppers. Due to the enormous benefits, artificial noise-aided security techniques have been widely used joint with multiantenna and cooperative diversity in order to increase the security performance of wireless networks, as detailed in the following sections.

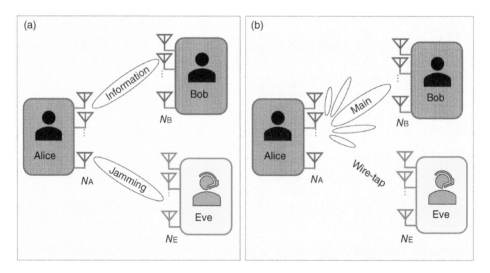

Figure 2 Schematics of MIMO wire-tap channels: (a) artificial noise generated from a multiantenna Alice and (b) beamforming.

Multiantenna Diversity

By exploiting the available spatial dimensions of wireless channels, multiple-input multiple-output (MIMO) systems can mitigate the effects of wireless fading while increasing the secrecy capacity in fading environments. The Gaussian MIMO wire-tap channel was studied by Khisti et al. (2007), where two cases were analyzed: (i) a deterministic case in which the CSIs of both the main and wire-tap links are fixed and known to all nodes and (ii) a time-varying Rayleigh fading scenario, where Alice has perfect CSI for the main channel and as statistical CSI knowledge for the wire-tap channel. For the deterministic case, a scheme based on the generalized singular value decomposition (GSVD) of the channel matrices is proposed, and it was shown that the secrecy capacity can be achieved at the high SNR. For the fading scenario, it was shown that secrecy capacity approaches zero if and only if the ratio of the number of eaves-dropper antennas to source antennas was larger than 2^1. The perfect secrecy capacity was analyzed in Oggier and Hassibi (2011) of the multiple antenna MIMO broadcast channel, by considering an arbitrary number of antennas at all nodes. Mukherjee and Swindlehurst (2011) proposed beamforming-based approaches for improving the secrecy of the wireless communications in MIMO channels as depicted in Figure 2a. The proposed schemes allocate power in order to attain a target SINR for the legitimate channel, and the remaining power is broadcasted as artificial noise in order to avoid the interception from the eavesdropper. It was considered that the CSI from the wire-tap link is not available while the availability of accurate CSI from the main channel is required. An analysis was also performed to quantify the effects of imperfect CSI. Results showed that the proposed schemes perform well for moderate CSI errors, but a large channel mismatch can eliminate the secrecy advantage of using artificial noise. The employment of maximal ratio combining (MRC) technique was analyzed by He et al. (2011), where it was proved that, through channel diversity, a target secrecy capacity can be attained with a reasonably low outage probability. Moreover, Alves et al. analyzed the outage performance of a transmit antenna selection (TAS) scheme in Alves et al. (2012) for the multiple-input single-output multiple-eavesdroppers (MISOME).

Cooperative Diversity

Relay-based wiretap scenarios have drawn intensive attention because, further from providing enhanced reliability and extended coverage, cooperative relaying techniques have proved beneficial for improving the secrecy performance of wireless networks against eavesdropping. Relays can play many different roles to counteract eavesdropping; some examples are depicted in Figure 3. For instance, they may act as traditional relays to assist the legitimate communication, or they may also act as both relay nodes and jammers by introducing AN in order to degrade the wire-tap channel. Moreover, the relays themselves can act as potential eavesdroppers of the confidential communication when they are untrusted. In the following sections, we discuss some relevant works on cooperative relaying techniques for providing secure transmissions.

Trusted Relays

Dong et al. (2010) introduced the use of cooperative relays to improve the secrecy performance of wireless networks. In that work, the widely known relaying protocols,

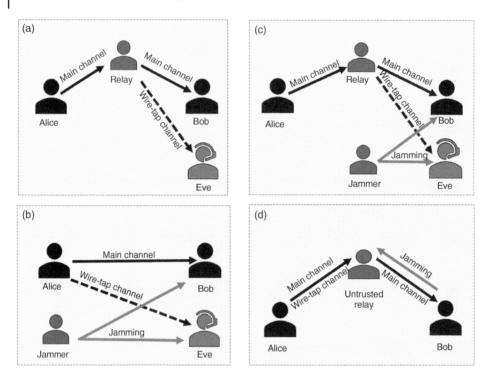

Figure 3 Schematics of cooperative diversity in the presence of an eavesdropper.

namely amplify and forward (AF) and decode and forward (DF), were evaluated by considering the secrecy rate maximization problem and power allocation subject to a power constraint, and the transmit power minimization problem subject to a secrecy rate. They also proposed the so-called cooperative jamming (CJ) technique, in which the relays contribute to provide secrecy by sending AN in order to interfere with the eavesdroppers. Krikidis et al. (2009) proposed an opportunistic relay selection scheme to increase security against eavesdroppers. Therein, it was considered that one relay is selected to assist Alice to send information to Bob using a decode-and-forward protocol. The second relay is used as a jammer to interfere the eavesdropper nodes. Moreover, a hybrid security scheme, which switches between jamming and nonjamming operation, is also discussed. Also, a collaborative beamforming relaying technique to maximize the secrecy rate was proposed in Zhang and Gursoy (2010), under the assumption of perfect CSI knowledge of all links and a total transmit power constraint, while the imperfect CSI knowledge case for multiple eavesdroppers was studied in Vishwakarma and Chockalingam (2012).

Untrusted Relays
The aforementioned works are based on the premise that the eavesdropper is an external node of the network. However, in practical scenarios, not all nodes might have the same rights to access information, even though they agree with participating on the communication process, as the case of untrusted relays. That scenarios have raised a

great interest once determining whether cooperation is beneficial or not is of critical importance for the network deployment (He and Yener 2009, 2010; Osorio et al., 2018; Wang et al., 2016). For instance, He and Yener (2010) considered a system where Alice and Bob rely on a relay node to assist their communication while keeping the information secret. Therein, the authors investigated whether cooperation with an untrusted relay node can ever be beneficial. For that purpose, the achievable secrecy rate was derived for the general untrusted relay channel, and two types of relay networks based on compress-and-forward protocol with orthogonal components were analyzed. The first model considered an orthogonal link from the source to the relay. The second model considered an orthogonal link from the relay to the destination. For the first model, it was found that the untrusted is not beneficial for the network. However, for the second model, by means of the achievable secrecy rate, it was demonstrated that a higher secrecy rate can be attained by relying on the untrusted relay to retransmit information than just treating the relay as an eavesdropper. Therefore, that work opened the path for a copious number of works that investigates different scenarios where untrusted relays are considered. Further, in He and Yener (2009), a positive secrecy rate was obtained by relying on the destination node or an external node to send a jamming signal in a two-hop compress-and-forward relaying network. This technique is referred to as destination-based jamming (DBJ). Moreover, in Osorio et al. (2018), the impact of the direct link on the secrecy outage probability was analyzed for a relaying network with multiple untrusted AF relays, where partial relay selection and DBJ are considered by means of a full-duplex destination. Furthermore, in Wang et al. (2016), a successive relaying scheme was proposed for a multirelay network, where the interrelay interference is used as jamming on the untrusted relays. Therein, the SOP was investigated, and it was shown that maximal secrecy diversity can be obtained by performing optimal relay selection.

Figure 4 illustrates the secrecy outage performance for the four cases depicted in Figure 3, where the advantages on introducing jamming to confuse the eavesdropper can be observed. By far, the worst case is the one where the relay only retransmits information without the help of AN. On the other hand, the best performance is obtained for the case that considers a relay and a jammer, as long as Eve channel is worse than the main channel. Otherwise, if relay-Eve channel is similar to relay-Bob channel, both are benefited from the retransmission of the relay, then the secrecy performance is poor. Also, the use of the untrusted relay joint with DBJ technique offers a good secrecy performance, and then the benefits on having a relay, even though being untrusted, are evidenced.

Physical-Layer Authentication and Secret-Key Generation

Authentication methods target to verify the identity of the legitimate parts, thus preventing two types of spoofing attacks, namely impersonation and substitution. In the former, the attacker sends messages to a legitimate receiver in order to be confused with other legitimate users, while in the latter, the attacker intercepts legitimate messages, modifies them, and then retransmits the altered messages to legitimate users. These methods, traditionally conducted at upper layers, may result in exorbitant latencies in large-scale networks, whereas the limited resources of a massive number

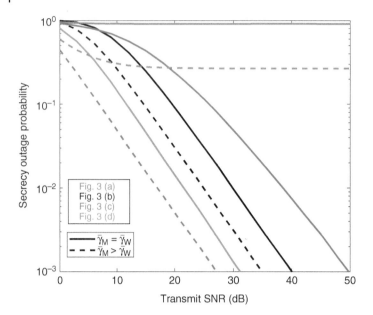

Figure 4 Secrecy outage probability versus transmit SNR for different scenarios with cooperative relays.

of heterogeneous devices from vertical industrial applications will demand robust and lightweight authentication alternatives for designing more secure mMTC networks. Moreover, because digital keys are generally used to identify and provide rights to users, attackers using unauthorized security keys cannot be efficiently detected in those scenarios, when physical-layer properties are overlooked. Therefore, physical-layer attributes of devices and environments, i.e. the so-called physical-layer device fingerprints, can be used to perform authentication with low computational power, energy, and overhead requirements, while being robust as those attributes are hard to be mimicked or predicted. This technique is referred to as physical-layer authentication (PLA) (Wang et al. 2016). Fingerprints can be of two types, channel-based fingerprints or analog front-end (AFE) imperfection-based fingerprints. Channel-based PLA exploits wireless channel parameters such as CSI, received signal strength (RSS), channel frequency response (CFR), and channel impulse response (CIR), as depicted in Figure 5a, in order to design the authentication of devices. As a downside, this approach requires significant channel monitoring, which is subject to imperfect estimates, thus being critical in highly dynamic environments as those of V2X communications. On the other hand, the AFE imperfection-based PLA relies on specific characteristics introduced during the fabrication of devices, including in-phase and quadrature-phase imbalance (IQI), digital-to-analog converter, carrier frequency offset, and power amplifier, among others. In practice, the reliability of estimating differences among the aforementioned attributes is deteriorated due to the noise and dynamic interference conditions.

The authentication process must be carried out periodically during the secret message transmission, within the coherence time of the channel, in order to guarantee a sufficient agreement of the channel signatures. Therefore, due to the time-varying attributes

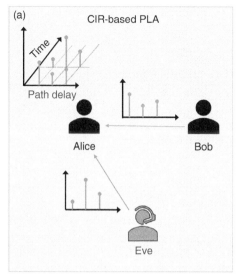
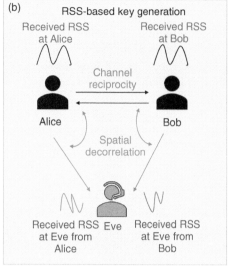

Figure 5 Schematics on (a) channel impulse response-based PLA and (b) receive signal strength-based secret-key generation.

and their imperfect estimation, PLA techniques may be difficult to design and standardize, thus presenting low reliability and accuracy. Although, multiattribute authentication techniques can be used to improve the robustness and accuracy of PLA, by combining a number of selected attributes according to the specific application scenario, thus attaining an increased level of security in the presence of attackers (Wang et al. 2016).

Physical-Layer Key Generation

Physical-layer key generation is based on three principles, namely temporal variation, channel reciprocity, and spatial decorrelation as illustrated in Figure 5b. Temporal variation is introduced by the movement of the transmitter, receiver, or any objects in the environment. Channel reciprocity implies that bidirectional wireless channel states are identical between two transceivers at a given moment in order to generate the same key, which is only valid for time-division duplex (TDD). Spatial decorrelation indicates that the properties of wireless channel are unique to the locations of the transceivers of the legitimate link, and then an eavesdropper at a position more than one-half wavelength away from the legitimate transceivers experiences a different and uncorrelated wireless channel. However, these assumptions may not be satisfied in all the environments. Therefore, physical-layer key generation faces some challenges to be overcome before their efficient use. For instance, there is a generation overhead as the key error-correction process (reconciliation) between the legitimate parties, which is generally attained using polar or low-density parity-check (LDPC) codes, demands a large number of extra bits, which consumes a significant amount of time overhead. Moreover, poor scattering or line-of-sight (LoS) channels, which present low randomness and variations, will present extremely low secret-key rates. Also, in wireless networks with multiple nodes, as those of the MTC scenarios, group key generation schemes are more efficient compared to one-by-one generation methods; however, the key generation process may suffer from a high complexity. Therefore, pursuing novel low-complex and efficient solutions for secret group-key generation schemes from physical-layer characteristics is an appealing research area for providing security in 5G-and-beyond networks.

Physical-Layer Security for 5G Technologies

The 5G new radio (NR) physical layer presents a flexible and scalable design to support diverse use cases with extreme and sometimes contradictory requirements, as well as a wide range of frequencies and deployment options. The key technology components of the NR physical layer are modulation schemes, waveform, scalable numerology, frame structure, multiantenna transmission, multiuser superposition and shared access, and channel coding (Zaidi et al. 2017).

Regarding modulation schemes, 5G NR will support the QPSK, 16-QAM, 64-QAM, and 256-QAM modulation formats. The 3rd Generation Partnership Project (3GPP) has included $\pi/2$-BPSK in the uplink to enable a further reduced peak-to-average power ratio and enhanced power-amplifier efficiency at lower data rates, which is important for mMTC services, for example. Since NR will cover a wide range of use cases, the supported modulation schemes may expand.

Also, the radio waveform is one of the most important specifications at physical layer. Then, the 3GPP agreed to adopt orthogonal frequency-division multiplexing (OFDM) with a cyclic prefix (CP) for both downlink and uplink transmissions, thus enabling low implementation complexity and low cost for wide bandwidth operations. NR also supports the use of discrete Fourier transform (DFT) spread OFDM (DFT-S-OFDM) in the uplink to improve coverage.

Besides, NR supports operation in the spectrum ranging from sub-1 GHz to millimeter wave bands. Scalable numerologies are key to support NR deployment in such a wide range of spectrum. NR adopts flexible subcarrier spacing of $2^\mu \cdot 15$ kHz ($\mu = 0, 1, ..., 4$). This scalable design allows support for a wide range of deployment scenarios and carrier frequencies.

Regarding the frame structure, NR supports time-division duplex (TDD) and frequency-division duplex (FDD) transmissions and operation in both licensed and unlicensed spectra. It enables very low latency, fast hybrid automatic repeat request (HARQ) acknowledgments, dynamic TDD, coexistence with long-term evolution (LTE), and transmissions of variable length (for example, short duration for URLLC and long duration for eMBB). A frame has a duration of 10 ms and consists of 10 subframes, which is also divided into an integer number of slots and/or minislots. Minislots are used to support transmissions with a flexible start position and a duration shorter than a regular slot duration, which facilitate very low latency for critical data as well as minimize interference to other links per the lean carrier design principle that aims at minimizing transmissions.

Moreover, NR will employ different antenna solutions. For lower frequencies, a low to moderate number of active antennas (up to around 32 transmitter chains) is assumed. For higher frequencies, a larger number of antennas can be employed in a given aperture, which increases the capability for beamforming and multiuser MIMO. For even higher frequencies (in the mmWave range), an analog beamforming implementation is typically required to be applied at both the transmitter and receiver ends to combat the increased path loss, even for control channel transmission.

Regarding multiple access, new schemes such as those based on nonorthogonal properties would introduce a "scheduling light" and/or "light initial access" mechanism to significantly reduce the control overhead and access latency in order to efficiently support mMTC. Also, NR employs low-density parity-check (LDPC) codes for the data channel and polar codes for the control channel.

Even though, these key technology components of 5G NR would bring advantages for providing security at the physical layer, as in the case of massive MIMO, new waveforms, or mmWaves. It is important to identify the challenges that these technologies face. In brief, we overview promising technologies for 5G and beyond and its related challenges with respect to PLS. Notice that our overview is by far not comprehensive; our goal is to illustrate the potential of combining PLS techniques into the design of future networks though select technologies.

Massive MIMO

Massive MIMO offers very directed beam patterns to locations of legitimate users, thus maintaining a reduced information leakage to undesired locations. Unlike traditional MIMO, massive MIMO introduces the following challenges: (i) the CSI estimation process is highly complex and (ii) channel models are correlated as the distances of antennas are very shorter than a half of the wavelength. Moreover, the pilot training period for the CSI estimation is vulnerable to attackers that can contaminate the uplink pilot sequences by generating identical pilots in order to modify the estimation, which is referred to as pilot contamination attacks, which is critical in MIMO systems as the eavesdropper can obtain a better SNR after beamforming. In these cases, secret keys extracted from channel estimates can be used to ensure that the eavesdropper will obtain the minimum information on the channel from the information exchange. However, solutions for practical systems need to be further investigated in order to efficiently prevent pilot contamination attacks.

mmWave Communications

The idea of mmWave is to take advantage of the unexploited range of high frequencies, from 3 to 300 GHz, to cope with future multi-gigabit-per-second mobile and multimedia applications. Compared to microwave networks, mmWave networks present some unique and special characteristics, namely larger system bandwidth, very short wavelength, different propagation laws, high directionality by using massive antenna arrays, and short-range transmissions, which can be exploited to further enhance the secrecy performance of future wireless networks.

Full-Duplex Communications

Full-duplex (FD) technology offers both opportunities and challenges for PLS. On the one hand, FD allows Bob to create AN to interfere Eve and receive the information at the same time. On the other hand, an FD Eve can actively attack the receiver in the transmission process while eavesdropping. Besides, FD communications can double the spectral efficiency with regard to the traditional half-duplex communications. Eventhough the management of the strong self-interference is a critical issue, recent advances have proved promising on making FD feasible.

Nonorthogonal Multiple Access

Nonorthogonal multiple access (NOMA) technology is a priority for enabling massive connectivity in 5G-and-beyond networks; thus, providing security to NOMA is of crucial importance. The security issues in NOMA context can be seen from two perspectives. First, by considering external eavesdroppers, the objective is to utilize NOMA transmission structure in a way for providing secrecy. This can be done by optimizing the transmission rates, the power allocation among users, the channel ordering of the NOMA users alongside their decoding order, and introducing artificial noise to enhance the secrecy performance. Second, by considering internal eavesdroppers, the main security concern lies in the fact that NOMA users have to decode other users' signals before being able to decode his own signal by following a process called successive interference cancellation (SIC). It can be noticed that, in such scenario, the transmission should be protected not only from external eavesdroppers, but also from the other internal multiplexed users.

Conclusions and Future Challenges

This article has tackled the main concepts and definitions regarding PLS from the information theory perspective. Secrecy notions as well as some of the most used secrecy performance metrics were described succinctly without the intention of performing an exhaustive study. Moreover, some of the most used PLS techniques and a brief description on the challenges and application of PLS on some important 5G technologies were also presented. In the following sections, some future challenges are tackled for the adoption of PLS techniques in 5G-and-beyond networks.

Practical Channel Models

Accurate channel models are crucial for the correct design of system parameters and system performance evaluation. In this sense, 5G brings huge challenges regarding the search for accurate channel models that efficiently fit with 5G environments. Therefore, future approaches on PLS techniques and metrics should be designed according to the challenges imposed by more practical and accurate channel models, which cover extremely wide frequency bands (Terahertz communications), visible light spectrum, and many new scenarios. Therefore, it is essential to revise PLS techniques and metrics regarding these new channel models. For instance, various PLS techniques are invalidated in poor scattering environments where a strong correlation between legitimate and wiretap channels can exist. Additionally, quasi-static and poor scattering channels can be challenging for secret-key generation.

Secure Waveforms

As previously mentioned, 5G is expected to support new waveforms that meet some specific requirements. Then, it is of notable interest to design new waveforms that are inherently secure. For instance, in Hamamreh and Arslan (2017), a new form of waveform, referred to as secure orthogonal transform division multiplexing (OTDM) waveform, is proposed in order to diagonalize the multipath channel matrix of only the legitimate receiver, while degrading eavesdropper's reception, thus providing an enhanced level of security at the physical layer. The design of secure waveforms is a critical aspect that can be exploited to either secretly transmit a message or extract a secret key between legitimate users. Thus, these kinds of solutions offer lightweight methods to provide security and deserve further investigation.

Physical-Layer Security in Terterahertz communications

Above 100 GHz, the directionality of transmitted signals is highly increased. These high-frequency, narrow angle broadcasts present a more challenging environment for eavesdroppers compared to that of lower frequencies. Even though high-frequency wireless data links are supposed to present increased security, the terahertz eavesdropping needs to be well characterized in order to have a comprehensive notion of security at those frequencies. In Ma et al. (2018), it was demonstrated that an eavesdropper can intercept signals in line-of-sight transmissions, even when they are transmitted at high frequencies with narrow beams. The eavesdropper's techniques are different from those for lower frequency transmissions, as they involve placing an object in the path of the transmission to scatter radiation toward the eavesdropper. One countermeasure involves characterizing the backscatter of the channel in order to detect some, although not all, eavesdroppers. Therefore, physical-layer security will be of extreme importance for terahertz wireless networks.

Physical-Layer Security for UAV-Aided Communications

Unmanned aerial vehicles (UAVs) are expected to play an important role for future mobile networks as enablers of new applications and services by offering several advantages, such as on-demand coverage, dynamic and cost-effective deployment, fast response to service demands, and mobility in three-dimensional (3D) space, and also due to its potential to improve the security of wireless networks by exploiting the characteristics of their flexible deployment and dominant LoS links compared to the ground base stations (BSs). On the other hand, the UAVs can also represent potential breaches to the terrestrial mobile network if they are misused by unauthorized agents for malicious purposes. Malicious UAVs can take advantage of their high mobility and flexibility to track their targets over time, thus overhearing or jamming their communications more effectively. In light of this, another challenging security problem arises regarding the protection of terrestrial communications, since the level of signal power received from the UAVs is higher than that received from ground BSs over a large area due to the dominant LoS links. This makes the prevention of terrestrial eavesdropping cumbersome. Moreover, resorting to the relatively high altitude of UAVs and their high mobility in 3D space, secure communications can be achieved by employing techniques from the PLS theory, such as UAV 3D beamforming, which can enhance the signal reception at the legitimate link, or by using ABS as aerial jammers to degrade the signal received at the eavesdroppers by sending an artificial noise.

Machine-Learning Techniques for Physical-Layer Authentication

By considering single-attribute-based PLA schemes, the performance is limited by the imperfect estimates of the considered attribute or variations of the physical layer, which can lead to low reliability and low robustness of the PLA. However, a multiple attribute-based authentication scheme is capable of achieving high security in the presence of adversaries, but this increases the challenges imposed on the legitimate users. For that purpose, adaptive and near-instantaneously PLA is more suitable for rapidly time-varying environments. Then, ML can be used to design learning-aided intelligent authentication approaches that can work with multidimensional attributes in order to provide security enhancement and more efficient management in 5G-and-beyond networks (Fang et al. 2018). However, some issues need to be considered for achieving effective solutions, namely (Krishna Sharma and Wang 2018): (i) the time consumed to the convergence of the selected learning technique may reduce the time for data transmission; then, this trade-off should be considered for the design, (ii) distributed implementation of the learning algorithm across multiple learning devices, (iii) parameters such as learning rate, discount rate, and exploration/exploitation trade-off should be dynamically adapted to enhance the performance of reinforcement learning algorithms in highly dynamic environments, and (iv) the heterogeneity of MTC devices, for instance, must be taken into account in terms of learning capability, cache size, delay tolerance, and data rate.

Acknowledgment

This work has been financially supported by the São Paulo Research Foundation (FAPESP) Proc. No 2017/20990-6, the Academy of Finland 6Genesis Flagship (grant 318927), the EE-IoT (grant 319008), and the Brazilian National Council for Scientific and Technological Development (CNPq) Proc. No 428649/2016-5.

Related Article

5G Security – Complex Challenges

Endnote

1 It is worthwhile to mention that a multiantenna eavesdropper can be seen as multiple single-antenna eavesdroppers.

References

3GPP (2017). Study on new radio (NR) access technology physical layer aspects, TR 38.802, March 2017.

Alves, H., Souza, R.D., Debbah, M., and Bennis, M. (2012). Performance of transmit antenna selection physical layer security schemes. *IEEE Signal Processing Letters* 19 (6): 372–375.

Barros, J. and Rodrigues, M.R.D. (2006). Secrecy capacity of wireless channels. 2006 IEEE International Symposium on Information Theory, July 2006, 356–360.

Csiszar, I. and Korner, J. (1978). Broadcast channels with confidential messages. *IEEE Transactions on Information Theory* 24 (3): 339–348.

Dong, L., Han, Z., Petropulu, A.P., and Poor, H.V. (2010). Improving wireless physical layer security via cooperating relays. *IEEE Transactions on Signal Processing* 58 (3): 1875–1888.

Fang, H., Wang, X. and Hanzo, L. (2018). Learning-aided physical layer authentication as an intelligent process. *IEEE Transactions on Communications*, 67 (3): 2260–2273.

Goeckel, D., Vasudevan, S., Towsley, D. et al. (2011). Artificial noise generation from cooperative relays for everlasting secrecy in two-hop wireless networks. *IEEE Journal on Selected Areas in Communications* 29 (10): 2067–2076.

Goel, S. and Negi, R. (2008). Guaranteeing secrecy using artificial noise. *IEEE Transactions on Wireless Communications* 7 (6): 2180–2189.

Hamamreh, J.M. and Arslan, H. (2017). Secure orthogonal transform division multiplexing (OTDM) waveform for 5g and beyond. *IEEE Communications Letters* 21 (5): 1191–1194.

He, X. and Yener, A. (2009). Two-hop secure communication using an untrusted relay. *EURASIP Journal on Wireless Communications and Networking* 2009 (1): 305146. doi: 10.1155/2009/305146.

He, X. and Yener, A. (2010). Cooperation with an untrusted relay: a secrecy perspective. *IEEE Transactions on Information Theory* 56 (8): 3807–3827.

He, F., Man, H. and Wang, W. (2011). Maximal ratio diversity combining enhanced security. *IEEE Communications Letters* 15 (5): 509–511.

He, B., Zhou, X. and Swindlehurst, A.L. (2016). On secrecy metrics for physical layer security over quasi-static fading channels. *IEEE Transactions on Wireless Communications* 15 (10): 6913–6924.

Khisti, A., Wornell, G., Wiesel, A., and Eldar, Y. (2007). On the Gaussian MIMO wiretap channel. 2007 IEEE International Symposium on Information Theory, June 2007, 2471–2475.

Krikidis, I., Thompson, J.S. and Mclaughlin, S. (2009). Relay selection for secure cooperative networks with jamming. *IEEE Transactions on Wireless Communications* 8 (10): 5003–5011.

Krishna Sharma, S. and Wang, X. (2018). Towards massive machine type communications in ultra-dense cellular IoT networks: current issues and machine learning-assisted solutions. *arXiv e-prints* arXiv:1808.02924.

Leung-Yan-Cheong, S. and Hellman, M. (1978). The gaussian wire-tap channel. *IEEE Transactions on Information Theory* 24 (4): 451–456.

Ma, J., Shrestha, R., Adelberg, J. et al. (2018). Security and eavesdropping in terahertz wireless links. *Nature* 563 (7729): 89–93.

Maurer, U. and Wolf, S. (2000). Information-theoretic key agreement: from weak to strong secrecy for free. In: Advances in Cryptology-EUROCRYPT 2000, Lecture Notes in Computer Science, vol. 1807, (ed. B Preneel), 351–368, Springer: Berlin.

Mukherjee, A. and Swindlehurst, A.L. (2011). Robust beamforming for security in MIMO wiretap channels with imperfect CSI. *IEEE Transactions on Signal Processing* 59 (1): 351–361.

Oggier, F. and Hassibi, B. (2011). The secrecy capacity of the MIMO wiretap channel. *IEEE Transactions on Information Theory* 57 (8): 4961–4972.

Osorio, D.P.M., Olivo, E.E.B., and Alves, H. (2018). Secrecy performance for multiple untrusted relay networks using destination-based jamming with direct link. 2018 IEEE 29th Annual International Symposium on Personal, Indoor and Mobile Radio Communications (PIMRC), September 2018, 1–5.

Poor, H.V. and Schaefer, R.F. (2017). Wireless physical layer security. *Proceedings of the National Academy of Sciences* 114 (1): 19–26. https://www.pnas.org/content/114/1/19.

Shannon, C.E. (1949). Communication theory of secrecy systems. *The Bell System Technical Journal* 28 (4): 656–715.

Vishwakarma, S. and Chockalingam, A. (2012). Decode-and-forward relay beamforming for secrecy with imperfect csi and multiple eavesdroppers. 2012 IEEE 13th International Workshop on Signal Processing Advances in Wireless Communications (SPAWC), June 2012, 439–443.

Wang, X., Hao, P., and Hanzo, L. (2016). Physical-layer authentication for wireless security enhancement: current challenges and future developments. *IEEE Communications Magazine* 54 (6): 152–158.

Wyner, A.D. (1975). The wire-tap channel. *The Bell System Technical Journal* 54 (8): 1355–1387.

Zaidi, A.A., Baldemair, R., Andersson, M. et al. (2017). Designing for the future: the 5G NR physical layer. Ericsson Technology Review. https://www.ericsson.com/en/ericsson-technology-review/archive/2017/designing-for-the-future-the-5g-nr-physical-layer.

Zhang, J. and Gursoy, M.C. (2010). Collaborative relay beamforming for secrecy. 2010 IEEE International Conference on Communications, May 2010, 1–5.

Zhou, X., McKay, M.R., Maham, B., and Hjorungnes, A. (2011). Rethinking the secrecy outage formulation: a secure transmission design perspective. *IEEE Communications Letters* 15 (3): 302–304.

Further Reading

Ahmad, I., Shahabuddin, S., Kumar, T. et al. Security for 5G and beyond. *IEEE Communications Surveys and Tutorials*. doi: 10.1109/COMST.2019.2916180.

Duong, T.Q., Zhou, X., and Poor, H.V. (ed.) (2017). *Trusted Communications with Physical Layer Security for 5G and Beyond*. The Institution of Engineering and Technology, ISBN-10: 1-78561-235-2; ISBN-13: 978-1-78561-235-0.

Wang, N., Wang, P., Alipour-Fanid, A. et al. (2019). Physical-layer security of 5G wireless networks for IoT: challenges and opportunities. *IEEE Internet of Things Journal* 6 (5): 8169–8181.

Wu, Y., Khisti, A., Xiao, C. et al. (2018). A survey of physical layer security techniques for 5G wireless networks and challenges ahead. *IEEE Journal on Selected Areas in Communications* 36 (4): 679–695. doi: 10.1109/JSAC.2018.2825560.

5

Security for Handover and D2D Communication in 5G HetNets

Alican Ozhelvaci and Maode Ma

Nanyang Technological University, Singapore, Singapore

Introduction

In recent years, many people know that 3G and 4G thanks to their smartphones. But, the technology will not stop there; therefore, we are now talking about the next generation of mobile communication technology: 5G. With every new generation of wireless mobile networks that start from 1G and continued with 2G, 3G, 4G, and 4G LTE have delivered faster speeds, more functionality, and services to the mobile devices. However, technological development does not stop there and even continues with much faster speed; therefore, we are going into a new era with the development of new wireless technology called the 5G, which means the fifth generation of mobile communication technology. In reality, the 5G is expected to bring solutions to the problem of the 4G and complete 4G technology. However, the truth is even experts couldn't tell us what 5G actually is because they do not even know yet as of in 2017. John Smee, VP of Engineering at Qualcomm, said in early 2017 that "Perhaps the most exciting thing about 5G is that we are designing a new wireless network fabric, unlocking the potential for a host of services that we have not even imagined yet." Thus, there are several reasons and needs to develop this new-generation technology such as 4G will reach its limit soon because more users and devices are connecting to the mobile network, demand on faster data rates by users, and connecting internet of things (IoT) devices to each other and work together over a wireless mobile network. Thus, the industry and academia work together to achieve the need for demand and the requirements of 5G. The expectation is that around 2020, these next-generation mobile networks will be deployed. Researchers are investigating how to serve 1000 or more times more devices and data traffic, and achieve 1 ms latency (Jungnickel et al. 2014). 5G will bring other new features, including integration of heterogeneous networks (HetNet), network security, provisioning, extreme reliability, more bandwidth, and 90% less energy consumption (Khodashenas et al. 2016). The official standards have been issuing by standardization institutions, such as WiMAX Forum, Third Generation Partnership Project (3GPP), ITU-R, or IEEE.

With the fast development of computer networks, mobile telecommunication networks, electronic technology, and control technology as well as the increased intelligent requirements in the modern industry, the need to become more compatible

The Wiley 5G REF: Security. Edited by Rahim Tafazolli, Chin-Liang Wang, Periklis Chatzimisios and Madhusanka Liyanage.

with information and the Internet on physical devices are quickly emerging (Agiwal et al. 2016). However, there is still a serious gap between the mobile world, where information is exchanged and transformed, and the physical world in which we live. In this regard, the evolution of mobile communication networks is getting faster by the new requirement for mobility, faster speeds, more performance, more energy efficiency, more coverage but less latency and lower energy consumption. To meet the requirements mentioned before, we are heading to the 5th generation mobile networks, shortly 5G, which are the next generation of mobile communication systems beyond 4G or 4G/LTE networks (Panwar et al. 2016). The next-generation wireless network is being developed and it is not only evolution of the legacy other mobile networks such as 3G and 4G but also with advanced wireless and networking technologies, it will be a system with many capabilities. What we expect from the new and next-generation wireless mobile communications network and what it will enable for us. Firstly, all research and developments aim to bring advanced technologies with 5G, such as software-defined networking (SDN), network function virtualization (NFV), and network slicing. Second, 5G will have higher capacity than 4G LTE, much denser in terms of the number of users and user equipment (UE), should support device-to-device (D2D) and massive machine-type communications (Zhang et al. 2019). 5G will be a better connection option for the implementation of the IoT in terms of very low latency and low energy consumption (Andrews et al. 2014). Moreover, 5G mobile network is expected to enable these eight high-level features: the speed of 5G is expected to be between 1 and 10 Gbps, the latency will be lower than 5 ms, have at least 1000 times bandwidth than 4G, number of connected devices will be 10 to 100 times higher, all-time availability, 100% coverage, reducing the usage of network energy, and more secure or the securest network compared with last generations (Gupta and Jha 2015). With the 5G era, new various network services will utilize what 5G enables for them. These services are vehicle-to-vehicle communication, machine-to-machine communication, smart grid, smart cities, smart homes, smart nations, blockchain-based services, mobile fog computing, eHealth services, unmanned aerial vehicle (UAV), and so on as shown in Figure 1. Hence, the expectation for the commercialization of 5G is around 2020, and after that, gradually these next-generation mobile networks will be deployed.

As mentioned above, to achieve the requirements of 5G in terms of performance, connection, coverage, latency, and security, there are technologies being developed such as millimeter waves (mmWave) and massive MIMO for performance, SDN, NFV, and network slicing in the backhaul (Zhang et al. 2015), lastly HetNets to support legacy networks. Also, this next-generation wireless communication system will enhance mobile broadband with massive IoT, D2D communication, and crucial services. Therefore, a general architecture of 5G mobile networks would be as shown in Figure 2. Furthermore, the next generation of mobile communication 5G, which will allow wireless and wired systems to monitor different environmental and physical conditions without human intervention, will be the major communication technology used in the most existing network system. Consequently, the 5G will have new architecture and its new technologies and new use-cases will make us face new challenges regarding security and privacy preservation (Liyanage et al. 2016). Therefore, security mechanisms should meet the following requirements to establish and have secure 5G HetNet communication (Chen et al. 2016a):

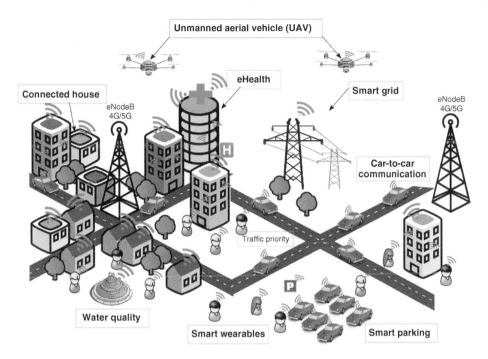

Figure 1 What will the 5G enable?

- *Confidentiality*: Confidentiality prevents private sensory data from being eaves-dropped by passive attackers, which can be achieved by using authentication and cryptographic encryption to ensure that the information transmitted in the 5G mobile communications can be viewed only by authorized parties.
- *Integrity*: Data transmitted in the 5G mobile communication system cannot be illegally altered by unauthorized participants before reaching the right recipient. Integrity, which can be achieved by employing cryptographic integrity checks, is critical in 5G mobile communications since tampering with the information could result in severe consequences.
- *Authentication*: Though integrity ensures that the data cannot be illegally altered in transmission, it cannot know if the recipient is authorized to access the private information. As a prerequisite for the secure 5G mobile communications, authentication allows the authenticator to identify whether the base stations and access points are trustworthy in the 5G HetNets.
- *Access control*: Access control mechanisms built on identification and authentication ensures that the accesses to critical and sensitive information are restricted to authorized 5G HetNet parties.
- *Availability*: Since wireless radio communication is often used in the 5G HetNet communication system, communication availability can be suffered from jamming attacks. To defend against jamming attacks, a certain mechanism should be applied in the system to provide availability, which ensures that data should be available whenever the authorized users' access the authenticator.
- *Privacy*: For some privacy-sensitive 5G applications, the protection of private information is important. For example, in e-healthcare systems, if the patients' health files

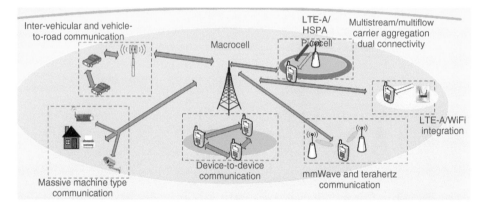

Figure 2 The general architecture of the 5G.

are illegally obtained by unauthorized third parties, it would lead to undesirable negative effects on patients' lives.

The security requirements mentioned above are also the requirements for the 5G HetNet communications to overcome the security vulnerabilities under various security threats, which have been specified by the 5G Infrastructure Public–Private Partnership (5GPPP) as follows (Alcaraz-Calero et al. 2017):

- Communication at 5G, data that will not be confined to individual customers, such as 3G and 4G, will be communicated to many vertical industries (IoT applications such as connected vehicles, supply chains, and smart cities will also be connected to 5G networks).
- 5G networks will process information (such as IMSI, MSISDN, IMEI) that contains confidential data and identities of users. This information can be accessed by many networked applications, potentially jeopardizing user privacy.
- Instead of the ad hoc security methods used in today's networks, suggested methods that provide end-to-end data security may be needed in 5G networks.
- Protocol attacks on the devices are the attacks on the protocol stack at the devices, including man-in-the-middle attack (MITM) on first network access, Denial of Service (DoS) attacks, compromising a device by exploiting weaknesses of the active network services, and attacks on over-the-air management (OAM) and its traffic.
- Attacks on the core network are the main threats to the SDN of ISPs, including the impersonation of devices, traffic tunneling among the impersonated devices, misconfiguration of the firewall in the modem/router/gateways, and DoS attacks against the core network. They may also include changing the device's authorized physical location in an unauthorized fashion or attacks on the radio access network by using a rogue device.
- User data and identity privacy attacks include eavesdropping of other users' or devices' data sent over the UTRAN or EUTRAN, masquerading as other users/subscriber devices, users' network ID or other confidential data revealed to unauthorized third parties.

- Especially with the development of IoT applications, remote access protocols will become widespread and more advanced authentication protocols will be needed to prevent unauthorized access to the devices.
- The high diversity of data to be transmitted over 5G networks will require that the encryption techniques to be used and the secure communication protocols to be designed/adapted to be contextually adaptable.

The requirements created by new areas of 5G technology (short latency, high bandwidth, long battery life, etc.) may conflict with high-security techniques. Besides the specification made by the 5GPPP security workgroup, the main security requirements of 5G mobile networks in different 5G application areas have been discussed in 3rd Generation Partnership Project (2018a) by 3GPP organizations, in which the main security threats, corresponding approaches, and future directions in the context of the reference stack of 5G wireless communication technologies have been also explored. Also, to ensure the security of 5G networks the standardization body of 3GPP has analyzed the security issues and the effect on the UE, NR (new radio) Node B (gNB), and the 5GC (Core) network entities using the postquantum era symmetric and asymmetric, and worked on the application of 256-bit key length encryption algorithms including with key derivation, AKA (Authentication and Key Agreement) key generation, distribution, refreshing, and size negotiation of key, lastly, handling the information of confidential CP (Control Plane)/UP (User Plane)/MP (Management Plane) in the release of 3GPP TR 33.841 (3rd Generation Partnership Project 2018b).

In recent literature, there are some investigations that have been done on security in 5G mobile networks in terms of security architecture, requirements, mechanisms, authentication, and privacy protection. Table 1 shows the comparison of our paper with previous survey papers regarding what areas they cover in 5G systems (5GSs) (Schneider and Horn 2015; Vij and Jain 2016; Ferrag et al. 2018; Ozhelvaci and Ma 2018; Fang et al. 2017). In Schneider and Horn (2015), the authors have addressed the potential security requirements and mechanisms on 5G mobile networks. The article mentions the confidentiality, integrity, and privacy of users' and devices' data or information and therefore, robustness against DoS attacks in 5G networks. Additionally, the security issues that can be happened by the new architecture using the virtualization and SDN are also analyzed. In the next survey (Vij and Jain 2016), the authors address how to maintain the security of the mobile computing environment better than LTE networks for the 5G. Also, the

Table 1 Comparison of areas covered by these papers.

Scheme	5G network security architecture	Handover security	D2D communication security	Heterogeneous networks
Schneider and Horn (2015)	Mentioned	No	No	No
Vij and Jain (2016)	Not mentioned	No	No	No
Ferrag et al. (2018)	Mentioned	Yes	No	No
Ozhelvaci and Ma (2018)	Mentioned	Yes	No	Yes
Fang et al. (2017)	Mentioned	Yes	Yes	Yes
Our Paper	Mentioned	Yes	Yes	Yes

security architecture is discussed for the future development of the 5G network by analyzing the security of the 5G network in five different features. In the survey (Ferrag et al. 2018), the authors are addressing the existing authentication and privacy-preserving schemes for 4G and 5G mobile networks and classifying the attack models into four categories, including privacy, authentication, availability, and integrity attacks and also providing the countermeasures with three types of categories that are cryptography, intrusion detection methods, and human factors. Moreover, there is a discussion and recommendation for the future direction of the 5G networks based on the current surveys. In the next paper (Ozhelvaci and Ma 2018), the authors have proposed a secure and efficient vertical handover authentication scheme for 5G HetNet by utilizing the SDN technology in the architecture because the phenomena of handover will occur hundred or more times than the existing legacy network because of the small cell deployment on the 5G networks. Therefore, the paper offers a lightweight and fast vertical handover scheme to meet the latency requirement of 5G networks. In Fang et al. (2017), the survey paper presents a comprehensive study on the security of 5G mobile networks comparing with the existing legacy networks. It starts with the potential attacks and security services which are summarized with the requirement and new use-cases of 5G mobile networks and investigates the security services into five categories, including authentication, availability, confidentiality, key management, and privacy. In addition to that, the new and existed technologies involving security are discussed, such as HetNet, D2D communications, massive MIMO, SDN, and IoT. Finally, they proposed a new security architecture for 5G networks with a handover procedure and a signaling load scheme.

We present a comprehensive security survey on 5G mobile networks, and analysis has been done on the architecture, attacks and security mechanism, and heterogeneous environment cases of 5G networks. The main contributions of this article are described as follows: (i) We start by drawing a general overview of the security architecture and its features in the 3GPP 5G networks. (ii) The main focus of the article is on the security directions in the 5G, including potential threats and attacks and the corresponding countermeasures which are categorized into five types, including authentication, cryptography, data confidentiality, availability, integrity, and key management. (iii) We analyze the security evaluation techniques and discuss the existing tools for the evaluation of security schemes or protocols. (iv) Finally, we give a security investigation on HetNets in terms of handover phenomena and D2D communications because of the heterogeneity of the 5G networks.

The remainder of this article is organized as follows. In the section titled "Security Architecture Overview" presents a general overview of the network and security architecture of 5G networks done by the 3GPP group. In the section titled "Security Attacks and Mechanisms in 5G" explains the potential and crucial attacks that need to be considered in 5G network security, security mechanisms that are used or going to be used in the 5G, and lastly, the tools that are used to evaluate and analyze the security protocols. In the section titled "Security in 5G Heterogeneous Networks", we provide security analyses in terms of features, functionalities, and vulnerabilities with the corresponding solutions and open research topics in the heterogeneous environment cases of 5G, such as handover phenomena and D2D communications. Finally, a conclusion will be drawn in the last section.

Security Architecture Overview

3GPP 5G Network Architecture

The 5G System (5GS) architecture contains mainly two parts: 5G Core Network (5GCN) and 5G Access Network (5G-AN), as shown in Figure 3. In the 5GS, there are entities that are Access and Mobility Management Function (AMF), Authentication Server Function (AUSF), Session Management Function (SMF), User Plane Function (UPF), and other new network functions. The UE communicates with the 5G base station gNB through the 5G radio access network (5G (R)AN), and to perform mutual authentication, the AMF will communicate with UE on behalf of the AUSF in the 5GC.

The 5GS architecture is defined to support data connectivity and services enabling deployments to use techniques, such as NFV and SDN. Thus, the 5GS has new network functions and entities called as 5G Network Function (NF) when comparing with the current LTE/LTE-A system. These 5G NFs are Structured Data Storage Network Function (SDSF), Unstructured Data Storage Function (UDSF), Network Exposure Function (NEF), Network Repository Function (NRF), Network Slice Selection Function (NSSF), Policy Control Function (PCF), Unified Data Management (UDM), Unified Data Repository (UDR), UE radio Capability Management Function (UCMF), Application Functions (AF), and CHarging Function (CHF). Exchanging information can be done between all network functions and entities when it is necessary. As seen in Figure 4, the 5GCN that integrates different Access Types, e.g. 3GPP access and non-3GPP access also support

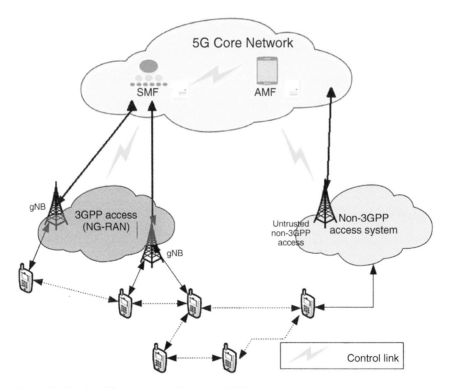

Figure 3 The simplified system architecture of 5G.

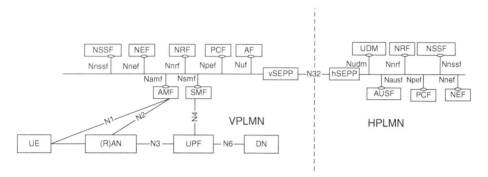

Figure 4 The interoperator security and the SEPP.

the connection of the non-3GPP access network, such as a Wireless Local Area Network (WLAN). If a UE connects to a 5GCN through a non-3GPP access network, the connection is done through the non-3GPP Inter-Working Function (N3IWF) in the 5GCN. The 5G networks are designed to support large-scale IoT, D2D, and V2X communications, for example, eMTC (enhanced machine-type communication) and NB-IoT (narrowband IoT) for IoT applications, for D2D communication; the proximity devices can communicate directly through the D2D communication channel without making a connection with the server, and for V2X communication, there is V2X control function and a V2X device can connect to the 5G network through this function. Also, supporting the communications of IoT, D2D and V2X improve resource utilization and network capacity thanks to modularizing the function design, which enables flexible and efficient network resource utilization.

3GPP 5G Security Architecture

5GS brings security enhancements compared to 4G/LTE systems and the major differences between 5G and 4G security are as follows: the trust model, key hierarchy, interoperator network security, privacy, and service-based architecture security. There are eight key points that are enhanced over 4G/LTE system security:

1) *Primary Authentication*: The mutual authentication between the UE and Service network in 5G is done by primary authentication. There are similarities between 4G and 5G, but essentially there are some differences. Basically, there are two authentication schemes which are mandatory in 5G Phase 1, namely 5G-AKA and Extensible Authentication Protocol (EAP) such as EAP–AKA. In addition, primary authentication can be done over non-3GPP technologies such as WLAN because it does not depend on radio access technology.
2) *Secondary Authentication*: This particular authentication is meant for optional EAP-based secondary authentication between the UE and a data network (DN) outside of the home domain. Thus, 5G supports different EAP-based authentication methods.
3) *Interoperator Security*: There are several security issues in the interoperator interface since the early generations of mobile communication. Thus, to resist against attacks, the SEPP is introduced as an entity during 5G phase 1 inside the PLMN that is shown in Figure 4.

4) *Privacy*: Unlike the early generation of mobile networks, 5G supports the privacy of permanent identifiers of users to protect against active attacks on the air interface as a solution to the issue of subscriber identity protection. To do privacy security, a home network public key is used.

5) *Service-Based Architecture*: The 5GCN is designed as a service-based architecture, which is never been used in the design of early generations, such as in 4G. Therefore, adequate security is supplied for the SBA by the 5G.

6) *Central Unit (CU)–Distributed Unit (DU)*: The CU and the DU are the logical split-tings in the base station of 5G and there is an interface that connects them to each other. So, there is a need for security in this interface in the 5G. Thus, the security is supplied by not giving access to the DUs when the confidentiality of user data pro-tection is enabled. The air interface security has remained the same in 5G with 4G, even the separation of CU–DU.

7) *Key Hierarchy*: The 5G hierarchy reflects the changes in the overall architecture and the trust model using the security principle of key separation. The primary security source is the long-term secret key (K) in the USIM and the 5G same as in the 4G networks. One of the main differences is the possibility of the integrity protection of the user plane between the 5G and the 4G systems.

8) *Mobility*: Overall the mobility remains the same as the 4G in the 5G networks; how-ever, one of the main differences is that mobility and security anchors can be sepa-rated in the core network depending on the carrier's requirement.

Security Attacks and Mechanisms in 5G

Security is the major concern of most 5G applications with high demands of appropri-ate security mechanisms to protect the 5G communication systems. Due to the flex-ible, open, and highly heterogeneity of the network with massive small cell deploy-ment and overlay coverage, it is harder to provide security features such as authenti-cation, integrity, and confidentiality. Therefore, new security schemes are supposed to be applied to a verity of new cases and the new networking paradigms, such as SDN, NFV, and network slicing (NOKIA 2017). Thus, in this section, we introduce the key security mechanisms and solutions currently available used and will be used in the 5G mobile communication network and a summary of the requiments of 5G in terms of security is shown as in Table 2.

Attacks in 5G

Attacks on security can be divided into two types which are passive attack and active attack. During a passive attack, attackers try to obtain or use information from legiti-mate users, but they do not want to attack the communication itself. There are mainly two types of popular passive attacks on a cellular network, namely eavesdropping and traffic analysis. Passive attacks are in the nature of monitoring and intended to violate data privacy, confidentiality, and user privacy. Unlike passive attacks, active attacks can involve interruption of legitimate communications or alternation of the data, such as replay attack, an MITM, masquerade, DoS attack, and distributed

Table 2 Security requirements for 5G HetNets.

Requirements with respect to 4G	Improve resilience and availability of the network against signaling-based threats including overload caused maliciously or unexpectedly
	Specific security design for use-cases, which require extremely low latency
	Comply with security requirements defined in 4G 3GPP standards. Need to apply especially to a virtualized implementation of the network
	Provide Public Safety and Mission Critical Communications (resilience and high availability)
Requirements from a radio access perspective	Improve system robustness against smart jamming attacks and security for 5G small cell nodes

denial of service (DDoS) attack. The security mechanisms used to address security attacks can be basically divided into two categories: physical layer security (PLS) approaches and cryptographic approaches with new network protocols. The most commonly used security mechanisms are cryptographic techniques and normally used in the upper layers of 5G wireless mobile networks with new network protocols. The modern cryptography mainly has been divided into two categories: symmetric (shared)-key cryptography and public (asymmetric)-key cryptography. Symmetric-key (shared) cryptography accredits to the encryption methods in which a secret key is shared between a sender and a receiver. Public-key (asymmetric) cryptography uses two different keys, where one is used as the public key for encryption and the other one is used as the secret key for decryption. A security service performance depends on the computational complexity of the algorithms and the key length. Figure 5 shows all the four attacks and each of them is discussed separately from the following three perspectives: the type of attack (passive or active), the security services provided to combat this attack, and the corresponding methods applied to intercept or prevent this attack. Our focus is on security attacks on the physical layer and the MAC layer, which is the main difference in security between wireless and wired networks.

Eavesdropping and Traffic Analysis

Eavesdropping is an attack that secretly listening to private communication and is used by an unintended receiver to intercept a message from others without their consent. This type of attack has become one of the major threats in wireless networks. As shown in Figure 5a, eavesdropping is a passive attack as normal communication is not affected by eavesdropping. Detecting eavesdropping is very difficult to detect due to it a passive attack. Encryption of the signals via a wireless connection is most extensively used to fight against the eavesdropping attack. Because of the encryption, the eavesdropper cannot directly intercept the received signal. Traffic analysis is another passive attack that an attacker uses to block information such as the location and identity of the communication parties by analyzing the traffic of the signal that is received by the intruder without understanding the content of the received signal. If we have encryption protection in place, a passive attacker might still able to observe the pattern of communication messages. In other words, traffic analysis can still be used to reveal the patterns of the communication parties even the signal is encrypted. The attacker could determine the location and identity of communicating hosts and could observe the frequency

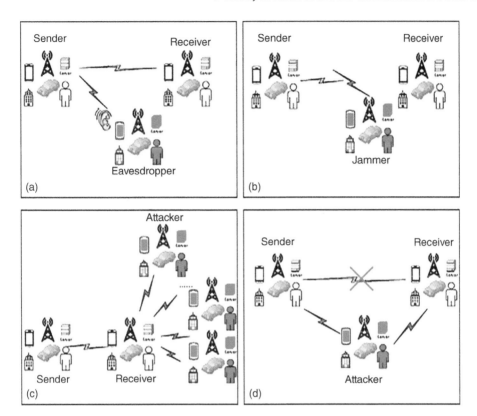

Figure 5 Attack in 5G mobile networks: (a) eavesdropping, (b) jammer, (c) DDoS, and (d) MITM.

and length of the messages being exchanged, where the information could be useful in guessing the nature of the communication that was taking place. Traffic analysis attack does not impact legitimate communications either. Moreover, because of some technologies applied to 5G HetNet wireless networks may further increase the difficulty to fight against eavesdroppers. Generally, the new characteristics of 5G mobile networks may cause many more complicated scenarios to cope with eavesdroppers, for instance, in Chen et al. (2016b), eavesdroppers with multiple antennas are considered.

Jamming
Unlike eavesdropping and traffic analysis, jamming can completely disrupt or block legitimate communications between authorized users. Figure 5b is an example of the jamming attack. This concept can be used to disrupt communications in wireless networks. Jamming occurs when the malicious node device generates intentional interference that can block or disrupt the data communications between authorized users. Jamming can also prevent authorized users from accessing network resources. The solutions for active attacks are normally detection based. These solutions are spread spectrum techniques including frequency hopping and direct sequence spread spectrum (FHSS and DSSS), and these solutions are widely used to fight against jamming and secure communication. But these methods may not be a solution for some applications in 5G networks to prevent jamming.

DoS and DDoS

An adversary can exploit the network resources by launching DoS attacks. DoS is a cybersecurity attack and a serious violation of the availability of the networks. One of the examples is jamming and it can be used to initiate a DoS attack. DDoS can be created when more than one distributed adversary exists. A DDoS model is shown in Figure 5c. DDoS can be launched from many different sources to flood the victim's traffic. DoS and DDoS are both active attacks that can be applied at different layers and it is almost impossible to stop the attack of DDoS by blocking simply a source because incoming traffic is from many different sources. In these days, the method to recognize DoS and DDoS attacks is detection, which is the most commonly used method. DoS and DDoS will likely become a serious threat for carriers with a high penetration of massive devices in 5G wireless networks (Alliance 2016). DoS and DDoS attacks can attack to block the access network via a very large number of connected devices in 5G wireless networks. Based on the attacking target, a DoS attack can be categorized into two types which is a network infrastructure DoS attack and a device/user DoS attack. If an attacker launches a DoS attack against the network infrastructure, it can strike the user plane, signaling plane, management plane, support systems, radio resources, and logical and physical resources (Alliance 2016). Also, if an attacker launches a DoS attack against device/user, it can target on battery, memory, disk, CPU, radio, actuator, and sensors.

Man in the Middle Attack

MITM is an attack that the attacker can secretly capture communication messages and most likely alter the messages between two entities. The most serious attach among the abovementioned attacks is the MITM. With MITM attack, the attacker secretly captures control of the communication channel between two legitimate entities. The MITM attacker can intercept, modify, and replace the communication messages between the two registered devices. To send new messages, the attacker must capture all connected messages going between two entities. Figure 5d is an illustration of a MITM attack model. MITM is an active attack that can be launched in different layers. Especially, MITM attacks target to compromise data confidentiality, integrity, and availability. The MITM attack is one of the most common security attacks according to Verizon's data investigation report (Baker et al. 2011). In the legacy cellular network, a malicious base station–based MITM is an attack that the attacker forces a registered user to create a connection with a malicious base transceiver station (Conti et al. 2016). Mutual authentication between the registered user's device and the base station is normally used to prevent MITM attacks coming from the malicious base station. Also, with the MITM attack, the attacker can initiate the abovementioned attacks to intercept keys and messages by unintended receivers.

Key Security Technologies in 5G

Software-Defined Networking (SDN)

SDN is a new network structure that separates or decouples the network's data forwarding and control planes and allows control of the network via a centralized controller (Dabbagh et al. 2015). SDN is based on an idea to centralize network management and to increase its agility and programmability to meet the requirements of emerging applications (Duan and Wang 2015). On the other hand, to reduce costs and boost performance,

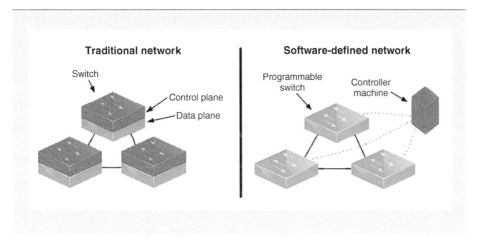

Figure 6 An architecture model of the SDN.

to operate and instantiate network and its services can be provided by emerging SDN and NFV (Guerzoni et al. 2014). In recent years, SDN has gained big attention as an emerging technology for the future wireless mobile network 5G. It has a different architecture than traditional network architecture, as mentioned above, the SDN network control plane is decoupled from data forwarding and has a feature of programmability. This feature of SDN has a big help for the simplification of network configuration and reconfiguration (Luo et al. 2015). Also, it brings very good opportunities to network and security management regarding flexibility and programmability.

Nowadays, large network producers like Cisco are announcing the network infrastructure that supports SDN (Inc 2013). Also, Google is using SDN to manage its data centers' network infrastructure (Jain et al. 2013).

NFV brings a new way to manage, implement, and design networking services. NFV allows network functions to work in software by separating them from proprietary hardware tools (Liyanage et al. 2015). NFV has advantages for overcoming specific DoS attacks (Schneider and Horn 2015). This decoupling control and data forwarding plane result in complicated traditional network routing devices into simple switches. These switches' jobs are to follow the policy that is produced by the logically centralized controller (as shown in Figure 6). This new structure is dissimilar from conventional networks (as shown in Figure 6), where switches and routers are performing both in data forwarding and control plane functions.

Data Forwarding Plane In this plane, there are basic switches. And these switches are interconnected to establish the physical network. Their role is just to forward incoming packets according to the routing policy of the control plane. These switches have a data forwarding table that is given by the control plane.

Control Plane The network's brain is in this plane and its responsibilities are to surveil the network, to make routing decisions, and to program how physical network behaves.

The advantages of SDN Security are as follows: The abstraction of functions allows coding top-level software for managing the network without considering how to configure the underlying physical network in the SDN. And that is one of the many advantages that SDN networking brings to the table. The next thing is that discussing each of these features by explaining how traditional networks are devoid of these features and exploiting each of them to increase network security.

View of the whole network; in the SDN, the controller has a view of the whole network. This is SDN's perhaps the biggest security advantages over conventional networks. This view of the whole network is accredited to centralization and to devices in the network, which are collecting and reporting statistics of traffic. The difference from conventional distributed networks whose devices need to exchange a lot of information and wait for a convergence time is this. In conventional network only, a few devices log traffic statistics, not like the SDN. The view of the whole network can allow the SDN controller to execute a network-wide intrusion system. To detect malicious traffic, IDS analyzes the network traffic statistics that are collected from all the network switches. This is also another difference between the conventional network and the SDN. The view of the whole network not only helps in detecting malicious behavior of switches in the network but also supports more efficient detection of attacks that are coming from malicious traffic. In the SDN, it is easier to identify misbehaving behavior because periodically switches are reporting to the control plane how many packets are received, forwarded, and dropped. The misacting switch or switches can be found after analyzing reports. The control plane's ability to record the view of the whole network over time, therefore, forensic analysis is facilitated. To find and analyze how an undetected intrusion has happened, one can trackback to the recorded traffic.

Mechanism of Self-Healing: For SDN, conditional rules are introduced in Shin et al. (2013) for this kind of mechanism. Conditional rules supply resiliency against DoS attacks, which are targeting links or the host of the network. If we compare the capabilities of control between the traditional networks and SDN networks, SDN networks have much more capability than traditional networks, because of the adoption of the flow-based scheme in SDN, where packet handling is defined by multiple header fields. But in traditional networks, this is relying on the destination address. Therefore, by determining which types of packets need to be moved in the network based on the source address, some other header filed value or, the payload type, the controller of SDN will have better access control. This can help to limit malicious traffic coming from any switch in the network of SDN.

Security challenges of SDN network are as follows: So far, we have explained that SDN's security advantages, now, in this section, security challenges that SDN is facing and its solutions will be explained. These challenges are based on which part of the SDN network they attack and these are data forwarding plane, control plane, and the connection between these two planes. While emphasizing the difficulties in developing these defense mechanisms, counterplot techniques that can be used to ameliorate, prevent, or mitigate some of these attacks are also explained.

The first security challenge in the SDN networks' data forwarding plane is the Switch DoS attack. Switches have limited storage capacity; that is why there is a reactive caching mechanism to handle storage issue. This mechanism is a vulnerable point against DoS attacks, where packets of large payloads that coming from different flows are flooded by a malicious user. Switch's buffer can be populated quickly, particularly if these packages have large payloads, leaving legitimate packages of new flows because of inefficient space on the buffer to store these packets. One of the solutions is proactive rule caching where switches cache as many rules that can fit their tables. Other solutions can be the increasing capacity of switch buffering, rule aggregation, and decreasing communication delay between switch and controller. The second one is tunnel bypassing and packet encryption. SDN networks can customize the treatment of packets with different payloads by the flow-based forwarding scheme. Because of dealing with encrypted packets which have invisible header fields, this can cause tunnel bypassing. A possible solution is to classify packet type based on traffic analysis (Fadlullah et al. 2010).

Our third security challenge is on the control plane. The control plane is sensitive to DDoS attacks which can cause multiple compromised hosts distributed over the network to simultaneously flood network switches with packets. Since all rules are not already present in the tables of switches, many requests will be generated and sent to the controller (Dabbagh et al. 2015). This will take advantage of the processing power of the controller, ultimately causing legitimate requests to be dropped or delayed. To prevent against this kind of attack, one solution is replication where there are multiple physical controllers rather than only one. Other solutions could be assigned a dynamic master controller and placement of the controller could also help. Other challenges that are the fourth challenge on the control plane is the attacker who can access the controller and can control all switches by the compromised controller. This problem is caused by SDN's centralization. To get over this problem, one solution is controller replication with diversity that can resist against such attacks.

The fifth security challenge is on the connection between data forwarding and control planes. If unencrypted communication messages sent on the connection of the data forwarding and control plane, frankly, this link will be sensitive to an MITM attack. An attacker may issue an audit policy by inferring to messages that exchanged on the link. This link must be secured to prevent those kinds of attacks. The solution is encryption. This can be used to prevent wiretapping. Also, to prevent replay attacks, a timestamp can be used in the encrypted messages. If not, a malicious user can collect encrypted rules and then, that user can send them back to switches to get back to use old policies.

In conclusion, old wireless mobile networks have been explained briefly to understand where we were and where we are going with the new technology of 5G. Also, what 5G is described and its emerging technologies with their solution to the requirements of 5G. There is an expectation that usage of SDN enables 5G architecture that will be progressed a lot more than these days, because of handling the challenges of 5G's reduced cell size and heterogeneity of network (Duan and Wang 2015). In summary, what SDN is, how it works, and its advantages and disadvantages are defined from the perspective of security. Even though SDN brings a solid solution to many security problems, at the same time, it is facing new security challenges on the data forwarding plane, control plane, and connection between these two planes. A few solutions have been explained to get over those security threats. There are many interesting topics that could be explored including the complexity of 5G network and security performance while there is or are a different attack(s) for SDN-enabled security mechanisms.

Heterogeneous Networks

5G will work in a highly heterogeneous environment characterized by the presence of a wide variety of access technologies, multilayer networks, multiple device flavors, and multiple user interaction diversity. Also, the 5G architecture has to meet the need for a more flexible core network than the existing technologies for managing heterogeneous access technologies and varying service components. This architecture must also have a dynamic structure that can meet the needs for different applications and traffic types.

The 5G network will become an intelligent communications network that manages M2M and IoT devices. 5G will describe the technological progress that the applied technologies can use together as a complement to all generation bands efficiently and harmoniously. HetNet is a promising technique for ensuring wireless coverage and high efficiency of blankets in 5G wireless networks. HetNets, which can be used for many different structures in very low frequency, very small cell (picocell, body cell, etc.) at very low frequency, can be used with existing (3G, 4G, WiFi) infrastructures as well. It is a multilayered system in which nodes in different layers have distinguishing features such as transmission power, coverage size, and radio access technologies. HetNet can achieve more capacity, wider coverage, and better performance with heterogeneous features. However, the requirements created by new areas of 5G technology (short latency, high bandwidth, long battery life, etc.) may conflict with high-security techniques. Moreover, with the density of small cells in HetNet, traditional handover mechanisms can experience significant performance problems due to handover frequency occurred very often between different cells (Duan and Wang 2016). The confidentiality problem on HetNet also faces a major problem. The location information becomes more precise because of the density of the small cells. Because of that, the traditional association mechanism can reveal information about location privacy (Farhang et al. 2015).

Coordinated multipoint transmission (CoMP) is applicable to improve communication coverage in HetNet (Xu et al. 2016). On the other hand, CoMP may cause the rise of the risk of being eavesdropped to authorized users. The knowledge of the cell to which a user is associated can easily reveal that user's location knowledge because of the high density of small cells. In Farhang et al. (2015), the authors examined the location privacy based on algorithms of physical association layers in 5G. Another way to improve security in 5G is to use the approach of the intrusion detection model. In Gai et al. (2016), attack and intrusion detection techniques for heterogeneous 5G and mobile cloud computing in 5G are introduced. Traditional password-based authentication and biometric authentication have been discussed to provide various levels of security.

Security Mechanisms in 5G

In 5G mobile communications systems, the new architecture, new technologies, and use-cases bring in new features and requirements of security mechanisms. In this section, we investigate types of security mechanisms into six categories.

Authentication

Authentication in 5G mobile networks is one of the most important security features and an authentication scheme is symmetric key encryption normally found in conventional cellular networks. Implementing the authentication scheme can introduce various security requirements on the table. Entity authentication and message authentication are two types of authentication. Both authentication and message authentication is important in 5G wireless mobile networks to prevent the above attacks. Asset authentication is used to make sure that the entity you communicate claims to be true. The mutual authentication between the UE and the mobility management entity (MME) is implemented before the two parties communicate with each other in conventional cellular networks. The most important security feature in the traditional cellular security framework is the mutual authentication between the UE and the MME.

4G LTE uses symmetric key cryptography in a challenge-response-based mechanism for AKA in cellular networks. However, at 5G, authentication is valid not only between the UE and HAM but also between other third parties, such as service providers. Since the trust model of 5G networks is different from that used in older cellular networks, hybrid and flexible authentication management such as Handover Authentication Mechanism (HAM) is required at 5G. The UE's hybrid and flexible authentication can be performed in three different ways: authentication by network only, authentication by the service provider only, and authentication by both the network and the service provider (Duan and Wang 2016). 5G wireless mobile networks are expected to be faster than usual in 5G authentication because of the very fast data rate and very low latency requirements. In addition, depending on the multilayer architecture of 5G, the network will have frequent hyperlinks and authentication between different layers or small cells in 5G. In Duan and Wang (2015), an SDN-enabled fast authentication scheme has been proposed that uses weighted secure content-information transfer to increase the effectiveness of authentication during handovers and meet the 5G latency requirement to overcome the difficulties of key management in HetNets and reduce unnecessary latency. Due to the favorites and authentication between the different layers, the centralized integration of SDN technology, while enabling the use of intelligence and programmable networking capabilities, has extended the Long-Term Evolution (LTE) hierarchical architecture provided by 3GPP to a 5G mobile network architecture. In the proposed architecture, an authentication handover module (AHM) based on SDN functions is proposed, which allows tracking the movement of the subscribed user and the next location. Thus, the AHM can identify potential target cells and, through this information, will begin the transfer procedure to reduce and optimize the induced signaling delay. In Eiza et al. (2016) and Zhang et al. (2017), it is proposed to provide more security services on a public key cryptography-based AKA 5G wireless network.

With 5G wireless networks, message authentication with various new applications and services is becoming increasingly important. In addition, with stringent requirements for low latency, spectrum efficiency (SE), and energy efficiency at 5G, message authentication faces new challenges. In Dubrova et al. (2015), an effective Cyclic Redundancy Check (CRC)–based message authentication for 5G has been proposed to enable the detection of both random and malicious errors without increasing the bandwidth. Mutual authentication is implemented between a base station and a network in third-generation (3G) cellular networks. After authentication, a password key and an integrity key are generated to provide data privacy and integrity between the mobile station and the base station.

Due to the low latency requirement of 5G networks, authentication schemes should be much more efficient at 5G than ever before. To take advantage of SDN, Duan and Wang in (2016) proposed a fast verification scheme in SDN that uses Secure Context-Information (SCI) transport, which is weighed as noncryptographic security technique to improve authentication efficiency in high-frequency handshakes in 5G HetNet to meet your low latency needs. Compared with digital cryptographic authentication methods, it is alleged that the proposed fiction is entirely in danger

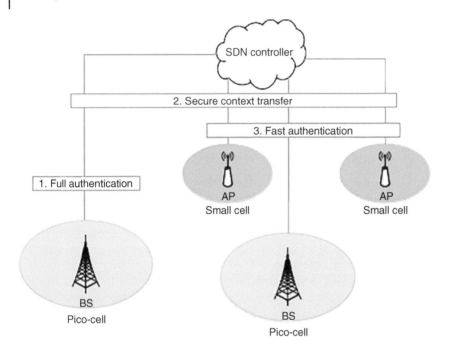

Figure 7 The SDN-enabled authentication model.

because it depends on the physical layer characteristics of the user. For applications requiring high levels of security, the SCI has multiple physical layer features that are used to improve authentication credentials.

The SDN-enabled authentication model is shown in Figure 7. The SDN controller applies a verification model to track and estimate the user location for the preparation of the respective cells prior to the user's arrival point. With this feature of the SDN controller, it can provide seamless transfer authentication. Physical layer properties are used to provide unique fingerprints to the user and simplify the authentication procedure. Three types of fingerprints are used as user-specific physical layer properties. Verified original attributes are obtained after a full authentication. Observations are continuously collected by sampling various physical layer properties from the packages received in the SDN controller. Both the original file and the observation results include the mean value of the attributes and the variance of the selected attributes. The average feature offset can then be calculated according to the authenticated original attributes and observed attributes. If the attribute offset is less than a predetermined threshold, the UE is considered legal.

Cryptography
Authentication for 5G mobile communication security is a prerequisite. Cryptography, which can be used for user authentication, is necessary to hide information that is transmitted over an untrusted medium. In addition to the PLS solutions described above, cryptographic methods are also used to implement data privacy by encrypting data with secret keys. Public (asymmetric) cryptography can be applied to key distributions. Symmetric (shared) cryptography has been adopted for data encryption to reduce the

cost of encryption. In Eiza et al. (2016), a participating vehicle may send a random symmetric key encrypted using the TA's common key. The symmetric key is used to encrypt the message between TA, DMV, and participating vehicles. A one-time encryption key is also encrypted with the public key. A one-time encryption key is used to encrypt the video. In Zhang et al. (2017), after establishing the client/server relationship, the initial symmetric (shared) session key is negotiated between the client and the server. The symmetric key is then used for data transmission between the client and the server.

Due to the large amount of information exchanged over the 5G network, the need for security should be paid attention to. Cryptographic techniques provide a secure foundation for exchanging messages over an insecure channel to guarantee the confidentiality and integrity of the exchanged information. For resource-constrained small cell access points, the computational cost of cryptographic algorithms should be less.

Confidentiality
To support the secure connection between legitimate users and the server, different authentication approaches have been proposed to support security and to meet the security requirements of 5G mobile communication systems. Authentication procedures can safeguard information transferred to legitimate users and protect 5G mobile communications from threats. Privacy consists of two aspects: data privacy and privacy. Data privacy protects data transmission from passive attacks by limiting data transfer only to targeted users and preventing access or disclosure to unauthorized users. Privacy protects and controls the information about legitimate users, for example privacy protects traffic flows from any analysis of an attacker, traffic patterns, and sender/receiver position. Such sensitive information can be used to diagnose. In a variety of applications in the 5G, there are tremendous data on user privacy coming from, for example, vehicle routing data, and health monitoring data.

Data encryption is commonly used to provide data confidentiality by preventing unauthorized users from extracting any useful information from the broadcast information. The symmetric key cryptography technique can be applied to encrypt and decrypt the data with a private key shared between the sender and the receiver. A secure key distribution method is required to share a key between the sender and the recipient. Traditional cryptography is based on the assumption that attackers have limited computing capabilities. For this reason, it is difficult to fight against attackers equipped with powerful computing capabilities. Instead of relying solely on general high-layer encryption mechanisms, the PLS can support privacy service (Chaudhry et al. 2015) against jamming and eavesdropping attacks. In addition to the data services of the 5G, users become aware of the importance of privacy protection services. The 5G privacy service attracts more attention than older cellular networks because of its large data connections. The anonymization service is a basic security requirement in the case of many users. In most cases, privacy breaches can cause serious consequences. For the examples, health monitoring data reveal sensitive personal health information (Zhang et al. 2017), vehicle routing data may reveal location privacy (Eiza et al. 2016), and 5G wireless networks cause serious concerns about privacy breaches. In the HetNets, due to the high density of small cells, the association algorithm can reveal the location latitude of users. In Farhang et al. (2015), a different proprietary algorithm is proposed to maintain location privacy. Eiza et al. (2016) proposed cryptographic mechanisms and schemes to provide secure, privacy-aware real-time video reporting services in vehicle networks.

Availability

Usability is an important metric to provide ultra-reliable communication at 5G. However, while wireless noise is spreading randomly, an attacker can significantly reduce the performance of mobile users and even prevent the availability of services. Jamming is one of the typical mechanisms used by DoS attacks. Most antijamming programs use a frequency hopping technique that users bypass from multiple channels to avoid jamming attacks and to ensure the availability of services.

In Li et al. (2011), the authors proposed a hidden adaptive frequency hopping scheme as a possible 5G technique against DoS based on a software-defined radio platform. Based on physical layer information, the proposed bit error rate (BER) estimator is applied to determine the frequency of blacklisting under a DoS attack. Because frequency hopping techniques require multiple channel access of users, dynamic spectrum access may not work efficiently for users due to the high switching rate and the possibility of high jamming.

To reduce the switching rate and the likelihood of jamming, a pseudorandom time-hopping anti-jamming scheme is proposed by Adem et al. (2015), to overcome a mock opposition plan to prevent jam-prevention attacks for 5G cognitive users. The impact of spectrum dynamics on the performance of mobile cognitive users is modeled by the presence of a cognitive jammer with limited resources. Analytical solutions of mixing probability, switching ratio, and error probability are presented. The probability of jamming is related to the delay performance and the probability of error. Jammer is less likely to be jammed when lacking access. The possibility of switching the time-lapse system performs better than the frequency-lagged system. With the same average symbol energy per joule, it has a lower error probability than a time skipped frequency, and the performance gain saturates at a certain symbol energy level. The authors noted that the proposed time-lapse technique is a strong candidate for D2D connectivity in 5G wireless networks because of its ability to provide jamming resistance with good EE and SE performance, as well as a small communication load. However, a preshared key is required for the time-delayed anti-jamming technique.

Integrity

Although message authentication allows the source of a message to be verified, there is no protection against duplication or modification of the message. 5G aims to bring the connectivity at any time, anywhere, and everywhere to support connections that are closely related to people's everyday life activities, such as planning the quality of drinking water, medicine, and transportation. Therefore, the integrity of the data is one of the most important security requirements in certain applications.

Integrity prevents the information from being altered or altered by unauthorized actives. It can be violated by insider malicious attacks, such as data integrity, message injection, or data modification. It is difficult to detect these attacks because internal attackers have valid identities. In use cases such as smart meters in intelligent networks or smart grid (Yan et al. 2012), data integrity service against manipulation must be provided. Compared to voice communication, data can be attacked and changed more easily (Alliance 2015). Integrity services can be provided using mutual authentication, which can create an integrity key. Personal health information integrity service is required (Zhang et al. 2017), and message integrity can be provided in authentication schemes (Eiza et al. 2016).

Key Management

Key management is the management of various cryptographic keys in a cryptosystem, which includes dealing with the generation, exchange, storage, use, and replacement of various keys. A novel collaborative protocol for the key establishment has been presented in Zhang et al. (2016) in a heterogeneous 5G environment. The efficient cooperative key establishment system could establish end-to-end secure communications between nodes with different resource capabilities. The scheme is designed to provide a secure and energy-efficient key establishment mechanism to make a highly resource-constrained node to be able to securely set up a session key with an external entity using the asymmetric cryptography primitives. Evolved from the collaboration scheme, a new approach has been proposed in Ma and Hu (2019) to enforce the session key derivation. A novel collaborative session key exchange method has also been proposed, by which, a high resource-constrained node can obtain the assistance from its more powerful neighbors when handling costly cryptographic operations. The approach does not require the resource-constrained nodes to perform heavy asymmetric operations. Instead, the assisting nodes in the neighborhood take charge of the heavy computational load in a distributed and cooperative way. Considering the security aspects of the cellular 5G communication architecture, a key agreement mechanism for secure communication between entities in the cellular 5G networks has been proposed in Prasad et al. (2018) to provide a key agreement function for the reliability and efficiency of the secure cellular 5G communication networks.

Key management is one of the major challenges in 5G network security. Symmetric key-based key management mechanisms are more popular than asymmetric key-based scheme due to less computation cost. With a large variety of devices deployed in the 5GS, group key management and hybrid key management which combines symmetric and asymmetric cryptography may be future trends.

Security Evaluations Techniques

Researchers in the Security and Privacy fields use the formal and informal techniques to analyze, prove, and verify the reliability of their proposed security scheme, and especially for schemes that are based on cryptography as a tool for achieving the authentication and privacy. Therefore, we divide these techniques into two classes, including (i) without an implementation tool and (ii) with an application tool, the unofficial and formal security analysis techniques used in the authentication and privacy protection schemes for 4G and 5G cellular networks as shown in Figure 8. For the first class, we classify in "without an implementation tool" eight techniques, including, Zero-Knowledge Proof, Mathematical difficulties, GNY logic, CK security model, Random oracle model, Game theory, Probabilistic functions, and BAN logic. To analyze the completeness of a cryptographic protocol, both schemes use the GNY logic. The scheme uses a random oracle model to show that there is an adversary A that can construct an algorithm to solve the CDH problem or the k-CAA problem separately. The scheme uses the BAN logic

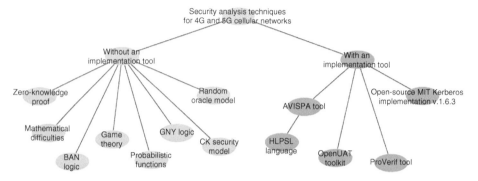

Figure 8 Classification of security evaluation techniques.

to demonstrate that the scheme is valid and practical. The mathematical difficulties are used by the scheme to achieve security and privacy using discrete logarithm and computational Diffie–Hellman problems. Furthermore, the game theory is used by the scheme (Yang et al. 2006) to prove the security of the bipartite protocol by designing a game that turns a CDH instance into the protocol. According to Manshaei et al., the game approach is related to the security problem to be solved, e.g. the Stackelberg game for jamming/eavesdropping, the static security cost game for Interdependent Security, and the static nonzero-sum game for Vendor Patch Management.

For the second class, we classify in "with an implementation tool" four techniques, including AVISPA tool, Open-source MIT Kerberos implementation, OpenUAT, and ProVerif. The Open-source MIT Kerberos is used especially to evaluate the performance of the enhanced Kerberos protocol such as the scheme (Pereniguez et al. 2011). The OpenUAT is used by the scheme (Mayrhofer et al. 2013) to implement some intuitive authentication methods in a common library. To verify the secrecy of the real identity and the resistance against known attacks, four schemes (Chaudhry et al. 2015; Fu et al. 2016; Gisdakis et al. 2015) use the ProVerif, which is an automatic cryptographic protocol verifier, in the formal model, called Dolev-Yao model. Specifically, the ProVerif takes as input a model of the protocol in an extension of the pi-calculus with cryptography. For more details about the ProVerif, we refer the reader to the work of Blanchet (2016). Therefore, five schemes (Ozhelvaci and Ma 2018; Hamandi et al. 2017; Fu et al. 2017; Abdelkader et al. 2010; Cao et al. 2012) use the AVISPA tool based on the HLPSL language to verify the security of these schemes against insider attacks and outsider attacks.

Security in 5G Heterogeneous Networks

Different technologies have been applied to 5G networks to achieve to serve 1000 or more times more devices and data traffic, and 1 ms latency. Additionally, 5G will bring other new features including integration of legacy networks, network security, provisioning, extreme reliability, more bandwidth, and 90% less energy consumption. One of the ways to meet the requirements is that using the concept of the technology of HetNets. So, using the concept of HetNets in 5G is so critical to achieving the requirements such as low cost, less energy consumption, 99.99 coverage, and high throughput.

HetNet can achieve wide coverage, high capacity, and better performance with characteristics of heterogeneity to support immanent coverage, and different cells are proposed for 5G, such as femtocells, microcells, and relays. In such a heterogeneous environment, security has become one of the crucial points of 5G networks. Therefore, security mechanisms should ensure the interaction among communication entities and protect against different attacks, and provide data integrity, confidentiality with having mutual authentication between the UE and the SN. In this section, we will provide two security aspects in terms of handover and D2D authentication in 5G HetNets.

Handover Authentication Security

The next-generation network is considered to be much more heterogeneous and due to the densified small cell deployment, the user will face more frequent handoffs, thus, the handover authentication should be fast and efficient enough to ensure low latency. Moreover, a handover authentication scheme should have low computational cost and overhead to have persistent connectivity and performance. Therefore, a seamless and secure handover authentication scheme is needed because users frequently move between different networks that can cause serious vulnerabilities against various attacks. Designing a fast and efficient authentication scheme to grant access to registered users in a foreign network is a tough task. However, for 5G mobile networks, there are few research studies to design specific architecture and security features. In Duan and Wang (2015), centrally integrating the SDN technology can make possible to use intelligence and programmable networking capabilities, and they proposed a 5G mobile network architecture by extending the LTE hierarchical architecture, which is provided by 3GPP. In the proposed architecture, an AHM is proposed based on SDN functions and that allows to monitor subscribed user's movement and next location. Thus, AHM can identify the potential target cells and by this information, it will start the handover procedure to reduce and optimize the induced signaling delay. In El Idrissi et al. (2012), the authors proposed an authentication method for LTE networks based on EAP–AKA using Elliptic Curve Diffie–Hellman (ECDH) and symmetric key cryptography. They overcome the drawbacks of the EAP–AKA mechanism by adding a local authenticator. Therefore, their proposed approach can protect user identity against various attacks and provide mutual authentication and data integrity. But with the huge number of users and small cells, the solution may not be efficient and scalable because it can add an additional delay in the environment of 5G networks.

In another proposed authentication scheme, in He et al. (2012) authors proposed a protocol based on bilinear pairing functions to have secure handover operation and reduce the overhead of computation and communication costs. But, because the authentication server in architecture is usually located remotely, the latency can be up to hundreds of milliseconds which make it unsatisfactory for 5G requirements due to the frequent handovers between the authentication server and small cell access points. The author in Alezabi et al. (2014) proposed an efficient authentication method to overcome various vulnerabilities of EAP–AKA, which is called Evolved Packet System AKA (EPS-AKA) in LTE networks by reducing computational overhead, authentication delay, and satisfying security requirements. EEPS-AKA is based on the Simple Password Exponential Key Exchange (SPEKE). The main aim of this method is to overcome the disclosure of the UE, reducing the size of messages exchanged and make the protocol faster by using a secret key method. Although the authentication procedure is simplified, latency can become higher due to a frequent number of inquiries in the environment of small cell 5G networks.

In Duan and Wang (2016), similarly with the handover procedure in Duan and Wang (2015), the protocol uses the weighted Secure-Context-Information (SCI) with several attributes to achieve fast authentication in 5G HetNets. For the handover authentication, instead of using complex and forgeable cryptographic exchange mechanisms, the SDN controller compares the observed attributes matrix through zero-mean white Gaussian noises with the validated original SCI attributes matrix and obtains the offset. Then, the SDN controller compares each attribute with the threshold. If the difference between them is less than the threshold, the authentication is successful. Otherwise, the authentication is failed. In this way, this scheme can significantly reduce handover latency. However, they forget to consider the false alarms and their impact on the latency. In Boujelben et al. (2015), they proposed a new handover self-optimization algorithm to minimize the number of handovers (HOs) and reduce energy consumption. In this algorithm, the velocity of the UE is added as a measurement parameter to select the appropriate target cell. Concretely, high-speed UEs can only be authorized to connect to the high loaded microcells, while low-speed UEs shall be directed to high loaded femtocells. This algorithm could greatly reduce the energy consumption. However, the algorithm only includes the measurement process during the whole handover procedure. In Table 3, there are comparisons of the aforementioned handover security schemes for 5G HetNets and including the technologies that they used, solved issues and what kind of security flaws that still exist in the schemes.

Mutual authentication between the UE and the network and key agreement used to provide keying material to protect the subsequent security procedures are the two most important security features in the 5G network. In the 5GS, a new AKA protocol named 5G AKA is supported by the 3GPP committee (3rd Generation Partnership Project 2018a,c), which enhances 4G AKA protocol, i.e. EPS AKA (3rd Generation Partnership Project 2018d) by offering the home network with the proof of successful authentication of the UE. Before 5G, after the Home Network (HN) sends the Authentication Vector (AV) to the Visited Network (VN), it does not participate in the subsequent authentication process, which easily leads to a security problem. That is, in the roaming scenario, the visiting operator obtains the complete authentication vector of the roaming user from the home operator, and the visiting operator falsifies the user location update information by using the authentication vector of the roaming user, thereby generating a roaming fee by forging the bill. As shown in Figure 9, to withstand this attack, 5G AKA protocol performs a one-way transformation on the authentication vector where the visited operator can only obtain the transformed authentication vector of the roaming user. The visited operator implements the authentication of the roaming user without acquiring the original authentication vector and sends the authentication result of the roaming user to the home operator, and thus the home operator enhances the authentication control for the visited operator (Table 4).

Except for the 5G AKA protocol, EAP–AKA's protocol is also supported to perform the mutual AKA in the 5G network as shown in Figure 10 (3rd Generation Partnership Project 2018a,c). In LTE/LTE-A network, the EAP–AKA or EAP–AKA' is only a complementary authentication approach and is only used for UE to connect to the 4G core network via non-3GPP access networks such as WLAN. In addition, it is implemented in

Table 3 Comparison of handover security in 5G HetNets.

Scheme	Technologies	Issues Solved	Security Flaws
Duan and Wang (2015)	SDN controller monitors and predicts the location of the UE	Handover latency reduced, resist against active attacks	The need of SDN and base station assistance
El Idrissi et al. (2012)	Using ECDH and symmetric key cryptography on EAP–AKA	Provides mutual authentication and data integrity	Ignore the impact of small cell deployment that can increase latency dramatically
He et al. (2012)	Utilize bilinear pairing functions	Greatly reduce the number of handovers	Ignore the impact of latency due to the located authentication server remotely
Alezabi et al. (2014)	Based on Simple Password Exponential Key Exchange on EPS-AKA	Greatly reduce the size of messages exchanged	Lack of consideration for handover security
Duan and Wang (2016)	Using secure-context-information and compare observed attributes matrix	Greatly reduce handover latency	Latency can be much higher due to including false alarms
Boujelben et al. (2015)	Select the base station depending on the UE's velocity	Energy consumption and the number of handovers are reduced	Only focus on measurement process during the whole handover process

a set of independent Network Elements (NEs) compared with the EPS–AKA protocol, such as Authentication, Authorizing, and Accounting (AAA) server. In the 5G network, a UE can execute the 5G AKA or the EAP–AKA' to accomplish the mutual authentication with the 5GS via 5G wireless access network. Here, the EAP–AKA' has been upgraded to the same position of 5G-AKA, and they use the same NEs. This means that 5G authentication NEs must support both two authentication methods on the standard. The 5GS also supports the non-3GPP access for the UE (3rd Generation Partnership Project 2018e). For an untrusted non-3GPP access network, the channel between the UE and the 5GCN is considered unsafe. To protect the communication between the UE and the 5GCN, the UE shall establish an IPSec tunnel by using IKEv2. Both the EAP–AKA' and 5G AKA are allowed for the authentication of UE via non-3GPP access during the IPSec tunnel establishment procedure (3rd Generation Partnership Project 2018f,g).

Consequently, the handover authentication process is one of the crucially important aspects of the new generation networks; therefore, it needs to be further improved not only to supply the required latency but also to stand against the various attacks that have been mentioned in the section titled "Security attacks and Mechanisms in 5G". Some handover authentication schemes have been proposed in Duan and Wang (2015, 2016), El Idrissi et al. (2012), He et al. (2012), Alezabi et al. (2014), and Boujelben et al. (2015) to meet the requirements. However, since the SDN technique is adopted in these schemes in Duan and Wang (2015, 2016) and Ma et al. (2017) to monitor the devices,

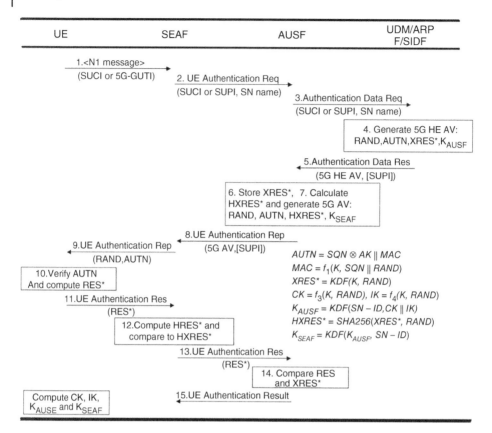

Figure 9 5G-AKA process.

once the SDN is compromised, the whole network will crash. In addition, these proposed schemes cannot be applied to all scenarios in the 5GS, especially, high-speed rail networks and satellite–terrestrial integration networks in the 5G environment. For the high-speed rail networks and satellite–terrestrial integration networks in the 5G environment, many UEs in the train or on the land must perform fast handover simultaneously to meet the users' lower latency requirements. If these existing schemes are adopted, it may incur a large amount of handover overheads in an instant when the train or satellite is moving quickly. It is serious in the ultra-dense small cell deployed 5G environment.

D2D Group Authentication Security

To build the next-generation network, there are various technologies to meet these requirements, for example, massive multiple-input multiple-output (massive MIMO), SDN, D2D communication, network slicing, and NFV (Sarraf 2019). As one of them, D2D communication is a promising technology for reducing communication delay, improve resource utilization, and enabling direct communication between two or more users in the proximity. Researchers have performed vivid studies on D2D communication in both academia and industry. In academia, resource allocation,

Table 4 Comparison of D2D communication security in 5G HetNets.

Scheme	Technologies	Issues solved	Security flaws
Wang et al. (2017)	Diffie–Hellman Key Exchange algorithm (DHKE) and MAC	Achieve mutual authentication and key agreement between two D2D devices under LTE network	The lack of identity protection and single point failure due to the involvement of VNs
Sun et al. (2019)	Identity-Based Prefix Algorithm and ECDH techniques	Achieve mutual authentication and key agreement, and resist several protocol attacks	Incur a lot of computation and communication costs in the scenario of involving many devices at a time
Hsu et al. (2017)	Identity-based k-anonymity secret handshake. Key-private encryption and Linear encryption, and zero-knowledge proof	Achieve network-covered and network-absent mutual authentication and key agreement among a group of D2D devices	Incur a lot of computation and communication costs and do not take roaming scenarios need to manage the D2D group
Wang and Yan (2017)	HMAC and identity-based signature, pseudonym management	Achieve mutual authentication and group key agreement for D2D group communication	Incur a lot of computation and communication costs and the scheme do not consider the roaming scenarios. The temporary identities of UEs are updated frequently, so, there a need to manage the D2D group communication
Xie et al. (2018)	Acoustics waves, acoustics channel response-based authentication	Achieve device discovery and bidirectional initial authentication	Lack of identity privacy protection and cannot provide strong security
Huang et al. (2013)	Beacon transmission pattern	Achieve distributed synchronization device discovery	Lack of mutual authentication and key agreement

interference control, key agreement (Wang et al. 2017; Sun et al. 2019), and mode selection (Wu et al. 2017) are some of the issues that have been studied by many researchers. On the other hand, in industry, use-cases of D2D communication are still underworking to develop new applications, such as content and information sharing, gaming, and social network service. Meanwhile, the standardization of D2D communication in 5G networks is still in progress and the 3GPP is the standardization body that has defined the D2D communication under the title of "Proximity-Based Services" (ProSe) (3rd Generation Partnership Project 2013, 2018h). 3GPP's standards specify many of new use-cases in 5G networks that will rely on the establishment of highly secure D2D group communication to achieve the low latency and multiconcurrent authentication.

Although D2D group communication has great potential in terms of usefulness and effectiveness for various scenarios in the network, there are a variety of security and privacy issues that need to be addressed. Also, these new application scenarios and

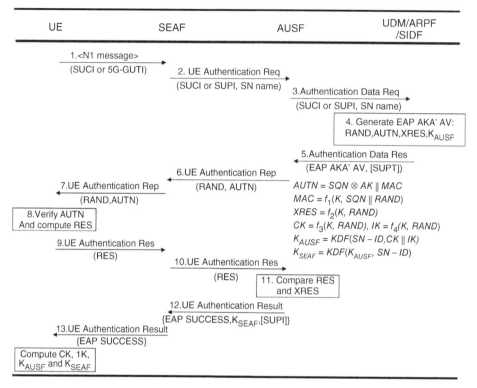

Figure 10 EAP–AKA process.

architecture can be open different type of active and passive attacks. Therefore, D2D communications need to supply the privacy of the user identity, mutual authentication, and additional security level for key agreement. Current solutions have been proposed mostly for two-user D2D communication scenarios. In Wang et al. (2017), the authors have proposed a D2D universal AKA scheme based on the DHEKE algorithm and the authentication code for the 4G/LTE networks. It can be applied to the D2D roaming and interoperator scenarios. In this scheme, the core network cannot acquire the final session key and only group members can share the final session key, therefore, it can provide privacy-preserving protection. But, due to a large amount of signaling cost and involving of the base station, the single point failure can be happened very likely. In Sun et al. (2019), the scheme is designed for D2D communication between two UEs. By the scheme, two UEs can have mutual authentication and as well as privacy protection during the device discovery phase. Also, this scheme can resist several protocol attacks thanks to the identity-based prefix encryption and ECDH techniques. However, in the scenario of D2D communication can involve more than 2 devices at a time to discover each other and build the communication; therefore, this scheme cannot achieve the ideal efficiency and could increase the latency dramatically under this circumstance. In Hsu et al. (2017), two protocols have been designed to provide anonymous D2D group communications. These protocols achieve anonymous protection of group information without addressing the protection of user's privacy. Besides, these protocols only support D2D communication between one announcing user and one monitoring user, which implies that it is still a data sharing and transmission method between two users. In Wang and Yan (2017), the authors have proposed two privacy-preserving key agreement protocols to set up an anonymous D2D group session. These approaches generate a group session key to protect user's privacy. However, each resource-constrained terminal needs to perform $(n + 2)$ bilinear pairing operations for setting up one D2D session between three users, which results in significant computational overhead. Moreover, the user's identity can be exposed in the process of registration, if there is an active attack such as the MITM attack. So, it can be said that these protocols are vulnerable to many different types of attacks. In Xie et al. (2018), the acoustic waves have been used to discover the proximity devices between two devices. Bidirectional initial authentication is done between two devices by calculating the physical response interval, and a key agreement is done by adopting a novel coding scheme between two devices. However, the scheme is not robust against strong attacks and can cause the issue of identity privacy protection. In Huang et al. (2013), the authors have proposed a distributed synchronization device discovery mechanism. The proximity devices can form a synchronization group and they can broadcast their existing by using this scheme. Even though the scheme is shortening the device discovery due to the signaling interference level could be so high, the scheme cannot provide security for D2D communication.

Based on the investigations above, it is clear to see that most recent studies do not emphasize on the heterogeneous environment of 5G and the security and efficiency of D2D communication. In D2D communication security, there should be a universal security device discovery, secure access, mutual authentication, and key agreement schemes and mechanisms to tackle the security pitfall points. Clearly, the abovementioned protocols can only meet one or two security requirements and cannot be put into application scenarios for 5G HetNets. Another reason is that the development of D2D communication is still in early-stage for 5G HetNets. Therefore, one of the key points is that there is a need to design a uniform AKA scheme to meet a variety of security requirements for D2D communication in 5G HetNets to make sure that there is no single point that is vulnerable to protocol attacks. Another key point is the balance between security level and the performance of the protocol for D2D communication in 5G HetNets. Since the high-level security protocols generally have huge computational and storage costs and transmission overheads, it is necessary to have some trade-off between performance and security features of the protocol. Hence, the 5G infrastructure will, also, support the IoT and its applications, and there should be enough performance to be applicable for the IoT devices because of their limited resources.

Since new use-cases and application scenarios will be more complicated such as communication between two devices on different networks, there is a need to have such protocol to provide or achieve the requirements of the D2D communications. The mobility characteristics of the D2D devices themselves, and the low latency and high heterogeneity of the 5G HetNets can easily cause more complex D2D group authentication scenarios with new security issues. Even though this is one of the crucial security problems of the D2D communication in the 5G, there are a few kinds of research that have been done on lightweight and fast handover authentication schemes with roaming scenarios in 5G HetNets. Thus, it is crucial to design a uniform, lightweight, and secure authentication protocol with privacy-preserving of user's communication for D2D communication in 5G HetNets.

Conclusions

The 3GPP is the standardization body revealing several standards to lead the development of 5G networks in terms of research and for industry needs. In this article, we have first introduced the background and the architecture of the 5G communication in the HetNets and addressed the security issues and vulnerabilities of 3GPP 5GSs. The SDN and the NFV standards which would be the fundamental technologies of 5G communication in the future are also introduced. We have then reviewed the state-of-the-art security solutions for the 5G mobile communications with the evaluation of the security mechanisms, security requirements, or vulnerabilities in heterogeneous scenarios. We have examined the security aspects in terms of handover authentication and D2D group communication in 5G HetNets. These security issues and corresponding solutions in the 5G mobile communications have paved the way for further research in this area. Moreover, with the rapid deployment of small cells technologies, the number of connected devices will increase quickly, so the further deliberated solutions of security issues are needed in HetNets. As a summary, with the development of the next generation of mobile technology, the existing security mechanisms are inadequate and need to be improved to make the 5G communication environment more secure. In the future, new security threats and solutions could also be discovered to cope with security and privacy issues, such as key management, privacy protection, data encryption, and integrity.

Related Articles

5G Mobile Networks Security Landscape and Major Risks
5G-Core Network Security

References

3rd Generation Partnership Project (2013). Technical specification group services and system aspects; feasibility study for proximity services (prose) (Release 12). 3GPP TR22.803.

3rd Generation Partnership Project (2018a). Technical specification group services and system aspects; security architecture and procedures for 5G system (Rel 15). 3GPP TS 33.501 V15.3.1 (December 2018).

3rd Generation Partnership Project (2018b). Technical specification group services and systems aspects; security aspects; study on the support of 256-bit algorithms for 5G (Rel 16). 3GPP TR 33.841 V16.0.0 (December 2018).

3rd Generation Partnership Project (2018c). Technical specification group core network and terminals; non-access-stratum (NAS) protocol for 5G system (5GS); stage 3 (Rel 15). 3GPP TS 24.501 V15.2.0 (December 2018).

3rd Generation Partnership Project (2018d). Technical specification group services and system aspects; 3GPP system architecture evolution (SAE); security architecture (Rel 15). 3GPP TS 33.401 V15.6.0 (December 2018).

3rd Generation Partnership Project (2018e). Technical specification group services and system aspects; system architecture for the 5G system; stage 2 (Rel 15). 3GPP TS 23.501 V15.4.0 (December 2018).

3rd Generation Partnership Project (2018f). Technical specification group core network and terminals; access to the 3GPP 5G core network (5GCN) via non-3GPP access networks (N3AN); stage 3 (Rel 15). 3GPP TS 24.502 V15.2.0 (December 2018).

3rd Generation Partnership Project (2018g). Technical specification group services and system aspects; procedures for the 5G system; stage 2 (Rel 15). 3GPP TS 23.502 V15.4.0 (December 2018).

3rd Generation Partnership Project (2018h). Technical specification group services and system aspects; proximity based services (prose) (Release 15). 3GPP TS23.303.

Abdelkader, M., Hamdi, M., and Boudriga, N. (2010). A novel advanced identity management scheme for seamless handoff in 4G wireless networks. GLOBECOM Workshops (GC Wkshps), 2010 IEEE, 2075–2080, IEEE.

Agiwal, M., Roy, A., and Saxena, N. (2016). Next generation 5G wireless networks: a comprehensive survey. *IEEE Communications Surveys & Tutorials* 18 (3): 1617–1655.

Alcaraz-Calero, J., Ioannis-Prodromos, B., and Jesus Bernardos Cano, C. (2017). Leading innovations towards 5G: Europe's perspective in 5G infrastructure public-private partnership (5G-PPP). Personal, Indoor, and Mobile Radio Communications (PIMRC), 2017 IEEE 28th Annual International Symposium on, 1–5, IEEE.

Alliance, N. (2015). Next generation mobile networks. 5G White Paper, 1–125.

Alliance, N. (2016). 5G security recommendations package. White Paper.

Adem, N., Hamdaoui, B., and Yavuz, A. (2015). Pseudorandom time-hopping anti-jamming technique for mobile cognitive users. Globecom Workshops (GC Wkshps), 2015 IEEE, 1–6, IEEE.

Andrews, J.G., Buzzi, S., and Choi, W. (2014). What will 5G be? *IEEE Journal on Selected Areas in Communications* 32 (6): 1065–1082.

Alezabi, K.A., Hashim, F., Hashim, S.J., and Ali, B.M. (2014). An efficient authentication and key agreement protocol for 4G (LTE) networks. Region 10 Symposium, 2014 IEEE, 502–507, IEEE.

Baker, W., Goudie, M., and Hutton, A. (2011). Data breach investigations report. Verizon RISK Team, 1–72. www.verizonbusiness.com/resources/reports/rp_databreach-investigationsreport-2011_en_xg.pdf (accessed 9 December 2019).

Blanchet, B. (2016). Modeling and verifying security protocols with the applied pi calculus and ProVerif. *Foundations and Trends® in Privacy and Security* 1 (1–2): 1–135.

Boujelben, M., Rejeb, S.B., and Tabbane, S. (2015). A novel green handover self-optimization algorithm for LTE-A/5G HetNets. 2015 International Wireless Communications and Mobile Computing Conference (IWCMC), 413–418, IEEE.

Cao, J., Li, H., Ma, M. et al. (2012). A simple and robust handover authentication between HeNB and eNB in LTE networks. *Computer Networks* 56 (8): 2119–2131.

Chaudhry, S.A., Farash, M.S., Naqvi, H. et al. (2015). An enhanced privacy preserving remote user authentication scheme with provable security. *Security and Communication Networks* 8 (18): 3782–3795.

Chen, S., Qin, F., Hu, B. et al. (2016a). User-centric ultra-dense networks for 5G: challenges, methodologies, and directions. *IEEE Wireless Communications* 23 (2): 78–85.

Chen, B., Zhu, C., Li, W. et al. (2016b). Original symbol phase rotated secure transmission against powerful massive MIMO eavesdropper. *IEEE Access* 4: 3016–3025.

Conti, M., Dragoni, N., and Lesyk, V. (2016). A survey of man in the middle attacks. *IEEE Communications Surveys & Tutorials* 18 (3): 2027–2051.

Dabbagh, M., Hamdaoui, B., Guizani, M., and Rayes, A. (2015). Software-defined networking security: pros and cons. *IEEE Communications Magazine* 53 (6): 73–79.

Duan, X. and Wang, X. (2015). Authentication handover and privacy protection in 5G hetnets using software-defined networking. *IEEE Communications Magazine* 53 (4): 28–35.

Duan, X. and Wang, X. (2016). Fast authentication in 5G HetNet through SDN enabled weighted secure-context-information transfer. 2016 IEEE International Conference on Communications (ICC), 1–6, IEEE.

Dubrova, E., Näslund, M., and Selander, G. (2015). CRC-based message authentication for 5G mobile technology. *Trustcom/BigDataSE/ISPA, 2015 IEEE* 1: 1186–1191.

Eiza, M.H., Ni, Q., and Shi, Q. (2016). Secure and privacy-aware cloud-assisted video reporting service in 5G-enabled vehicular networks. *IEEE Transactions on Vehicular Technology* 65 (10): 7868–7881.

El Idrissi, Y.E.H., Zahid, N., and Jedra, M. (2012). Security analysis of 3GPP (LTE)—WLAN interworking and a new local authentication method based on EAP-AKA. 2012 International Conference on Future Generation Communication Technology (FGCT), 137–142, IEEE.

Fadlullah, Z.M., Taleb, T., Vasilakos, A.V. et al. (2010). DTRAB: combating against attacks on encrypted protocols through traffic-feature analysis. *IEEE/ACM Transactions on Networking (TON)* 18 (4): 1234–1247.

Fang, D., Qian, Y., and Hu, R.Q. (2017). Security for 5G mobile wireless networks. *IEEE Access* 6: 4850–4874.

Farhang, S., Hayel, Y., and Zhu, Q. (2015). PHY-layer location privacy-preserving access point selection mechanism in next-generation wireless networks. 2015 IEEE Conference on Communications and Network Security (CNS), 263–271, IEEE.

Ferrag, M.A., Maglaras, L., Argyriou, A. et al. (2018). Security for 4G and 5G cellular networks: a survey of existing authentication and privacy-preserving schemes. *Journal of Network and Computer Applications* 101: 55–82.

Fu, A., Song, J., Li, S. et al. (2016). A privacy-preserving group authentication protocol for machine-type communication in LTE/LTE-A networks. *Security and Communication Networks* 9 (13): 2002–2014.

Fu, A., Qin, N., Wang, Y. et al. (2017). Nframe: a privacy-preserving with non-frameability handover authentication protocol based on (t, n) secret sharing for LTE/LTE-A networks. *Wireless Networks* 23 (7): 2165–2176.

Gai, K., Qiu, M., Tao, L., and Zhu, Y. (2016). Intrusion detection techniques for mobile cloud computing in heterogeneous 5G. *Security and Communication Networks* 9 (16): 3049–3058.

Gisdakis, S., Manolopoulos, V., Tao, S. et al. (2015). Secure and privacy-preserving smartphone-based traffic information systems. *IEEE Transactions on Intelligent Transportation Systems* 16 (3): 1428–1438.

Guerzoni, R., Trivisonno, R., and Soldani, D. (2014). SDN-based architecture and procedures for 5G networks. 5G for Ubiquitous Connectivity (5GU), 2014 1st International Conference on, 209–214, IEEE.

Gupta, A. and Jha, R.K. (2015). A survey of 5G network: architecture and emerging technologies. *IEEE Access* 3: 1206–1232.

Hamandi, K., Abdo, J.B., Elhajj, I.H. et al. (2017). A privacy-enhanced computationally-efficient and comprehensive LTE-AKA. *Computer Communications* 98: 20–30.

He, D., Chen, C., Chan, S., and Bu, J. (2012). Secure and efficient handover authentication based on bilinear pairing functions. *IEEE Transactions on Wireless Communications* 11 (1): 48–53.

Hsu, R.-H., Lee, J., Quek, T.Q., and Chen, J.-C. (2017). GRAAD: group anonymous and accountable D2D communication in mobile networks. *IEEE Transactions on Information Forensics and Security* 13 (2): 449–464.

Huang, P.-K., Qi, E., Park, M., and Stephens, A. (2013). Energy efficient and scalable device-to-device discovery protocol with fast discovery. 2013 IEEE International Conference on Sensing, Communications and Networking (SECON), 1–9, IEEE.

Inc, C. (2013). Software defined networking: why we like it and how we are building it. White Paper.

Jain, S., Kumar, A., and Mandal, S. (2013). B4: experience with a globally-deployed software defined WAN. *ACM SIGCOMM Computer Communication Review* 43 (4): 3–14.

Jungnickel, V., Habel, K., Parker, M. (2014). Software-defined open architecture for front-and backhaul in 5G mobile networks. Transparent Optical Networks (ICTON), 2014 16th International Conference on, 1–4, IEEE.

Khodashenas, P.S., Aznar, J., and Legarrea, A. (2016). 5G network challenges and realization insights. 2016 18th International Conference on Transparent Optical Networks (ICTON), 1–4, IEEE.

Li, Y., Kaur, B., and Andersen, B. (2011). Denial of service prevention for 5G. *Wireless Personal Communications* 57 (3): 365–376.

Liyanage, M., Ahmed, I., and Ylianttila, M. (2015). Security for future software defined mobile networks. 2015 9th International Conference on Next Generation Mobile Applications, Services and Technologies, 256–264.

Liyanage, M., Abro, A.B., Ylianttila, M., and Gurtov, A. (2016). Opportunities and challenges of software-defined mobile networks in network security. *IEEE Security & Privacy* 14 (4): 34–44.

Luo, S., Wu, J., Li, J. et al. (2015). Toward vulnerability assessment for 5G mobile communication networks. 2015 IEEE International Conference on Smart City/SocialCom/SustainCom (SmartCity), 72–76, IEEE.

Ma, T. and Hu, F. (2019). A cross-layer collaborative handover authentication approach for 5G heterogeneous network. *Journal of Physics: Conference Series* 1169 (1): 012066.

Ma, T., Hu, F., and Ma, M. (2017). Fast and efficient physical layer authentication for 5G HetNet handover. 2017 27th International Telecommunication Networks and Applications Conference (ITNAC), 1–3, IEEE.

Mayrhofer, R., Fuß, J., and Ion, I. (2013). UACAP: a unified auxiliary channel authentication protocol. *IEEE Transactions on Mobile Computing* 12 (4): 710–721.

NOKIA (2017). Security Challenges and Opportunities for 5G Mobile Network, White Paper .

Ozhelvaci, A. and Ma, M. (2018). Secure and efficient vertical handover authentication for 5G HetNets. 2018 IEEE International Conference on Information Communication and Signal Processing (ICICSP), 27–32, IEEE.

Panwar, N., Sharma, S., and Singh, A.K. (2016). A survey on 5G: the next generation of mobile communication. *Physical Communication* 18: 64–84.

Pereniguez, F., Marin-Lopez, R., Kambourakis, G. et al. (2011). PrivaKERB: a user privacy framework for Kerberos. *Computers & Security* 30 (6–7): 446–463.

Prasad, A.R., Arumugam, S., Sheeba, B., and Zugenmaier, A. (2018). 3GPP 5G security. *Journal of ICT Standardization* 6 (1): 137–158.

Sarraf, S. (2019). 5G emerging technology and affected industries: quick survey. *American Scientific Research Journal for Engineering, Technology, and Sciences (ASRJETS)* 55 (1): 75–82.

Schneider, P. and Horn, G. (2015). Towards 5G security. *Trustcom/BigDataSE/ISPA, 2015 IEEE* 1: 1165–1170.

Shin, S., Yegneswaran, V., Porras, P., and Gu, G. (2013). Avant-guard: scalable and vigilant switch flow management in software-defined networks. Proceedings of the 2013 ACM SIGSAC Conference on Computer & Communications Security, 413–424, ACM.

Sun, Y., Cao, J., Ma, M. et al. (2019). Privacy-preserving device discovery and authentication scheme for D2D communication in 3GPP 5G HetNet. 2019 International Conference on Computing, Networking and Communications (ICNC), 425–431, IEEE.

Vij, S. and Jain, A. (2016). 5G: evolution of a secure mobile technology. 2016 3rd International Conference on Computing for Sustainable Global Development (INDIACom), 2192–2196, IEEE.

Wang, M. and Yan, Z. (2017). Privacy-preserving authentication and key agreement protocols for D2D group communications. *IEEE Transactions on Industrial Informatics* 14 (8): 3637–3647.

Wang, M., Yan, Z., and Niemi, V. (2017). UAKA-D2D: universal authentication and key agreement protocol in D2D communications. *Mobile Networks and Applications* 22 (3): 510–525.

Wu, D., Zhou, L., and Cai, Y. (2017). Social-aware rate based content sharing mode selection for D2D content sharing scenarios. *IEEE Transactions on Multimedia* 19 (11): 2571–2582.

Xie, P., Feng, J., Cao, Z., and Wang, J. (2018). GeneWave: fast authentication and key agreement on commodity mobile devices. *IEEE/ACM Transactions on Networking (TON)* 26 (4): 1688–1700.

Xu, M., Tao, X., Yang, F., and Wu, H. (2016). Enhancing secured coverage with CoMP transmission in heterogeneous cellular networks. *IEEE Communications Letters* 20 (11): 2272–2275.

Yan, Y., Qian, Y., Sharif, H., and Tipper, D. (2012). A survey on cyber security for smart grid communications. *IEEE Communications Surveys and Tutorials* 14 (4): 998–1010.

Yang, C.-C., Chu, K.-H., and Yang, Y.-W. (2006). 3G and WLAN interworking security: current status and key issues. *International Journal of Network Security* 2 (1): 1–13.

Zhang, J., Xie, W., and Yang, F. (2015). An architecture for 5G mobile network based on SDN and NFV. 6th International Conference on Wireless, Mobile and Multi-Media (ICWMMN 2015).

Zhang, K., Mao, Y., and Leng, S. (2016). Energy-efficient offloading for mobile edge computing in 5G heterogeneous networks. *IEEE Access* 4: 5896–5907.

Zhang, A., Wang, L., Ye, X., and Lin, X. (2017). Light-weight and robust security-aware D2D-assist data transmission protocol for mobile-health systems. *IEEE Transactions on Information Forensics and Security* 12 (3): 662–675.

Zhang, P., Yang, X., Chen, J., and Huang, Y. (2019). A survey of testing for 5G: solutions, opportunities, and challenges. *China Communications* 16 (1): 69–85.

Further Reading

Abd-Elrahman, E., Ibn-Khedher, H., and Afifi, H. (2015). D2D group communications
security. 2015 International Conference on Protocol Engineering (ICPE) and
International Conference on New Technologies of Distributed Systems (NTDS), 1–6,
IEEE.

Alam, M.J. and Ma, M. (2017). DC and CoMP authentication in LTE-advanced 5G HetNet.
GLOBECOM 2017-2017 IEEE Global Communications Conference, 1–6, IEEE.

Al-Shaikhli, A., Esmailpour, A., and Nasser, N. (2016). Quality of service interworking over
heterogeneous networks in 5G. 2016 IEEE International Conference on
Communications (ICC), 1–6, IEEE.

Boubakri, W., Abdallah, W., and Boudriga, N. (2017). Access control in 5G communication
networks using simple PKI certificates. 2017 13th International Wireless
Communications and Mobile Computing Conference (IWCMC), 2092–2097, IEEE.

Chopra, G., Jha, R.K., and Jain, S. (2018). Security issues in ultra dense network for 5G
scenario. 2018 10th International Conference on Communication Systems & Networks
(COMSNETS), 510–512, IEEE.

Dubrova, E., Näslund, M., and Selander, G. (2015). CRC-based message authentication for
5G mobile technology. 2015 IEEE Trustcom/BigDataSE/ISPA, Vol. 1, 1186–1191, IEEE.

Gohil, A., Modi, H., and Patel, S.K. (2013). 5G technology of mobile communication: a
survey. 2013 international conference on intelligent systems and signal processing
(ISSP), 288–292, IEEE.

Meneses, F., Guimares, C., Corujo, D., and Aguiar, R.L. (2018). SDN-based mobility
management: handover performance impact in constrained devices. 2018 9th IFIP
International Conference on New Technologies, Mobility and Security (NTMS), 1–5,
IEEE.

Yoo, T. (2016). Network slicing architecture for 5G network. 2016 International Conference
on Information and Communication Technology Convergence (ICTC), 1010–1014,
IEEE.

Zhang, A. and Lin, X. (2017). Security-aware and privacy-preserving D2D communications
in 5G. *IEEE Network* 31 (4): 70–77.

6

Authentication and Access Control for 5G

Shanay Behrad[1], Emmanuel Bertin[1], and Noel Crespi[2]

[1] *Orange Labs Caen, Caen, France*
[2] *Institut Mines-Telecom, Telecom SudParis, Paris, France*

Introduction

Authenticating users and controlling their access to network services is one of the first procedures in cellular networks. This procedure is mandatory for providing suitable connectivity services to the network's subscribers, preventing the network from being abused, and protecting the subscribers' privacy and their information. In other words, the AAC (Authentication and Access Control) mechanisms provide secure network services for network subscribers. Despite the progress of the cellular networks in each generation to fulfill a broader range of requirements and use cases, the similar progress has not been made in the way of their AAC mechanisms.

The first generation of cellular networks (1G, Nordic Mobile Telephony in Europe and Advance Mobile Phone System in the United States) was based on analog technologies and supports only voice call services. Due to its analog nature, its radio links did not support any encryption and an attacker just needed a radio scanner to intercept the calls.

The second generation of cellular networks, GSM (2G, Global System for Mobile Communications) was introduced in 1991 as the first digital communication system. In addition to provide messaging services, it also introduces AAC for its users. The AAC is done through the SIM (Subscriber Identity Module). A SIM card is a well-known secure element that is provided by the operator to its subscribers and contains the subscriber's permanent identity calls IMSI (International Mobile Subscriber Identity) and a long-term secret key used for encryption and establishing a secure connection between the subscriber and the network. However, its lack of mutual authentication has led to active attacks against subscribers (e.g. an attacker can impersonate itself as a valid network to subscribers) (Behrad et al. 2018, 2019a).

In UMTS (3G, Universal Mobile Telecommunications system), the data application and mobile internet services were introduced. In terms of AAC, the 3GPP (3rd Generation Partnership Project) defined AKA-based (Authentication and Key-Agreement) protocols (3GPP TS 33.102 2018) with mutual authentication feature to address the security issues raised in 2G.

In 2010, LTE (4G, Long-Term Evolution) system was introduced to support higher data transmission speed (up to 100 Mbps at the early stage) all-IP architecture. In LTE,

The Wiley 5G REF: Security. Edited by Rahim Tafazolli, Chin-Liang Wang, Periklis Chatzimisios and Madhusanka Liyanage.
© 2021 John Wiley & Sons Ltd. Published 2021 by John Wiley & Sons Ltd.

the AKA-based protocols are used for the AAC purpose as in the 3G systems. The AKA mechanism in LTE (evolved packet system, EPS–AKA) is a complementary form of the AKA mechanism in 3G (UMTS–AKA), with a few differences (3GPP TS 33.401 2019).

In the fifth generation of the cellular networks, 5G, the aim is to fulfill the increasing demand for the higher throughput, the low latency, and the better quality of service. Some additional concepts have also been included in the scope of 5G, such as handling the connectivity for the massive number of Internet of Things (IoT) devices, providing network slices to specific customers or vertical sectors, and managing heterogeneous network access (e.g. addressing Wi-Fi and cellular access networks from a converged network) (3GPP TS 23.501 2019). All of these requirements and concepts affect the whole network and the associated security needs. Although the different security requirements of the new use cases, the way of AAC, and main protocols provided to fulfill its requirements (e.g. 5G-AKA), still remains the same as the two previous generations (3GPP TS 33.501 2019). The standards enhance the AAC mechanisms in 5G from the security point of view only with the central role of the connectivity provider (operator). The main focus is on detecting the shortcomings of the pre-5G generation's AAC methods (e.g. EPS–AKA) and solve their security issues in designing the AAC methods for 5G. The need of more open and flexible network in presence of new actors in the 5G environment is not considered in the mentioned enhancements as well as in the AAC mechanisms proposed in the literature.

This article reviews the AAC procedures proposed by the 3GPP for the 5G systems as well as the AAC mechanisms in the literature. The primary contributions of this article are the following:

1) A survey of the vulnerabilities of the AAC mechanism in 5G, the possible attacks against them, and the clarification of their goals in the section titled "Authentication and Access Control in 5G."
2) A study of the new needs arising from the new 5G use cases and a discussion about the abilities of the proposed AAC in fulfilling the security requirements of these new use cases in the sections titled "5G-Specific Use Cases and Requirements" and "AAC Proposal in Cellular Networks."

The section titled "Basics of Authentication and Access Control" explains the basics of the AAC mechanisms. The section titled "Overall architecture of the AAC in 3G, 4G, and 5G" gives an overall view of the AAC mechanisms in the recent 3 cellular network generations and the main entities evolving in these mechanisms. The section titled "5G Network Architecture" details the networks functions of the 5G architecture. The section titled "New Concepts in 5G" is an introduction of the new concepts and issues raised in 5G and depicts the reasons of why the enhancements just in the security part of the AAC are not enough in the 5G systems.

Basics of Authentication and Access Control

The general concepts of AAC mechanisms in different cellular systems are similar although their associated entities may vary. The main purposes of these mechanisms are protecting the subscribers and the network and applying billing rules for the usage of network resources (Wong et al. 2017; Koutsopoulou et al. 2004). The objective of

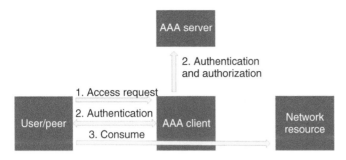

Figure 1 Functions of an AAC system. Source: Behrad et al. (2019b) Reproduced with permission of IEEE.

the authentication is to verify the user's identity (to know who the user is) by checking its credentials (Velte and Velte 2006). It enables authorization which defines rules to access specific resources and services (specify what a user can do) (Metz 1999). Access control is the method to control and enforce the access rights of users on the resources, such as data or even to IoT objects (Gusmeroli et al. 2012).

The architecture of the AAC mechanisms consists of two main entities: an AC (Access Control) server and an AC client. The AC server consists of a database storing the users' data and its responsibility is to manage the AC processes according to this database. When users try to access the network, they send their request to the AC client and it queries the AC server according to these requests (Yegin and Watanabe 2005). In the different systems, different physical entities play the role of the AC servers and AC clients. Figure 1 depicts an AAC system used in cellular networks. First, the user (peer) sends its access request to the AC client. Then, according to this request, the AC client authenticates the user considering the UE's related data stored in the AC server. Finally, the user accesses the network resources.

Overall Architecture of the AAC in 3G, 4G, and 5G

From the AAC perspective, a cellular network (e.g. 3G, 4G, and 5G) consists of three main parts: UEs (User Equipment) or devices, a SN (Serving Network), and a HN (Home Network). The home network is the network which the UEs have subscriptions with. The serving network is the network which the UEs served at and it is changed in the roaming scenarios. Considering the AAC entities described in the section titled "Basics of Authentication and Access Control," the SN is the AC client and the HN is the AC server. Figure 2 depicts the AAC model in the cellular networks.

Following is the summary of the AAC systems used in the cellular networks (the details of the architectural entities and the AAC protocols in 5G are described in the sections titled "5G Network Architecture" and "Authentication and Access Control in 5G," respectively). These AAC systems rely on a long-term secret key shared between:

- a hardware security module in the form of an UICC (Universal Integrated Circuit Card) running an USIM application (Universal Subscriber Identity Module, the counterpart of the SIM in 2G) inside the UE.
- a database holding the UEs' subscription data: the HLR (Home Location Register) in 3G, the HSS (Home Subscriber Server) in 4G, and the UDM/ARPF (Unified Data

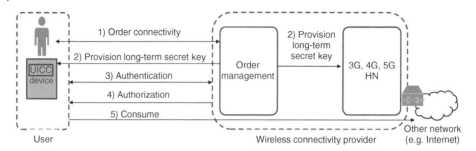

Figure 2 AAC model in the cellular networks.

Management/Authentication Repository and Processing Function) in 5G, which is capable of authenticating the UICC.

The subscription information of the UEs is stored in the HN at the ordering time. After confirming the identity of the UE by authenticating the UICC, the operator as the cellular network provider retrieves the UE's subscription information to determine the cellular services that the UE is authorized to use. The cellular network provider then bills the UE for the services it consumes.

The main mechanisms to fulfill the AAC requirements in the cellular networks are based on AKA protocols that are challenge-response protocols with mutual authentication feature. It means that in addition to the authentication of the subscribers in the network, the subscribers also authenticate the network. These AAC mechanisms are called UMTS–AKA, EPS–AKA, EAP–AKA (Extensible Authentication Protocol), EAP–AKA', 5G-AKA, and EAP–TLS. All of these mechanisms (except the EAP-TLS) are based on symmetric cryptography and secret keys shared between the UEs and the HN. The UMTS–AKA, the EPS–AKA, and the 5G-AKA are used to authenticate the subscribers connected across 3GPP access networks to the core network of 3G, 4G, and 5G, respectively.

The UMTS–AKA involves the USIM in the subscriber's mobile equipment, the VLR/SGSN (Visitor Location Register/Serving GPRS Support Node, responsible for mobility management) in the SN, and the HLR in the HN. Authentication of the subscribers is based on their unique identity, IMSI, and a shared secret key K that is stored both inside the USIM and the HLR. This identity is provisioned by an order management module from the mobile network operator Information System while the subscriber buys a UICC from the operator (Figure 2) (3GPP TS 33.102 2018).

As with the UMTS–AKA, the EPS–AKA operates between the USIM, the MME (Mobility Management Entity, the main control node of the network) in the SN, and the HSS in the HN. EPS–AKA is also based on IMSI and a shared secret key between the USIM and the HSS (previously provisioned by the order management module). One of the important differences between the EPS–AKA and the UMTS–AKA protocols is that the EPS–AKA uses the serving network's identity in deriving the further keys in the key hierarchy (from the shared secret key K), to secure the connections between the network elements. The binding of the keys to the serving network identity reduces the probability of a serving network impersonation fraud (Behrad et al. 2019a). The 5G-AKA in 5G network also operates between the UEs and the network (details are in the section titled "Authentication and Access Control in 5G") with the same AAC manner as the AAC in the pre-5G cellular network generations (3GPP TS 23.501 2019).

The EAP–AKA and the EAP–AKA′ are responsible for the authentication of subscribers when they try to access a 3GPP core network via a non-3GPP access network (e.g. via a public or private Wi-Fi network). These two protocols belong to the EAP framework. In this framework, we have "authenticators" (the AC client) and "EAP servers" (the AC server). In the EAP–AKA and the EAP–AKA′, the authentication process is based on NAI (Network Access Identifier, derived from IMSI) and a shared secret key as in UMTS–AKA, EPS–AKA, and 5G-AKA. It is performed between USIM or any other application with a similar functionality (this part is left unspecified in 3GPP specifications because of the use of non-3GPP access networks) and the network. The EAP–AKA′ is more secure than the EAP–AKA as it uses serving network identity in key derivation processes like the EPS–AKA (3GPP TS 23.501 2019; 3GPP TS 33.401 2019).

5G Network Architecture

3GPP has provided a technical specification to define the architecture of 5G systems and to specify the main nodes and their responsibilities (3GPP TS 23.501 2019). In this architecture, control planes and user planes are separated as much as possible to achieve more flexible and scalable deployment. Instead of Network Entities grouping many functions, 3GPP attempted to define NFs (Network Function) with more atomistic roles (i.e. one specific responsibility per function). However, most of these NFs are somehow a mapping of existing 4G entities. Two representations are possible for NF interactions: one of them is based on the SOA (service-oriented architecture) viewpoint and the other is based on traditional reference points. In service-based representation, an NF exposes a set of services it offers to other NFs, and it uses the services provided by other NFs. All interactions are carried by the same protocol for API invocations. Each time a new NF needs to be plugged in, only its new API should be declared to the other components. In the reference point representation, specific protocol links are kept between pairs of network functions.

The following NFs can be seen as an evolution of the HSS:

- AUSF (Authentication Server Function) provides a unified framework for authentication issues (for 3GPP access as well as non-3GPP access); and
- UDM (Unified Data Management) contains data that is related to the HSS (i.e. user data). The UDM stores only some part of the data (such as a user's subscription data) and not all of it. It also supports authentication credential processing, user identification handling, and access authorization.

Two other NFs can be seen as a division of the 4G MME:

- AMF (Core Access and Mobility Management Function) has different functionalities, including access authentication and authorization, registration management, and mobility management. Since different access technologies will be used, 5G needs a common framework for access management, as well as for handling mobility between different types of access. Therefore, the AMF will support both 3GPP access networks and non-3GPP access networks. Unlike 4G (where the MME is used for 3GPP access and the AAA server for non-3GPP access), the structure of the core network will be common for 3GPP access and non-3GPP access in the 5G system.

- SMF (Session Management Function) is responsible for session management and some other functionality, such as the allocation of IP addresses and the control of the policy enforcement and QoS (establishment of a session is totally separated from mobility management in 5G).

A new function is also dedicated to network slicing:

- NSSF (Network Slice Selection Function) determines the serving AMF for the UE and selects network slice instances for it (in addition to the network slicing concept, network slice instances provide specific services to different enterprises).

There are also other network functions in the 5G network architecture, but we mentioned the network functions that involve in the UEs/devices network attachment and AAC procedures.

Authentication and Access Control in 5G

The 5G architecture comes with some new design choices for the authentication and access control, but also brings much continuity. The most important continuity concerns the symmetric key-based authentication through a secure element. In 5G specifications release 16, it is decided to keep a secure element in the UE or device (like the UICC in 4G and 3G and the SIM card in 2G) to process subscription credentials (3GPP TS 33.501 2019), which could also be an ESIM (Embedded Subscriber Identity Module) provided by device makers and with which operators can provision their profile over-the-air at the subscription time. The authentication methods introduced in 5G specifications are 5G-AKA, EAP–AKA′, and EAP–TLS (Transport Layer Security).

5G defines a new type of identifier, the SUPI (Subscriber Permanent Identifier), which is somehow equivalent to the IMSI but with a more global footprint, as it can be used not only for cellular service subscribers but also for different environments like the IoT. The SUPI can have different formats: IMSI and NAI (Network Access Identifier). The NAI is more flexible than the IMSI and it can include different identifiers (including the IMSI). To protect user privacy, the MSIN part of the identifier will be encrypted with the public key of the subscriber's home network (limiting the IMSI disclosure vulnerability). This choice can be justified as follows: if all parts of the identifier were encrypted, the decryption would have to be done in the serving network to route the messages to the right home network. This would impose the need for a global mechanism to distribute and manage certificates as well as to control multiple public keys for different serving networks. The SUCI (Subscription Concealed Identifier) contains the concealed SUPI. The public key of the home network could be stored in the secure element of the UE. The 5G-GUTI is also used as the temporary identifier, like the GUTI in the 4G systems. Sending the SUCI as the encrypted form of the SUPI over the radio links is the major security improvement of 5G in comparison with the former generations. It prevents the UEs' or devices' permanent identifiers to be sent in a clear text over the radio (3GPP TS 33.501 2019).

5G-AKA Protocol

Without the loss of generality, we focus on the 5G-AKA protocol (and its differences with the EPS–AKA protocol) as it is the main authentication and key agreement protocol to fulfill the AAC requirements in 5G. Figure 3 depicts the detailed message flow in 5G-AKA procedures (3GPP TS 33.501 2019). As mentioned in the section titled "Authentication and Access Control in 5G," the authentication mechanisms in 5G systems will be done along with the same principle as in 4G systems with some minor differences. These differences in AKA mechanisms will be from the network perspective only, and not from the UE perspective. AKA mechanisms in 5G systems, like those in 4G systems, use a "serving network name" (like SNid in 4G) to derive the anchor key (K_{SEAF}); thus, the anchor key will belong to the specific serving network and this serving network cannot pretend to be another serving network. Different from the EPS–AKA, there are four actors in the 5G-AKA, which are explained further in this section. Another difference in AKA mechanisms for 5G systems is that the anchor key (K_{SEAF}), which is derived in a 3GPP access, can also be used in a non-3GPP access without a new authentication process. As mentioned in the section titled "Overall Architecture of the AAC in 3G, 4G, and 5G," the 4G systems use EPS–AKA for 3GPP access and EAP–AKA for non-3GPP access, but in 5G systems both 5G-AKA and EAP–AKA' can be used in 3GPP access and non-3GPP access. The NAS context (e.g. K_{AMF} which is derived from K_{SEAF}) is needed for 5G-AKA, which is not present for non-3GPP access, and so, at the beginning of non-3GPP access, only EAP–AKA' is foreseen (Dehnel-Wild and Cremers 2018).

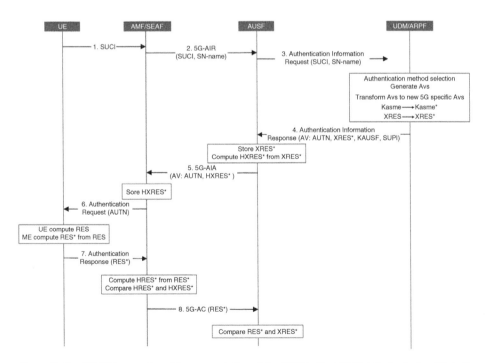

Figure 3 The 5G-AKA procedure. The computation of the RES* in the ME (Mobile Equipment) is in the same way as the computation of the XRES* in the ARPF. The computation of the HRES* in the SEAF is in the same way as the computation of the HXRES* in the AUSF.

The authentication process will involve the UE, the SEAF (Security Anchor Function) in the serving network, the AUSF in the home network, and the UDM/ARPF (Authentication Repository and Processing Function) also in the home network (3GPP TS 33.899 2018). The SEAF will be included in the AMF and interact with the AUSF to obtain authentication data from the UDM. It accomplishes UE authentication for different access networks. The ARPF stores subscribers' profiles and the information related to security. It also selects an authentication method (e.g. 5G-AKA, EAP–AKA', EAP–TLS) based on the subscriber's identity and computes the keying materials for the AUSF. According to Figure 3, the call flow of the 5G-AKA is as follows (Khan and Martin 2019):

1) At the beginning of the authentication process, the UE sends its SUCI to the SEAF.
2) After receiving the signaling message from the UE, the SEAF sends the 5G-AIR (Authentication Initiation Request) message to the AUSF. The 5G-AIR contains the SUCI or SUPI of the UE and the name of the serving network. This message also indicates that the UE is using a 3GPP access or a non-3GPP access.
3) After verifying the authorization of the serving network that asks for the authentication service, the AUSF sends the AIR message to the UDM/ARPF. If the AUSF sends the SUCI in this message, the SIDF (Subscription Identifier De-concealing Function) which is collocated with the UDM/ARPF decrypts the SUCI to obtain the SUPI. After receiving the authentication information request from the AUSF, the UDM/ARPF generates an AV as in 4G, and then transforms them to new AVs that are specific to 5G systems. This transformation will be different according to the chosen authentication method.
4) The UDM/ARPF sends the AVs containing the AUTN, XRES*, KASUF, and the decrypted SUPI to the AUSF in the Authentication Information Response message. On receiving this message, the AUSF computes the HXRES*, which is the hash of the XRES* and stores the KAUSF.
5) The AUSF sends the 5G-AIA (Authentication Information Accept) including the HXRES* to the SEAF. This message does not include the SUPI and the AUSF (in the home network) sends the SUPI to the SEAF (in the serving network) only after the successful UE authentication.
6) After storing the HXRES*, the SEAF sends the AUTN token in the Authentication Request message to the UE. The UE checks the validity of the AUTN (using its secret key shared with the home network). If the AUTN is valid, the network authentication in the UE is successful. If the AUTN is not valid, the UE sends the MAC Failure message to the SEAF (Message Authentication Code). Next, as in the ESP–AKA procedure, the UE checks the sequence number (SQN) derived from the AUTN to control the freshness of the AUTN. If this verification fails, the UE sends the Synchronization Failure message to the SEAF. The UE also computes the RES*.
7) The UE sends the RES* to the SEAF in the Authentication Response message. The SEAF checks the validity of the RES* by computing the HRES* and comparing it with the HXRES*.
8) For making the final decision about the UE's authentication by the home network, the SEAF sends the 5G-AC (Authentication Confirmation) message including the RES* to the AUSF. The AUSF checks the validity of the RES* by comparing it with the XRES*. Sending the 5G-AC message from the SN to the HN is a prevention against the possible billing cheat raised by the SN (Jover 2019).

In terms of success, the AUSF computes the K_{SEAF} and sends it to the SEAF along with SUPI. The further keys for securing the radio connections are derived from the K_{SEAF}. In addition to use of the encrypted form of the UEs or devices' permanent identifies, 5G-AKA differs from the EPS–AKA in the following areas:

- In the 5G-AKA, the AUSF, which is a part of the home network, makes the final decision on the UE's authentication. But in the EPS–AKA, the HN (HSS) only generates the authentication vectors and does not make decision on the UE's authentication. This property in 5G-AKA reduces the level of trust the 5G system has to put into the SNs. Thus, the SN cannot send fake authentication information requests to the HN for the UEs not attached to one of its gNBs (5G base station).
- The key hierarchy in the 5G-AKA is different from the key hierarchy in the EPS–AKA. In addition to the K_{SEAF} which operates like the K_{ASME} in EPS–AKA (the anchor key), the 5G-AKA also introduces the K_{AMF} as another intermediate key.

The 5G system also supports the EAP–AKA′ and the EAP–TLS methods. The EAP–AKA′ method is also based on the symmetric cryptography and it has the same security characteristics as the 5G-AKA with some differences related to the message flows, the role of the SEAF, and the derivation of the K_{AUSF}. The EAP–TLS method is different from the 5G-AKA and EAP–AKA′ and it can be used in some private networks and IoT use cases. The mutual authentication in this method is based on the certificates. Although the EAP–TLS eliminates the need of storing the long-term keys in the home network, it increases the overhead of the system as it has to manage the certificates (Zhang et al. 2019).

Security Flaws in 5G-AKA

Although the 5G-AKA is not in the operational stage yet, some security flaws have already been recognized. This section summarizes the 5G-AKA vulnerabilities found so far. As in the 4G network, the communications between the network functions within the 5G core network is done through the secure channels (the communications between the AMF/SEAF, AUSF, and UDM/ARPF) (3GPP TS 33.501 2019; Dehnel-Wild and Cremers 2018). But the communications between the UEs or devices and the AMF/SEAF are subject to passive and active attacks (Dehnel-Wild and Cremers 2018; Basin et al. 2018). The vulnerabilities of the 5G-AKA and the possible attacks against it are as follows:

- According to step 6 of the 5G-AKA procedure in the section titled "5G-AKA Protocol," the UE or device sends the failure messages in clear text. This vulnerability can cause the "Linkability Attack." The attacker can capture the authentication request message, which is sent from the SEAF to the UE (or device) and replay it after. If the UE (or device) answers with the Synchronization Failure message, the attacker determines the presence of the target UE (or device) in a particular area (Basin et al. 2018; Liu et al. 2018; Braeken et al. 2019; Koutsos 2019). In Borgaonkar et al. (2019), the authors introduce the "Location Confidentiality Attack," which is against the user location confidentiality but as it is mentioned in Khan and Martin (2019). This attack is an extension of the "Linkability Attack." The proposed solutions for addressing the mentioned vulnerability and the attacks are based on the encryption of the failure messages with the public keys of the connectivity providers (operators). But in this case there is a need for a global PKI (Public Key Infrastructure) among all the operators, which is not feasible (Liu et al. 2018). The authors in Borgaonkar et al. (2019) also introduce another attack called "Activity Monitoring Attacks," which is also caused by the transmission of the Synchronization Failure message in clear (Braeken et al. 2019). They claimed that an attacker can break the confidentiality of the SQN and monitor the activity of the target UE or device and learn its typical service consumption from the difference between the SQNs at two different times. But as it is mentioned in (Khan and Martin 2019), the prerequisite of this attack is the compromise of the identity confidentiality and the location confidentiality of the target UE, which is difficult to obtain (especially with using the SUCI instead of the SUPI).

- The pre-authentication messages, such as the RRC (Radio Resource Control) messages (e.g. RRC Connection Request), the NAS messages (e.g. Attach Request), and some other messages (e.g. Paging), are transmitted in clear. All the following procedures between the UEs or devices and the network are based on these messages that may come from fake base stations or fake UEs (Jover 2019; Jover and Marojevic 2019). This vulnerability also exists in the EPS–AKA procedure and can cause the same attacks in 5G such as the DoS attacks against UEs or their location confidentiality. In 5G-AKA, only the disclosure of the UE's permanent identity which is related to this vulnerability is addressed.

As it is mentioned in the section titled "Authentication and Access Control in 5G," the main improvement in the AAC procedure in 5G is the encryption of the UE's/device's identity. Consequently, the attacks that are related to the disclosure of the IMSI in the first attachment of the UE to the network in 4G are addressed in 5G. But the other attacks against the 4G network still remain in the 5G network (Behrad et al. 2018).

New Concepts in 5G

The fifth generation of mobile communications has a number of goals, such as achieving low latency, high data rates, increased convergence, accessibility, and dense connectivity. 5G will also support IoT services and address the needs of different vertical markets, such as health care, automotive, and transport. The 5G-PPP (Fifth Generation Public Private Partnership) has defined several different use cases for 5G, including enhanced mobile broadband and critical communications (Alliance 2015).

These different goals and use cases have important impacts on the security aspects of the system, and service-specific security requirements should be considered when designing appropriate authentication and access control mechanisms for 5G networks (e.g. fast communications need fast AAC procedures) (Schneider and Horn 2015). As another example, in the IoT, numerous devices may access the network at the same time, and so the network should have the ability to control this large amount of signaling traffic and authenticate the devices correctly to avoid DDoS (Distributed Denial of Service) attacks. The IoT devices have low power capacity and cannot support strong authentication procedures. In addition, they are usually able to connect to the network via non-3GPP access options (some of them will not have 5G radio access and will use Wi-Fi or Bluetooth) (Naslund et al. 2017). In light of these limitations, some solutions based on group-based authentications with an IoT gateway have been proposed to decrease the number of full AKA procedure executions (Li et al. 2013; Su et al. 2016). But these group-based AKA solutions have their own weaknesses. While some of these include the traditional AKA weaknesses mentioned in the previous section, some are specific to the group-based nature of these approaches. For example, an attacker can pose as a member of a group and get access to the network (Giustolisi and Gerhmann 2016).

The aforementioned requirements of 5G have also produced new concepts, and thus new security issues:

- Network slicing is a solution to meet heterogeneous requirements from different vertical markets (Foukas et al. 2017). Networks slices are logical networks relying on a single physical network (Chatras et al. 2017). Each network slice is composed of various network functions to provide specific capabilities and to satisfy a specific type of usage (Chatras et al. 2017). For example, in some IoT cases (e.g. a smart factory), mobility will not be very high, so it may not need mobility handling functions (Chatras et al. 2017). There can be different approaches in providing network slicing (for example, we can have a slice per service or a slice per vertical market). Different technologies like SDN (Software-defined Network), NFV (Network Function Virtualization), and automation as with ONAP (Open Network Automation Platform) will be used to deploy slicing. Ordonez-Lucena et al. (2017), Katsalis et al. (2017), and Rost et al. (2017) present some proposals for network slicing architecture and implementations. Concerning security, network slicing also adds some issues such as slice isolation to prevent threat propagation through slices, authentication, and integrity protection of input data, and access control between slices (Naslund et al. 2017).
- Heterogeneous network access, as different radio technologies might be used to access 5G networks. As we mentioned before, one of the 5G goals is to provide a better accessibility to users; therefore, when users do not have 5G connectivity, they may connect to 5G network through other types of accesses, e.g. satellite access. In IoT case, devices may also use different radio access technologies. In these situations, the enterprises or satellite providers may have their own AAA servers and the management of the connection between different AAA servers, especially in roaming scenarios which is very important (Naslund et al. 2017; Nasland et al. 2016). It is also important to prevent the network against unauthorized access in this heterogeneous infrastructure (Bisson and Waryet 2017).

5G-Specific Use Cases and Requirements

The fifth generation of mobile cellular networks, 5G, is designed to support a set of new use cases and requirements. The purpose of addressing these use cases and the derived requirements not only involves the 5G network operators (connectivity providers) and the end users but also brings different 3rd parties in the 5G environment. By emerging the different 3rd parties and business actors in the 5G environment, the concepts like "wholesale wireless connectivity" is gaining more and more attention. With wholesaling wireless connectivity, network operators (connectivity providers) sell connectivity to different 3rd parties which in turn provide them to their own users, in a

B2B2C business model (Business to Business to Consumer). Therefore, the wholesaling of wireless connectivity appears as a key issue, especially for the IoT use cases targeting vertical sectors that are involving end-users (e.g. connected car occupants).

The connectivity providers are trying to address these different use cases and their requirements by using network slicing architecture. The standards categorized the different use cases in four groups and defined four types of slices for each of them in the 5G specifications release 16: eMBB (Enhanced Mobile Broadband), URLLC (Ultra-Reliable and Low Latency Communications), MIoT (Massive Internet of Things), and V2X (Vehicle to Everything). But these network slices only consider the different QoS requirements (e.g. bandwidth, latency, etc.) of the use cases while they have other types of requirements as well. As it is mentioned in the section titled "Introduction," the standards provide the same AAC model for the different use case although they have different AAC requirements.

Motivated Use Cases

Three typical use cases are described below for deriving the requirements to address on the end-users' side, on the 3rd party organizations' side, and for the 5G network operators (Behrad et al. 2019c).

1) Alice buys a device with cellular connectivity to stay connected everywhere (e.g. a connected vehicle). She wants to have wireless connectivity embedded inside her device. That means she does not want to have an additional subscription with a wireless carrier and the need to set up an accounting plan with that carrier.
2) Alice lives in a smart home with a smart light system, a smart energy usage control system, a smart entertainment system, and a smart lock system. The IoT devices of these systems are connected to the outside world through a 5G network. The security of the data issued by the different elements of Alice's smart systems is important to her, but the leakage of some of them would cause more serious problems to her than the others' (malicious access to the smart lock system is more dangerous than malicious access to the entertainment system) (Geneiatakis et al. 2017). On the other hand, most of her devices are constrained devices with low energy and processing power and they are not able to support strong security algorithms (Naslund et al. 2017).
3) Alice works as factory manager at Acme Corporation. She wants to better automate the production of her factory. Alice subscribes to a 5G network slice, so her factory robots can access this slice through 5G connectivity. Acme only trusts itself to provide security policies, accounting, and configurations data for its factory robots (Nasland et al. 2016). So Alice wants to manage the identities and credentials of the robots, as well as their life cycles (from enrollment to decommissioning). She does not want to rely on the 5G network operator for installing each new robot or for uninstalling and eliminating a robot's profile and credentials from the network.

Derived Requirements

As can be inferred from the use cases, there are a number of requirements for slice-specific AAC mechanisms. These requirements are summarized as follows:

- R1: Provide embedded connectivity inside devices. Future connected devices such as connected vehicles and future things for automation and assisted living are now believed to be best retailed when connectivity is directly commercialized with the device, for a better customer experience. In these cases, a connectivity provider (i.e. the 5G network operator) sells connectivity to different verticals, which in turn provide them to their own users in a B2B2C business model. The 3rd parties (verticals) should then be able to manage the identities and credentials of their provided devices to control their subscriptions and connectivity usage.

- R2: Allow 3rd parties to choose their own AAC methods. The security requirements in each of the use cases are distinct. In other words, the sensitivity of the signaling and data messages between the devices and the network is not the same for all types of devices (nor for all use cases). Therefore, the network should have the ability to allow the 3rd parties to choose the appropriate AAC mechanisms according to the security requirements of their proposed services.

- R3: Allow 3rd parties to manage the lifecycles of their devices. The fleet of devices belonging to a specific 3rd party is not static. New devices are regularly added to this fleet and old one uninstalled. The network should offer 3rd parties the ability to control the whole lifecycle from their devices, from enrollment to disenrollment processes.

- R4: Provide AAC mechanisms for constrained devices. The devices involved in each use case are different in terms of computational power and restricted in their energy supply. The network should give 3rd parties the ability to apply the most suitable AAC mechanisms for each type of constrained devices.

- R5: Support for a massive number of devices. A massive number of devices attempting to simultaneously connect to the 5G network operator's core network (by sending attachment and AAC requests) may cause congestions in the core network and bring latency. Therefore, the network should be able to give the ability to the 3rd parties to manage the AAC of their provided devices to avoid the congestions in the 5G network operator's core.

AAC Proposal in Cellular Networks

Today, the use of eSIM (more precisely, eUICC) that means an embedded SIM instead of a plastic SIM card is gaining more and more attention. Through eSIMs, users can choose which operators they would like to subscribe to. Over-the-air activation methods are proposed to provision the needed credentials to the eSIMs in a secure manner (GSMA 2016). Although it is possible to add embedded connectivity features to some devices through the eSIMs, identity management and connectivity usage control of these devices are still done under the responsibility of the operator and not of the device providers (3rd parties). Therefore, the 3rd parties are not able to choose their AAC mechanisms according to their security requirements and manage the lifetime of their devices (They rely on the connectivity provider in the AAC level.). The AAC mechanisms in eSIMs are also based on the AKA protocols. However, AKA protocols used in cellular networks (e.g. 5G-AKA) are not fully suitable for constrained devices, as these devices may not be able to compute with the required cryptographic algorithms. Moreover, when a massive number of devices is simultaneously attaching to the network, these protocols increase the computations overhead on the operator's network side as well (Ferrag et al. 2017; Parne et al. 2018).

To overcome the shortcomings of AKA protocols in the presence of massive-constrained device, group-based AAC mechanisms have been proposed (3GPP TS 22.368 2017). The general process of these mechanisms is the following one:

- to form a group of devices based on their local communication areas, applications, or behaviors;
- to choose a leader device for the group based on its computational and battery capacity;
- to forward the signaling messages (authentication requests) of the group members to the network through this group leader (Parne et al. 2018; Li et al. 2013; Yao et al. 2016; Cao et al. 2012).

In Lai et al. (2013), the devices form a group also, but they do not choose a group leader. The authentication is done between the first device who attempt to connect to the network and then continued locally with the remaining members of the group (more precisely, the remaining devices and the serving network). These group-based AAC mechanisms address the requirements of constrained devices and solve the network congestion problems caused by a massive number of authentication requests. However, as the management of joining and leaving the devices in the group is done locally in the serving network, the core network is not aware of each individual device's behavior. It means that, although the core network provides services to each member of the groups, it is not able to control their connectivity usage and security issues (Giustolisi and Gerhmann 2016). For example, it is also not possible to provide different services to each member of the group (including AAC services) although their requirements would be different.

There are also some AAC mechanisms designed for preserving the privacy of the UEs (devices) when trying to connect to a service provider network, or foreign serving networks in the roaming scenarios. In Ni et al. (2018), the authors propose an authentication procedure between the UEs and the IoT service providers, in addition to the existing 5G-AKA between the UEs and the 5G network provider. They try to protect the service data and UEs' privacy (UEs are able to anonymously ask for services) against the intermediate nodes like gNBs (i.e. 5G base station) and inhibit them to capture sensitive information about UEs. Lai et al. (2014) and Liu et al. (2014) also provide anonymity when the UEs visit a serving network that is different from its home network. Although these mentioned papers show that it is possible to design AAC mechanisms based on the service providers or the visited serving networks' security requirements, the network does not provide the ability of choosing the AAC mechanisms in a dynamic way. They are not suitable for authenticating the massive number of constrained devices as well.

The different AAC methods proposed for cellular networks and their compatibilities with the different requirements mentioned in the section titled "Derived Requirements" are summarized in Table 1. As we can see in this table, cellular AKA and service-oriented and anonymity-based methods fully meet none of the requirements; eSIM method just addresses embedded connectivity inside the device; while group-based AAC methods address the AAC requirements of the constrained devices and the mass number of devices' simultaneous connectivity request.

Table 1 Different AAC mechanisms and their compatibility with the different requirements (Behrad et al. 2019c).

AAC method	R1	R2	R3	R4	R5
Cellular AKA	−	−	−	−	−
eSIM	+	−	−	−	−
Group based	−	−	−	+	+
Service oriented and anonymity based	−	+/−	−	−	−

Summary

Since 1G, the cellular networks have made a significant progress in technologies used (from the analog communications in 1G to the all-IP networks in 4G) and the use cases addressed (from voice calls in 1G to the high-speed data transmissions in 4G) in the pre-5G networks. This progress in each generation is the continuation of the previous generation in terms of the technologies used and the use cases addressed. However, this is not the case for 5G. The 5G network is not just a continuity of the 4G network and it aims to address much broader use cases in many different vertical domains (than any other pre-5G networks) using the virtualization technologies and providing new virtual network functions. Considering the use cases that the 5G aims to address, the 5G environment does not only consist of the connectivity providers (operators) and the end uses (mobile devices) there are also other business actors such as different 3rd parties (who are not network operators) in the 5G environment to address the broad range of use cases. These different 3rd parties require different connectivity services (e.g. in terms of latency) and network functionalities from the connectivity providers according to their addressed use cases (e.g. smart home, smart factory). The security requirements of the 3rd parties are also different and depend on the sensitivity of their provided services, and the predicted effects of the security flaws on their network and customers (the 3rd parties need a trade-off between their required security level and the processing power of their network and provided devices).

Despite the different security requirements in the different use cases, the 5G network only provides one way of AAC of the devices in the network, which is same as the AAC in the 3G and 4G networks with few enhancements (AKA-based AAC). In this AAC model, the connectivity provider (operator) plays the central role and it is only responsible for the device identities and credentials. The need for supporting the device identities and credentials owned by an entity (3rd party) separated from the connectivity provider (operator) is mentioned as a key issue in the 3GPP technical report, "Study on enhanced support of Non-Public Networks," release 17 (3GPP TR 23.700-7 2019). In another 3GPP technical report, "Study on Security Aspects of Enhanced Network Slicing," release 16 (3GPP TR 33.813 2019), the authentication for access to a specific network slice is mentioned as a key issue. But in the proposed slice-specific authentication and authorization solution, it is mandatory to have a primary authentication between the devices and the 5G network (using 3GPP credentials), then doing the slice-specific authentication with the corresponding slice as the secondary authentication (using slice-specific identities and credentials).

Related Articles

5G-Core Network Security
5G Security – Complex Challenges

References

3GPP TR 23.700-7 (2019). Study on enhanced support of non-public networks. TR 23.700-7, Tech. Report. 0.2.0.

3GPP TR 33.813 (2019). Study on security aspects of enhanced network slicing. TR 33.813, Tech. Report. 0.7.0.

3GPP TS 22.368 (2017). Service requirements for machine-type communications (MTC). TS 22.368, Tech. Spec. 14.0.1.

3GPP TS 23.501 (2019). System architecture for the 5G system. TS 23.501, Tech. Spec. 16.0.2.

3GPP TS 33.102 (2018). Security architecture. TS 33.102, Tech. Spec. 15.1.0.

3GPP TS 33.401 (2019). Security architecture. TS 33.401, Tech. Spec. 15.7.0.

3GPP TS 33.501 (2019). Security architecture and procedures for 5G system. TS 33.501, Tech. Spec. 15.4.0.

3GPP TS 33.899 (2018). Study of security aspects of the next generation system. TS 33.899, Tech. Spec. 1.3.0.

Alliance, N. (2015). Next generation mobile networks. White paper.

Basin, D., Dreier, J., Hirschi, L., et al. (2018). A formal analysis of 5G authentication. Proceedings of the 2018 ACM SIGSAC Conference on Computer and Communications Security, ACM, 1383–1396.

Behrad, S., Bertin, E., and Crespi, N. (2018). Securing authentication for mobile networks, a survey on 4G issues and 5G answers. 2018 21st Conference on Innovation in Clouds, Internet and Networks and Workshops (ICIN), IEEE, 1–8.

Behrad, S., Bertin, E., and Crespi, N. (2019a). A survey on authentication and access control for mobile networks: from 4G to 5G. *Annals of Telecommunications* 74: 593–603.

Behrad, S., Tuffin, S., Bertin, E., and Crespi, N. (2019b). Network access control for the IoT: a comparison between cellular, Wi-Fi and LoRaWAN. 2019 22nd Conference on Innovation in Clouds, Internet and Networks and Workshops (ICIN), IEEE, 195–200.

Behrad, S., Bertin, E., Tuffin, S., and Crespi, N. (2019c). 5G-SSAAC: slice-specific authentication and access control in 5G. 2019 IEEE Conference on Network Softwarization (NetSoft), IEEE, 281–285.

Bisson, P. and Waryet, J. (2017). 5G PPP Phase1 security landscape. 5G PPP Security Group White Paper.

Borgaonkar, R., Hirschi, L., Park, S., and Shaik, A. (2019). New privacy threat on 3G, 4G, and upcoming 5G AKA protocols. *Proceedings on Privacy Enhancing Technologies* 2019: 108–127.

Braeken, A., Liyanage, M., Kumar, P., and Murphy, J. (2019). Novel 5G authentication protocol to improve the resistance against active attacks and malicious serving networks. *IEEE Access* 7: 64040–64052.

Cao, J., Ma, M., and Li, H. (2012). A group-based authentication and key agreement for MTC in LTE networks. 2012 IEEE Global Communications Conference (GLOBECOM), IEEE, 1017–1022.

Chatras, B., Kwong, U.S.T., and Bihannic, N. (2017). NFV enabling network slicing for 5G. 2017 20th Conference on Innovations in Clouds, Internet and Networks (ICIN), IEEE, 219–225.

Dehnel-Wild, M. and Cremers, C. (2018). Security vulnerability in 5G-AKA draft. Tech. Rep. Department of Computer Science, University of Oxford.

Ferrag, M.A., Maglaras, L.A., Janicke, H. et al. (2017). Authentication protocols for internet of things: a comprehensive survey. *Security and Communication Networks* 2017: doi: 10.1155/2017/6562953.

Foukas, X., Patounas, G., Elmokashfi, A., and Marina, M.K. (2017). Network slicing in 5G: survey and challenges. *IEEE Communications Magazine* 55: 94–100.

Geneiatakis, D., Kounelis, I., Neisse, R. et al. (2017). Security and privacy issues for an IoT based smart home. 2017 40th International Convention on Information and Communication Technology, Electronics and Microelectronics (MIPRO), IEEE, 1292–1297.

Giustolisi, R. and Gerhmann, C. (2016). Threats to 5G group-based authentication. *13th International Conference on Security and Cryptography (SECRYPT 2016)* (26–28 July 2016). Madrid, Spain: SciTePress.

GSMA (2016). Remote provisioning architecture for embedded UICC technical specification. ver 3, 1–297.

Gusmeroli, S., Piccione, S., and Rotondi, D. (2012). IoT access control issues: a capability based approach. 2012 Sixth International Conference on Innovative Mobile and Internet Services in Ubiquitous Computing, IEEE, 787–792.

Jover, R.P. (2019). The current state of affairs in 5G security and the main remaining security challenges. arXiv preprint arXiv:1904.08394.

Jover, R.P. and Marojevic, V. (2019). Security and protocol exploit analysis of the 5G specifications. *IEEE Access* 7: 24956–24963.

Katsalis, K., Nikaein, N., Schiller, E. et al. (2017). Network slices toward 5G communications: slicing the LTE network. *IEEE Communications Magazine* 55: 146–154.

Khan, H. and Martin, K.M. (2019). On the efficacy of new privacy attacks against 5G AKA.

Koutsos, A. (2019). The 5G-AKA authentication protocol privacy. 2019 IEEE European Symposium on Security and Privacy (EuroS&P), IEEE, 464–479.

Koutsopoulou, M., Kaloxylos, A., Alonistioti, A. et al. (2004). Charging, accounting and billing management schemes in mobile telecommunication networks and the internet. *IEEE Communications Surveys & Tutorials* 6: 50–58.

Lai, C., Li, H., Liang, X. et al. (2014). CPAL: a conditional privacy-preserving authentication with access linkability for roaming service. *IEEE Internet of Things Journal* 1: 46–57.

Lai, C., Li, H., Lu, R., and Shen, X.S. (2013). SE-AKA: a secure and efficient group authentication and key agreement protocol for LTE networks. *Computer Networks* 57: 3492–3510.

Li, J., Wen, M., and Zhang, T. (2015). Group-based authentication and key agreement with dynamic policy updating for MTC in LTE-A networks. *IEEE Internet of Things Journal* 3: 408–417.

Liu, J.K., Chu, C.-K., Chow, S.S. et al. (2014). Time-bound anonymous authentication for roaming networks. *IEEE Transactions on Information Forensics and Security* 10: 178–189.

Liu, F., Peng, J., and Zuo, M. (2018). Toward a secure access to 5G network. 2018 17th IEEE International Conference On Trust, Security And Privacy In Computing And Communications/12th IEEE International Conference On Big Data Science And Engineering (TrustCom/BigDataSE), IEEE, 1121–1128.

Metz, C. (1999). AAA protocols: authentication, authorization, and accounting for the internet. *IEEE Internet Computing* 3: 75–79.

Naslund, C.S., Stahl, P., Innov, I.T. et al. (2017). Deliverable D2. 7 security architecture (final).

Nasland, M., Selander, G., Phillips, S. et al. (2016). 5G-ENSURE-D2. 1 use cases.

Ni, J., Lin, X., and Shen, X.S. (2018). Efficient and secure service-oriented authentication supporting network slicing for 5G-enabled IoT. *IEEE Journal on Selected Areas in Communications* 36: 644–657.

Ordonez-Lucena, J., Ameigeiras, P., Lopez, D. et al. (2017). Network slicing for 5G with SDN/NFV: concepts, architectures, and challenges. *IEEE Communications Magazine* 55: 80–87.

Parne, B.L., Gupta, S., and Chaudhari, N.S. (2018). SEGB: security enhanced group based aka protocol for M2M communication in an IoT enabled LTE/LTE-A network. *IEEE Access* 6: 3668–3684.

Rost, P., Mannweiler, C., Michalopoulos, D.S. et al. (2017). Network slicing to enable scalability and flexibility in 5G mobile networks. *IEEE Communications Magazine* 55: 72–79.

Schneider, P. and Horn, G. (2015). Towards 5G security. 2015 IEEE Trustcom/BigDataSE/ISPA, IEEE, 1165–1170.

Su, W.-T., Wong, W.-M., and Chen, W.-C. (2016). A survey of performance improvement by group-based authentication in IoT. 2016 International Conference on Applied System Innovation (ICASI), IEEE, 1–4.

Velte, T. and Velte, A. (2006). *Cisco: A Beginner's Guide*. McGraw-Hill.

Wong, S., Sastry, N., Holland, O. et al. (2017). Virtualized authentication, authorization and accounting (V-AAA) in 5G networks. 2017 IEEE Conference on Standards for Communications and Networking (CSCN), IEEE, 175–180.

Yao, J., Wang, T., Chen, M. et al. (2016). GBS-AKA: group-based secure authentication and key agreement for M2M in 4G network. 2016 International Conference on Cloud Computing Research and Innovations (ICCCRI), IEEE, 42–48.

Yegin, A.E. and Watanabe, F. (2005). Authentication, authorization, and accounting. Next generation mobile systems 3G and beyond 315–343.

Zhang, J., Wang, Q., Yang, L., and Feng, T. (2019). Formal verification of 5G-EAP-TLS authentication protocol. 2019 IEEE Fourth International Conference on Data Science in Cyberspace (DSC), IEEE, 503–509.

Further Reading

Alawe, I., Hadjadj-Aoul, Y., Ksentini, A. et al. (2018). On the scalability of 5G Core network: the AMF case. 2018 15th IEEE Annual Consumer Communications & Networking Conference (CCNC), IEEE, 1–6.

Alliance, N. (2016). 5G security recommendations package# 2: network slicing. *NGMN Alliance* 1–12.

Ahmad, I., Kumar, T., Liyanage, M. et al. (2018). Overview of 5G security challenges and solutions. *IEEE Communications Standards Magazine* 2: 36–43.

Ahmad, I., Kumar, T., Liyanage, M. et al. (2017). 5G security: analysis of threats and solutions. 2017 IEEE Conference on Standards for Communications and Networking (CSCN), IEEE, 193–199.

Panwar, N., Sharma, S., and Singh, A.K. (2016). A survey on 5G: the next generation of mobile communication. *Physical Communication* 18: 64–84.

7

5G-Core Network Security

Ijaz Ahmad[1], Jani Suomalainen[1], and Jyrki Huusko[2]

[1] *VTT Technical Research Centre of Finland, Espoo, Finland*
[2] *VTT Technical Research Centre of Finland, Oulu, Finland*

Introduction

The core network in fifth generation (5G) has changed dramatically, compared to the core network in prior generations such as LTE. The major changes include the cloudification of the core network elements, virtualization of network functions, and separation of the data and control plane parts in the core network. The core network of the LTE system is called the evolved packet core (EPC), which is the latest evolution of the 3GPP core network architecture.

To begin with the GSM, the architecture relied on circuit switching in which a circuit is established between the calling and the called party through the access, the fixed, and the core networks of the operator. In GPRS, packet switching is added to the circuit switching for data transmission, whereas voice is carried through circuits. In 3G system, this dual system (circuit and packet switched) is kept on the core network side. With the evolution of the 3G system, the 3GPP community decided to move toward an all IP-based core network. Hence, the IP (Internet Protocol) is adopted as the key protocol to transport all services. Thus, the EPC, to remind, is evolved further from the packet-switched architecture of the prior generations.

With this evolution, the security landscape also evolved from simple security challenges such as phone tapping to advance IP-based attacks such as denial of service (DoS) attacks. During the first generation, i.e. 1G, enabling voice communication over the air was the main priority overlooking the challenges of security and privacy. Hence, the circuit-switched nonencrypted plain voice calls could be tapped or heard by unwanted third parties. Encryption was added in the second generation (2G); however, spamming became common. With the introduction of third generation, IP-based attacks, already common on the Internet, started targeting mobile networks. In the fourth generation (4G), cyberattacks on mobile device and networks remained a major issue. In the 5G networks, all IP-based communication among almost all connected things, security is a major concern. The core network, overlooking the operation of the rest of the network, the user profiles and billing, and network management residing in it, is a major security concern.

The Wiley 5G REF: Security. Edited by Rahim Tafazolli, Chin-Liang Wang, Periklis Chatzimisios and Madhusanka Liyanage.
© 2021 John Wiley & Sons Ltd. Published 2021 by John Wiley & Sons Ltd.

In this article, first we provide a brief description of the 5G core network (5GCN), and its design principles, and then we describe the possible security challenges, the potential solutions and what needs to be done more, i.e. future research directions.

5G Design Principles and the Core Network

The 5G design principles drastically changed due to the emergence of radically new services (Agyapong et al. 2014). The 5G design principles has its basics revolving around flexibility and support for new value creation. The design principles outlined by Next Generation Mobile Networks (NGMN) Alliance (2015) has strong emphasis on creating a common composable core with the following characteristics:

- Minimize the number of entities and functionalities
- Control-user plane separation and lean protocol stack
- No mandatory user plane functions
- Minimal legacy networking
- Radio access-agnostic core
- Fixed and mobile convergence

According to the NGMN Alliance, the operations and management must be simplified with automation and self-healing, etc. New approaches that enable flexible functions and capabilities such as network slicing, and network function virtualization (NFV) and software-defined networking (SDN) should be embraced. Having these characteristics and technologies such as SDN and NFV will drastically change the core network of 5G.

The demanding service and network requirements in 5G required a fundamental change in the core networks compared to the previous generations. The EPC in 5G is using, and is thus based on, the recent technological developments such as cloud computing, SDN, NFV, and slicing. The core network components in 5G are mostly network functions (NFs) implemented in software and deployed in cloud platforms that can scale up and down dynamically based on service requirements. Therefore, the 5GCN architecture definitions calls it a "service-based architecture (SBA)" framework (3rd Generation Partnership Project (3GPP) 2019), where the architecture elements are "NFs" rather than traditional network entities. All NFs are connected via interfaces to enable services of one NF to be accessed by other authorized NFs or any authorized consumers. The 5G system architecture, showing the core NFs, is presented in Figure 1.

The 5G system is composed of the user equipment (UE), the radio access network (RAN), the user plane function (UPF), the data network (DN), and the core NFs. These NFs described in the latest 3GPP release 15 (3rd Generation Partnership Project (3GPP) 2019) and include:

- Network repository function (NRF)
- The network exposure function (NEF)
- The unified data management (UDM)
- The unified data repository (UDR)
- The unstructured data storage function (UDSF)
- Network slice selection function (NSSF)
- Authentication server function (AUSF)
- Policy control function (PCF)

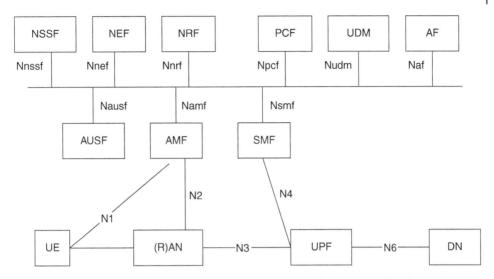

Figure 1 The 5G system architecture by 3GPP. Source: European Telecommunications Standards Institute (ETSI) (2017a). Reproduced with permission of ETSI.

- Access and mobility management function (AMF)
- Session management function (SMF)

The last two functions handle the mobility of UEs. The AMF supports user mobility and is in charge of signaling which is not specific to user data. Mobility can be hidden from the application layer to avoid service delivery interruptions. The SMF is in charge of signaling related to user data traffic such as session establishment, etc. The SBA approach of 5GCN utilizes the concepts of cloud computing beyond centralized clouds, for example edge computing, and virtualization, and slicing. Hence, the security of each of these approaches must also be taken into consideration. It is highly possible that the security requirements of each of these technologies change due to its use in a different domain, i.e. wireless networks. Therefore, specific care must be taken due to different requirements of security of each of these technologies. Furthermore, inter-operator and geographical roaming need the security policies and practices to be in place with roaming of user. One of the challenge for network operators is maintaining the same level of security due to poor security policies of other operators, both having agreements for user roaming. It is also necessary that a user gets the same security and privacy assurance while roaming among operators.

In the following section, we provide a brief overview of the security architecture of 5G. Subsequently, the security challenges and possible solutions of the core network are discussed.

Security of Core Network

Overview of 5G Security Architecture

The 5G security architecture has been defined in the latest technical specification release of 3GPP (release 15) (3rd Generation Partnership Project (3GPP) 2018a) with different

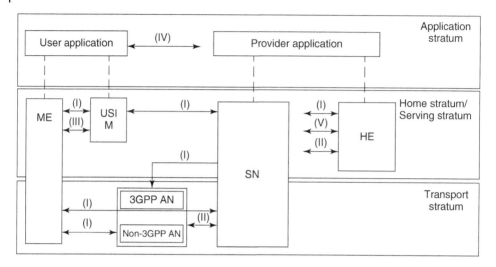

Figure 2 5G security architecture. Source: 3rd Generation Partnership Project (3GPP) (2018a). Reproduced with permission of ETSI.

domains. The security architecture is picturized in Figure 2 and has the following main security domains.

Network Access Security (I). A set of security features that enables a UE to securely authenticate and access network services. It includes security of 3GPP and non-3GPP access technologies, and delivery of security context from SN to the UE.

Network Domain Security (II). A set of security features that enables network nodes to securely exchange signaling and user plane data.

User Domain Security (III). Security features that enable secure user access to UE.

Application Domain Security (IV). Security features that enable user and provider domain applications to securely exchange messages.

Service-Based Architecture (SBA) Domain Security (V). Security features for network element registration, discovery, and authorization, as well as security for service-based interfaces.

The 5G security architecture is a comprehensive response to security threats that have been identified in mobile networks. The architecture incorporates security solutions that are either inherited from the previous generations with some modifications or defined newly for 5G. The LTE security concepts are the starting points, and considered as benchmarks for security of future wireless networks (Dan Forsberg et al. 2012). Yet security architectures of the previous generations will not suffice mainly due to new disruptive technologies. For example, network virtualization, multi-tenancy, and slicing were not there before; cloud platforms were not containing the most important network entities or functions as in 5G. As a consequence, there are new security considerations, and thus, there are new security designs for 5G systems.

There are three high-level visions (Nokia 2017) for 5G security; it will be build-in, flexible, and automated.

- *Built-in security* provides increased robustness against cyber threats that existed in previous generations and that are arising due to new technologies and use cases. As in the previous generations, 5G addresses remote threats against the operators and the users. But 5G also addresses insider threats – the lack of full trust between different cooperating parties – e.g. by filtering inter-network threats, by isolating different trust zones, and by application slicing. More focus is also given for enhancing privacy and for security assurance, security monitoring, and verification.

- *Flexibility of security* mechanisms enables tailoring and optimization of cyber-defense features for different end users. Unlike earlier generations, where operators provided only one type of security service and then end users optionally applied own end-to-end application-layer security, 5G enables varied flexibly in security services. 5G will provide and enable alternative authentication and security mechanisms suitable for different use cases and devices, e.g. some with resource restrictions or some with high security requirements.
- *Automation of security* enables operators to easily customize and adapt security controls according to dynamic needs and situations. Security orchestration and management enables operators to deploy security functions that provide holistic security through the whole end-to-end connection chains. Automation enables operators to deploy intelligent security controls that detect and mitigate ongoing attacks and evolving threats.

5G security architecture provides a framework that is fulfilled by security controls – security functions – that address threats and security needs in different domains and strata. The 5G-ENSURE project analyzed (Arfaoui et al. 2018) composition of 5G security architecture and categorized security controls into 10 classes: identity and access control, authentication, nonreputation, confidentiality, integrity, availability, privacy, audit, trust and assurance, as well as compliance. The key functions and their roles are briefly summarized in Table 1. The table lists functions with explicit or implicit security role, their locations in the architecture, as well as highlights the scope of 3GPP specifications in security-related standardization.

Security of Core Network Elements

There are a number of security challenges that have been discussed in the literature regarding the security of 5G. Due to the centralized role of the core network, the security challenges in the core network are pretty serious since the whole network can be compromised due to security lapses. The NGMN has highlighted several potential security challenges in Alliance (2015), where the most pertinent ones related to core network are:

Flash Network Traffic. Since the number of users and connected devices in 5G are projected to increase exponentially, mainly due to the conglomeration of diverse sets of IoT, traffic surges will be no uncommon. Flash traffic can occur due to sudden connectivity of massive number of IoT devices due to known or unknown abrupt activities. The result will be a flash of signaling traffic that will directly affect the core network. Hence, the 5GCN must efficiently handle large swings in traffic, and provide resilience whenever such surges of traffic occur (Ahmad et al. 2019).

Denial of Service (DoS) Attacks. DoS attacks cause exhaustion of resources through specifically crafted traffic flows or requests, usually targeted toward central control points. Since, the core network in 5G is logically centralized into cloud platforms, DoS and distributed DoS (DDoS) attacks can be threatening. Since the number of

Table 1 Key functions and roles in 5G core network security.

Security functions, which are defined by 3GPP for 5G core networks	
Authentication credential repository and processing function (ARPF)	A home network entity, which keeps authentication credentials and generates authentication vectors for the authentication server function
Authentication server function (AUSF)	A home network entity, which authenticates user equipment and provides keying material for the serving network
Subscription identifier de-concealing function (SIDF)	A home network entity which decrypts user's permanent identifiers, which are encrypted with public key cryptography in 5G to increase privacy of users' location
SEcurity anchor function (SEAF)	A serving network entity that re-authenticates devices that are moving to different access network. Minimizes signaling costs with the home network
Security edge protection proxy (SEPP)	A firewall that filters inter-operator network traffic (particularly, between home and serving network domains)
3GPP functions with a central role in security	
Access and mobility management function (AMF)	A serving network entity with a central role in access control. AMF provides identification and authentication, network selection, service authorization, access barring, and policy control by interacting with AUSF, network slice selection function (NSSF), and policy control function (PCF). Integrity and confidentiality of communication between AMF and UE is protected with cryptography
Session management function (SMF)	An entity distributed both to serving and home network in order to control QoS flows and policies (coming from PCF). Thus plays an important role in assuring availability. SMF provides also lawful intercept
Non-3GPP security functions (essential for 5G network security but not specified by 3GPP)	
Security tunneling gateways	Secure IPsec tunneling e.g. between access and serving networks and between serving networks and user equipment in non-3GPP access network
Firewalls	External security threat filtering between different domains and network segments
Identity and access management for control plane	Solutions for controlling who can configure and manage core network functions
Integrity protection for infrastructure	Security solutions in physical, hardware, operating system and virtualization levels for protecting and attesting integrity, authenticity, and trustworthiness of software and functionality that is running in the cloud platform and in the core network
Security monitoring	Security monitoring solutions consist of the following different functional elements: probes, which are distributed to different domains and layers to collect information; inter-domain and intra-domain distribution of information on monitored events; intrusion detection engines based on data correlation, signature matching, and machine learning; as well as auditing
Security orchestration	Automated deployment and control of security functions and resources to fulfill end-user-specific end-to-end security requirements and to respond to dynamic security events and attacks

devices grows at enormous rates, mainly due to massive IoT, distinguishing a DoS or DDoS attack from normal traffic surges can also become challenging. Thus, one of a big resilience challenge will be to timely and appropriately distinguish flash traffic from DoS or DDoS attacks (Ahmad et al. 2019).

The core NFs, which are now centralized in cloud platforms, can also be vulnerable to security challenges. These challenges are multi-faceted due to the use of different technologies such as cloud computing, virtualization, and the SDN. Each of these technologies will introduce their own vulnerabilities into the core network and thus the solutions will be based on not only direct threats to these elements but also the technologies involved. For example, the AMF is involved in mobility-related functions. The mobility can be multi-pronged, for example physical mobility and slice-to-slice mobility. Since, the cell sizes are shrinking, physical mobility handovers will increase. Changing mobility parameters, such as received signal strength values in mobile terminals through malicious code can cause an increase in handovers and cause a signaling storm on the AMF. Therefore, proper security of NFs must be ensured from such scenarios, specifically due to increase in the number of IoT devices which can be easily compromised (Zhang et al. 2014). A large-scale analysis of firmware of constrained devices has been conducted by Costin et al. (2014) that reveal that the firmware of most low-capacity/resource devices is ripe with security vulnerabilities. Hence, such devices can trigger bearer activation, attach and detach signaling attacks on the NFs.

There have been security challenges involved in handover between 3GPP and non-3GPP systems in 4G heterogeneous networks as highlighted in Qachri et al. (2012). One of the main challenges with non-3GPP access systems is in compliance with 3GPP security procedures and principles. Hence malicious access points exposing keys exchanged between the core network elements must be properly investigated. Similarly, security vulnerabilities related to the control of signaling constitute another major research area. For instance, the signaling procedures such as attach/detach, authentication, bearer activation, and location updates will increase massively in the realm of massive number of IoT devices connected to 5G. Such signaling procedures can also result in non-access stratum (NAS) signaling storms (Agiwal et al. 2016). The overall challenge is handling the signaling of massive number of devices, and effectively distinguishing from resource exhaustion attacks. Two approaches are discussed regarding the massive increase in signaling traffic in 5G PPP Security WG (2017). First is using lightweight authentication and key agreement (AKA) protocols for massive IoT communication. The second approach is using protocols that allow grouping devices together, for example group-based AKA protocols. Hence, various grouping approaches have been proposed. For example, group-based authentication scheme, that group narrow-band IoT devices with similar attributes together, is proposed in Cao et al. (2018). A group leader is chosen that gathers sensitive information together and sends it to the corresponding node in the core network. A similar group-based authentication scheme has been proposed in Lai et al. (2017) for SDN-based vehicle-to-anything (V2X) communication to minimize the signaling costs in the network.

Security of Interfaces/Protocols

Access to core functions and communication within and through core network is protected with different authentication, authorization, and communication security solutions. 5G introduces some new security protocols and functions as well as adopts and hardens some protocols from the earlier generations.

Security in the *home and serving stratum* – supporting communication to and between core NFs – is heavily impacted by two new 5G core technologies: SBA and exposure API, enabling third parties to access 5G services (Mayer 2018). 5G adopts HTTP/2 for communication between and with the core network entities (Hu et al. 2018). For instance, DIAMETER-based communication for authentication is replaced with web APIs in 5G. The web-based interfaces are secured, typically with Transport Layer Security (TLS)-based security protocols as well as with authorization and identity management solutions. Further, function discovery and registration require confidentiality, integrity and replay protection as well as authorization and mutual authentication.

Inter-operator security is hardened with new security edge protection proxy (SEPP) function. It filters control communication between home and visited network. SEPP is motivated by the changed trust models and inter-operator threats that were present in the previous generations (Prasad et al. 2018): SS7 key theft and rerouting attacks as well as DIAMETER node impersonation and source spoofing attacks.

Operators will protect their networks from threats from different external networks with proprietary firewall solutions. As in previous generations, firewalls are deployed to e.g. between core and Internet, between core and IP multimedia subsystems, between core and different access networks, and between different operator networks. Operators may also utilize firewalls to segment different parts of their network, including protecting their assets in cloud and edge platforms. Network softwarization and virtualization provide new opportunities for firewall implementations.

Communication security between access network and serving network in *transport stratum* is based on IPsec VPN tunneling and public key infrastructure. 5G introduces some latency and throughput-related requirements (ultra-reliable low-latency communication) that the IPsec security gateways in the network-side must guarantee. Security solutions must also work with new 5G features; SDN-based traffic routing, and EDGE computing. For instance, SDN transport layer needs 5G awareness in order to route packets with encrypted headers.

Security in the application stratum relies heavily on application layer end-to-end security protocols. 5G end users and user companies may apply own end-to-end security mechanisms such as TLS, IPsec, and HTTPS when implementing e.g. virtual private networks or accessing secure cloud services. 5G network supports end-to-end security by enabling the use of 3GPP credentials, keys, and authentication framework in third-party network access control (Prasad et al. 2018). Different end-user applications and verticals have nevertheless their own security needs and functions in the core network, for instance:

- Proximity service-specific security protocols and network-assisted key management mechanisms (3rd Generation Partnership Project (3GPP) 2018d) provide confidentiality and authenticity for direct (device-to-device) communication as well as authentic and authorized discovery of proximity services (and nearby devices).
- Multimedia broadcast/multicast service (MBMS) security (3rd Generation Partnership Project (3GPP) 2018b) provides group key management and delivery with efficient rekeying.

- Mission-critical security framework (3rd Generation Partnership Project (3GPP) 2018c) addresses end-to-end confidentiality, authentication, and key management of group push-to-talk and data applications, which are reliability and delay critical.
- Lawful interception provides e.g. law-enforcement officials' (genetive) deep packet inspection capabilities to access headers or payloads.

UE and Network Authentication

In the LTE architecture, evolved packet system-authentication and key agreement (EPS-AKA) methodology was used to perform mutual authentication between the UE and the network (Li and Wang 2011; Kien 2011). AKA started in GSM, evolved with next generations, and is still considered the most suitable authentication and authorization mechanism in 5G (Zhang and Fang 2005; Arkko 2006). AKA runs in subscriber identity module (SIM) and is based on symmetric keys. Extensible Authentication Protocol (EAP)-AKA method was developed by 3GPP for 3G networks to support identity privacy and fast re-authentication. The EPS-AKA uses multiple keys in different contexts, features renewal of keys without involving the home network for every time. The EPS-AKA has some challenges such as latency and computation and communication overhead (Alezabi et al. 2014), but has no visible vulnerabilities demonstrated so far (Schneider and Horn 2015). Essential changes that 5G introduce for authentication – with new 5G-AKA, EAP-AKA', and EAP-TLS – include (Prasad et al. 2018; Cablelabs 2019):

- New service authentication (secondary authentication) enables UE to use EAP authentication and credentials to gain access to third-party DNs that are not controlled by the mobile operator.
- Deeper key hierarchy and more complex key management will address architectural changes and new trust models of 5G
- Unified authentication framework – security context (i.e. secret session keys) can be shared across different access networks in order to make roaming between 3GPP and non-3GPP networks more fluent.
- Alternative EAP-TLS provides certificate-based mutual authentication that is targeted for private networks and IoT devices.

For trusted non-3GPP access network, a UE is authenticated through the authentication, authorization and accounting (AAA) server using AKA. For untrusted non-3GPP access networks, the UE uses the evolved packet data gateway IPsec tunnels to connect to the 5GCN (Cao et al. 2014). These mechanisms have many benefits including short message size, and needs only one handshake between the UE and SN, and between the serving and home networks (Schneider and Horn 2015).

Security of Emerging New Network Technologies

Since 5G is using new technologies such as SDN, NFV, and the concepts of cloud computing beyond the centralized servers like mobile edge computing, the security implications due to these technologies must be considered. Hence, in the following sections we describe the security challenges and their solutions for each of these technologies.

Security of Cloud Platforms

The surging data rates, massive number of connected things, and increasing number of services have given cloud platforms a cardinal stage in communication networks. The concepts of cloud computing and storage are used in the 5GCNs in different ways. Thus the security of cloud platforms in general will have implications on the security of the core network. In 5G, the core network elements are implemented as software functions and deployed in cloud platforms, and therefore, use the concepts of cloud computing and storage in wireless networks. These platforms can be either centralized in high-end servers, or decentralized and located in the wireless network edge, for instance, in base stations. The decentralized cloud systems implemented in the form of multi-access edge computing (MEC) platforms can bring the core NF near the wireless edge, or in other words, the users. However, both of these methodologies will have their own security implications. The decentralized platforms must also secure the interfaces that connect different core NFs, or instances of the same function distributed in different edge nodes for availability or scalability reasons.

The security challenges of cloud platforms can be due to multiple reasons. For instance, virtualization of cloud resources will be common. Hence, security challenges related to virtual systems will be important to consider. Similarly, concentrating important functions into centralized systems make it an attractive choice of DoS attacks. In the 5GCNs, both of these challenges will be crucial. In relation to virtualization, core NFs will be deployed as VNF (virtual network functions) leveraging NFV. Hence, proper isolation of different functions, protecting resource theft or overuse, and protecting one function from other (compromised) functions will be highly important. Regarding DoS attacks, the centralization of key functions into cloud platforms makes it more attractive. Furthermore, due to the massive number of connected things, differentiating signaling traffic from resource exhaustion or DoS attacks will be further challenging. Signaling storms and flash network traffic must be recognized early enough to distinguish those from cyberattacks. Enhancing availability through scalability, and distributing work load of the core NFs among multiple distributed nodes will be of paramount importance in this regard. Leveraging the concepts of MECs, live service or NF migration from dense nodes to lightly loaded nodes will be highly important. Furthermore, strict access control policies and isolation through slicing will improve the resilience of the platforms.

Security of SDN

SDN offers interesting features and capabilities for the core network. Through separation of the control and data planes, SDN logically centralizes the network control plane and renders the data plane simplified. Programmable interfaces between the control and data planes empower the control plane to adjust the behavior of the data forwarding devices at run-time. Controlling the underlying network from a central vintage point highly simplifies the network management, is cost-effective in updating the network, and eases deployment of new NFs. The potential of SDN has been already demonstrated with its most used implementation, i.e. the OpenFlow (McKeown et al. 2008). The OpenFlow architecture implements SDN in a three-tier architecture comprising the application, control, and data planes. The OpenFlow protocol interfaces the control and data plane. The control plane is called the OpenFlow controller, the data plane is made of OpenFlow switches, and the application consists of OpenFlow applications.

Even though, OpenFlow has gained a lot of momentum, it has also exposed the security challenges that SDN can face. These challenges are not limited to the OpenFlow implementation, but are related to the very concepts of SDN, namely, the centralization (logically though) of the control plane, simplification of the data plane, and the recognizable communication between the control and data planes (Ahmad et al. 2015).

The centralized control plane can be targeted for DoS and DDoS attacks (Kandoi and Antikainen 2015). In SDN, the data plane is simple, and it acts on the directions from the controller. The data plane switches contain flow tables and consult those for forwarding packets to specific ports. The controller populates the flow tables. In OpenFlow variant of SDN, the OpenFlow protocol is used for the purpose. When a data plane switch receives packets from a new host, it sends the first packets to the controller, and the controller then installs flow rules in the switch flow tables. This mechanism of sending flow setup requests, and then flow rules installation in the data plane can be used to fingerprint the controller and attack the controller. For example, changing packet header fields or specifically crafted flows can be sent to the switches, for which the switch will not be able to find flow rules and forward them to the controller. In this way, the controller can be flooded with flow setup requests mainly to exhaust its resources. Hence, the SDN controller is one target for DoS and DDoS attacks, or resource exhaustion attacks (Ahmad et al. 2015).

The solutions for securing the control plane are also diverse, ranging from dividing control plane functionalities among multiple physical or logical controllers to increasing the capacities of the controllers. Hence, there are various proposals to increase the security of the control plane through increasing the control plane resources, distributing control functions among multiple nodes, and enhancing the access control procedures. There are also different approaches that devolve some of the control plane functions back to the data plane. For example, the DevoFlow framework (Mogul et al. 2010), preserves the local routing decisions within switches to avoid controller involvement in every packet forwarding decision, specifically the micro flows. This improves the controller resilience and reliability and ensures its availability for the setup of crucial macro-flow rules in the data plane.

In principle, SDN can separate different users from each other and reserve resources so that DoS attacks performed in one data flow do not affect other users. Also, end-user's packet flows are not visible for other users. However, isolation requires that all functions and resources are dedicated for particular SDN, when functions or resources are shared attacks on one SDN may have effects on others. Also, SDN fingerprinting techniques (timing of control plane operations) have been shown to reveal information from other slices (Cui et al. 2016). This information that is acquired from the shared control plane may be used to intensify DoS attacks that affect every slice (i.e. cause excessive processing in the common control plane). Fingerprinting may also reveal privacy-sensitive information (when and how intensively some application is used).

There are also other security challenges in SDN. For example, authorization of applications to interact with the controller or generate flow rules. In SDN, different NFs can be implemented as applications. For instance, a load balancing application can retrieve the network statistics through the control plane from the data plane switches. In OpenFlow implementation of SDN, the OpenFlow switches maintains flow tables which contain packet or byte counters associated with different flows. A load balancing application can fetch those counter values to decide if a switch is congested or will be congested after a certain threshold. Hence, important NFs can be implemented as SDN applications. However, proper authentication and authorization of application is required before access to the application is given. There are some frameworks such as FortNox (Porras et al. 2012), SEFloodlight (Ahmad et al. 2019) that secures the control plane from malicious access. The ROSEMARY (Shin et al. 2014) controller platform applies a permission system on applications to secure the controller from malicious applications. The VeriFlow (Al-Shaer and Al-Haj 2010) platform verifies flow rules before being installed in the data plane. Moreover, a detailed study on the security of SDN is presented in Ahmad et al. (2015) that highlights the potential security challenges, their solutions, and interesting research directions for further improving the security of SDNs beyond the existing literature.

Security of Network Virtualization

NFV brings new advantages for security. Operators can *orchestrate* – i.e. create, automate, control, and tailor – security services according to customers' needs more easily when the services are based on software rather than hardware. Operators may, for instance, automate dynamic scaling of resources to ensure availability, update security functions rapidly, and enforce use-case-specific security policies. Use of custom security functions on end-user-specific slices enables e.g. monitoring functions to apply application-specific algorithms and react quicker by focusing on homogenous data flows. Further, software functions, which are dedicated for end-user-specific network slices, are also more isolated from the disturbances and e.g. DoS attacks occurring in other function instances.

On the other hand, virtualization (i) changes trust models – enables new (not fully trusted) players to enter the game – and (ii) the added complexity – arising from new interfaces as well as from variety of possible security configurations – introduces new security challenges. With virtualization, the same hardware infrastructure is shared between different operators, i.e. tenants, operating the software functions. Software and hardware functions may originate from different sources and vendors with diverse security practices. Operators and service providers, with own motivations, may cooperate to provide end-to-end services for the end users. A misbehaving function or service provider may inflict serious damages for the operator or the user. Hence, cooperating entities need means to ensure that misbehaving functions or operators do not compromise security and that infrastructure components can be trusted. Further, shared hardware must provide strong isolation guaranty to protect their integrity, confidentiality, and availability of software functions and data.

Security solutions for NFV have been specified by ETSI (European Telecommunications Standards Institute (ETSI) 2017a, 2015, 2017, 2018, 2014b). The high-level security management framework is illustrated in Figure 3. Virtualization layer's security functions provide protected isolation for virtual resources and network and security functions. To protect network layer under virtual functions physical security functions like firewalls are needed. Physical security functions are managed by security element managers and security functions in NFV infrastructure are managed by network function virtualization Infrastructure security manager. NFV security manager provides overall

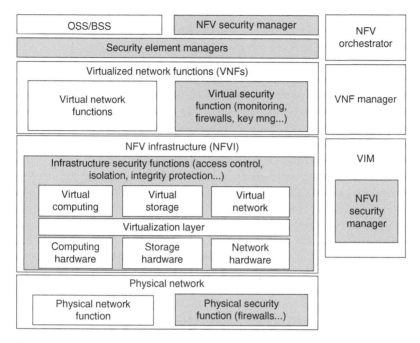

Figure 3 NFV security management framework. Source: Adapted from European Telecommunications Standards Institute (ETSI) (2017a).

security management over the NFV framework. Orchestration of security functions, that serve end users, is a responsibility of NFV orchestrator who may cooperate with orchestrators in other domains to provide end-to-end services.

Essential security controls required by NFV architecture include:

- Identity and access management for controlling who can use and manage VNF.
- Security monitoring – i.e. correlating and analyzing data collected from the user data, and management and control planes – is an essential enabler for automated security orchestration but it is also necessary for detecting intrusions against the NFV framework and functions.
- Function isolation depends on operating system and hypervisor level mechanisms as well as on physical controls.
- Communication security protocols are needed to protect the authenticity and confidentiality of communication between the elements of the framework.
- Firewalling, zoning, and topology hiding solutions are needed to protect virtualization framework against external threats.
- Trusted computing technologies, such as secure boot and attestation, are needed to ensure and attest integrity and trustworthiness of the software that is running in shared hardware and in NFs.

When NFV is applied in an end-to-end 5G network with multiple cooperating operators, solutions for cross-domain trust management are needed. Trust management between different cooperating domains can be based on service level agreements, monitoring, and auditing. As an interesting research direction, the use of blockchain technologies has also been proposed (Backman et al. 2017; Alvarenga et al. 2018; Scheid et al. 2019) as a mean to enable more autonomous auditing.

Discussion and Conclusion

The 5GCN is very different than the previous ones. New technologies are used in the core network. There are also new technologies throughout the network, e.g. in backhaul network and access networks. For instance, the functional split is happening nearly throughout the network, and the main technologies used are SDN and NFV. SDN separates the network control from the data planes, and NFV facilitates placing networking functions (e.g. network control function) in different network perimeters. Hence, the security of the core network is even more crucial due to these technologies. For instance, if the core network is compromised, the backhaul and access network having its control plane in the core network will be compromised. Similarly, the increasing number of devices enabled by IoT and diverse services require the involvement of the core network on a much higher scale than the previous generations. All these changes necessitate new thinking from the security perspectives with new cautionary preparations to secure not only the core network but to avoid security lapses in the rest of the network due to security lapses in the core network. Since the core network is overseeing the rest of the network as a control point, a compromised core network means that the rest of the network is also compromised. Therefore, the security of the core network is of paramount importance which has become more crucial due to the recent technologies like SDN and NFV.

This article has outlined the potential security challenges and the possible solutions to those challenges. The security challenges that could arise due the new technological concepts such as SDN and NFV are also highlighted along with the proposed solutions for those technologies. Since 5G is yet to be deployed on a large scale, there is a possibility of new security vulnerabilities. However, the existing security solutions are secure enough to avoid major security risks. Since machine learning is gaining momentum in wireless networks, security is one such interesting area in which machine learning can be very useful. Therefore, proactive security measures using learning-based systems leveraging machine learning must also be investigated to enhance security of the core network. Furthermore, improving availability of the core network through enhancing scalability or redundancy with minimum costs is another important research area for future networks with secure core network. MEC platforms having core NFs to improve resiliency and availability of the core network need further research in this direction.

Related Article

Security in Network Slicing

References

3rd Generation Partnership Project (3GPP) (2018a). Technical specification group services and system aspects (SA3);TS 33.501: security architecture and procedures for 5G system, Release 15. *3rd Generation Partnership Project (3GPP), Tech. Rep. 33.501.*

3rd Generation Partnership Project (3GPP) (2018b). Technical Specification Group Services and System Aspects; 3G Security; Security of Multimedia Broadcast/Multicast Service (MBMS).

3rd Generation Partnership Project (3GPP) (2018c). Security of the mission critical service. TS 33.180.

3rd Generation Partnership Project (3GPP) (2018d). Proximity-Based Services (ProSe); Security aspects. TS 33.303.

3rd Generation Partnership Project (3GPP) (2019). Technical specification group services and system aspects. Release 15 Description; TR 21.915 v1.0.0 (2019-03). *3rd Generation Partnership Project (3GPP), Technical Report*.

5G PPP Security WG (2017). 5G PPP phase1 security landscape. White paper. 5GPPP Papers.

Agiwal, M., Roy, A., and Saxena, N. (2016). Next generation 5G wireless networks: a comprehensive survey. *IEEE Communications Surveys Tutorials* 18 (3): 1617–1655, thirdquarter.

Agyapong, P.K., Iwamura, M., Staehle, D. et al. (2014). Design considerations for a 5G network architecture. *IEEE Communications Magazine* 52 (11): 65–75.

Ahmad, I., Namal, S., Ylianttila, M., and Gurtov, A. (2015). Security in software defined networks: a survey. *IEEE Communications Surveys & Tutorials* 17 (4): 2317–2346.

Ahmad, I., Shahabuddin, S., Kumar, T. et al. (2019). Security for 5G and beyond. Proceedings of the IEEE Communications Surveys & Tutorials.

Alezabi, K.A., Hashim, F., Hashim, S.J., and Ali, B.M. (2014). An efficient authentication and key agreement protocol for 4G (LTE) networks. Proceedings of the 2014 IEEE Region 10 Symposium, (April 2014), 502–507.

Alliance, N.G.M.N. (2015). 5G white paper. Next generation mobile networks, white paper, 1–125.

Al-Shaer, E. and Al-Haj, S. (2010). FlowChecker: configuration analysis and verification of federated OpenFlow infrastructures. Proceedings of the 3rd ACM Workshop on Assurable and Usable Security Configuration, ser. SafeConfig'10. ACM, 37–44.

Alvarenga, I.D., Rebello, G.A.F., and Duarte, O.C.M.B. (2018). Securing configuration management and migration of virtual network functions using blockchain. Proceedings of the NOMS 2018-2018 IEEE/IFIP Network Operations and Management Symposium, IEEE.

Arfaoui, G., Bisson, P., Blom, R. et al. (2018). A security architecture for 5G networks. *IEEE Access* 6: 22466–22479.

Arkko, J. (2006). Extensible authentication protocol method for 3rd generation authentication and key agreement (EAP-AKA). IETF RFC 4187, 1–78.

Backman, J., Yrjola, S., Valtanen, K., and Mammela, O. (2017). Blockchain network slice broker in 5G: slice leasing in factory of the future use case. Proceedings of the 2017 Internet of Things Business Models, Users, and Networks, IEEE.

Cablelabs (2019). A comparative introduction to 4G and 5G authentication. https://www .cablelabs.com/insights/a-comparative-introduction-to-4g-and-5g-authentication.

Cao, J., Ma, M., Li, H. et al. (2014). A survey on security aspects for LTE and LTE-A networks. *IEEE Communications Surveys Tutorials* 16 (1): 283–302.

Cao, J., Yu, P., Ma, M., and Gao, W. (2018). Fast authentication and data transfer scheme for massive NB-IoT devices in 3GPP 5G network. *IEEE Internet of Things Journal* 6 (2): 1561–1575.

Costin, A., Zaddach, J., Francillon, A., and Balzarotti, D. (2014). A large-scale analysis of the security of embedded firmwares. Proceedings of the 23rd USENIX Conference on Security Symposium, ser. SEC'14. Berkeley, CA, USA: USENIX Association, 95–110. http://dl.acm.org/citation.cfm?id=2671225.2671232.

Cui, H., Karame, G.O., Klaedtke, F., and Bifulco, R. (2016). On the fingerprinting of software-defined networks. *IEEE Transactions on Information Forensics and Security* 11 (10): 2160–2173.

Dan Forsberg, Horn, G., Moeller, W.-D., and Niemi, V. (2012). *LTE Security*. Wiley.

European Telecommunications Standards Institute (ETSI) (2014a). GS NFV-SEC 003. NFV Security; Security and Trust Guidance.

European Telecommunications Standards Institute (ETSI) (2014b). GS NFV-SEC 001. Network Functions Virtualisation (NFV); NFV Security; Problem Statement.

European Telecommunications Standards Institute (ETSI) (2015). GS NFV-SEC 002. Network Functions Virtualisation (NFV); NFV Security; Cataloguing security features in management software.

European Telecommunications Standards Institute (ETSI) (2017a). GS NFV-SEC 013. Network Functions Virtualisation (NFV) Release 3; Security Management and Monitoring specification.

European Telecommunications Standards Institute (ETSI) (2017b). GS NFV-SEC 012. Network Functions Virtualisation (NFV) Release 3; Security; System architecture specification for execution of sensitive NFV components.

Hu, X., Liu, C., Liu, S. et al. (2018). Signalling security analysis: is HTTP/2 secure in 5G core network?. Proceedings of the 2018 10th International Conference on Wireless Communications and Signal Processing (WCSP), Hangzhou, 1–6.

Kandoi, R. and Antikainen, M. (2015). Denial-of-service attacks in OpenFlow SDN networks. Proceedings of the 2015 IFIP/IEEE International Symposium on Integrated Network Management (IM), IEEE, 1322–1326.

Kien, G.M. (2011). Mutual entity authentication for LTE. Proceedings of the 2011 7th International Wireless Communications and Mobile Computing Conference (July 2011), 689–694.

Lai, C., Zhou, H., Cheng, N., and Shen, X.S. (2017). Secure group communications in vehicular networks: a software-defined network-enabled architecture and solution. *IEEE Vehicular Technology Magazine* 12 (4): 40–49.

Li, X. and Wang, Y. (2011). Security enhanced authentication and key agreement protocol for LTE/SAE network. Proceedings of the 2011 7th International Conference on Wireless Communications, Networking and Mobile Computing (September 2011), 1–4.

Mayer, G. (2018). RESTful APIs for the 5G service based architecture. *Journal of ICT Standardization* 6 (1): 101–116.

McKeown, N., Anderson, T., Balakrishnan, H. et al. (2008). OpenFlow: enabling innovation in campus networks. *ACM SIGCOMM Computer Communication Review* 38 (2): 69–74.

Mogul, J.C., Tourrilhes, J., Yalagandula, P. et al. (2010). Devoflow: cost-effective flow management for high performance enterprise networks. Proceedings of the 9th ACM SIGCOMM Workshop on Hot Topics in Networks, ACM, 1.

Nokia (2017). Security challenges and opportunities for 5G mobile networks.

Porras, P., Shin, S., Yegneswaran, V. et al. (2012). A security enforcement kernel for OpenFlow networks. Proceedings of the First Workshop on Hot Topics in Software Defined Networks, ser. HotSDN'12. ACM, 121–126.

Prasad, A.R., Arumugam, S., Sheeba, B., and Zugenmaier, A. (2018). 3GPP 5G security. *Journal of ICT Standardization* 2018: 137–158.

Qachri, N., Markowitch, O., and Dricot, J.-M. (2012). Vertical handover security in 4G heterogeneous networks: threat analysis and open challenges. In: *Future Generation Information Technology* (ed. T.-h. Kim, Y.-h. Lee and W.-c. Fang), 7–14. Berlin, Heidelberg: Springer Berlin Heidelberg.

Scheid, E.J., Rodrigues, B.B., Granville, L.Z., and Stiller, B. (2019). Enabling dynamic SLA compensation using Blockchain-based smart contracts. Proceedings of the 2019 IFIP/IEEE Symposium on Integrated Network and Service Management (IM), IEEE.

Schneider, P. and Horn, G. (2015). Towards 5G security. Proceedings of the 2015 IEEE Trustcom BigDataSE ISPA, Vol. 1, (August 2015), 1165–1170.

Shin, S., Song, Y., Lee, T. et al. (2014). Rosemary: a robust, secure, and high-performance network operating system. Proceedings of the 2014 ACM SIGSAC Conference on Computer and Communications Security, ser. CCS '14. New York, NY, USA: ACM, 78–89. [Online]. http://doi.acm.org/10.1145/2660267.2660353.

Zhang, M. and Fang, Y. (2005). Security analysis and enhancements of 3GPP authentication and key agreement protocol. *IEEE Transactions on Wireless Communications* 4 (2): 734–742.

Zhang, Z.K., Cho, M.C.Y., Wang, C.W. et al. (2014). IoT security: ongoing challenges and research opportunities. Proceedings of the 2014 IEEE 7th International Conference on Service-Oriented Computing and Applications, (November 2014), 230–234.

Further Reading

Ahmad, I., Kumar, T., Liyanage, M. et al. (2017). 5G security: analysis of threats and solutions. Proceedings of the 2017 IEEE Conference on Standards for Communications and Networking (CSCN), Helsinki, 193–199.

Ahmad, I., Kumar, T., Liyanage, M. et al. (2018). Overview of 5G security challenges and solutions. *IEEE Communications Standards Magazine* 2 (1): 36–43.

Dehnel-Wild, M. and Cremers, C. (2018). Security vulnerability in 5G-AKA draft: 3GPP TS 33.501 draft v0.7.0. *3rd Generation Partnership Project (3GPP), Tech. Rep.* [Online]. https://www.cs.ox.ac.uk/5G-analysis/5G-AKA-draft-vulnerability.pdf.

Namal, S., Ahmad, I., Gurtov, A., and Ylianttila, M. (2013). Enabling secure mobility with OpenFlow. 2013 IEEE SDN for Future Networks and Services (SDN4FNS), Trento, 1–5.

8

MEC and Cloud Security

Aaron Yi Ding

Delft University of Technology (TU Delft), Delft, The Netherlands

Introduction

The fifth-generation cellular network technology (5G) is generating new opportunities for various applications in Internet of Things (IoT), multimedia, smart grid, and mobility domains (Ding and Janssen 2018). Given its role as a critical infrastructure for both industry and society, 5G security has always been a top priority (Liyanage et al. 2018b). Besides conventional confidentiality, integrity, and availability (CIA) issues, 5G is facing new security challenges coming from new technologies, services (e.g. industrial Internet of Things (IIoT), autonomous driving), regulations, and change of user demands.

Besides embracing cloud solutions, 5G is rapidly integrating multi-access edge computing (MEC), which is standardized by European Telecommunications Standards Institute (ETSI). MEC aims to bridge the gap between cloud and IoT by providing high bandwidth and low latency access to 5G resources. Those advantages will facilitate mobile operators to open their networks to a new edge-driven ecosystem and corresponding value chains. In this regard, cloud computing enables 5G providers to flexibly outsource storage and processing functionalities. MEC further fills the gap between centralized cloud and distributed computing resources in terms of scalability, location awareness, and mobility. MEC allows to filter extremely large amounts of data to support efficient data processing at scale. Together with cloud intelligence, MEC and cloud can accelerate decision making based on both the locally processed data with context awareness and centralized management overview. This advantage is vital for various 5G enabled services such as fully autonomous driving, remote surgeries, and HD mobile video conferencing, which demand reliability, availability and ultra-low latency. In addition, MEC offers localized caching and storage, which are necessary to support fine-grained offloading in terms of data traffic and computational load (Ding et al. 2015; Cozzolino et al. 2017).

Although MEC and cloud offer several appealing features to 5G ecosystem, the security aspects of those technologies must be fully conceived. As shown in studies on D2D (Haus et al. 2017) and IoT communications (Hafeez et al. 2017), security must be enforced at the start of system development and deployment. For MEC and cloud in 5G, security concerns include both technical and societal aspects, such as trustworthiness of MEC and cloud, IoT-oriented security threats, secure infrastructure access, and

The Wiley 5G REF: Security. Edited by Rahim Tafazolli, Chin-Liang Wang, Periklis Chatzimisios and Madhusanka Liyanage.
© 2021 John Wiley & Sons Ltd. Published 2021 by John Wiley & Sons Ltd.

distributed security logging at scale. Since 5G is going to serve as a fundamental infrastructure, its security, safety, reliability, and resilience are of critical importance to our future digitalized society.

Given the rapid process of consolidating both edge and cloud to better support 5G systems (Morabito et al. 2018), this article focuses on the integrated view of applying MEC and cloud to 5G. Therefore, the term "5G MEC-Cloud" or "MEC-Cloud" is utilized to represent a unified discussion for edge and cloud security in the 5G context. The article is organized as follows. First, an overview is presented which covers features and synergies of 5G MEC-Cloud. Second, in-depth discussions on MEC-Cloud security are presented, which include threat models, key security challenges, considerations, and future directions. Finally, the article concludes with an outlook on key concerns that are pragmatic and valuable to 5G engineers, researchers, service developers, and policy makers from industry, academia, and government.

5G MEC-Cloud Overview

Edge and Cloud Computing for 5G

For 5G, edge and cloud are complementary technologies that can promote the operational efficiency, computational capability, service diversity, and low-latency access to 5G radio network resources. The advantages of integrating both edge and cloud will allow 5G operators to open up their infrastructure and service offering towards a more advanced and dynamic ecosystem. By utilizing edge and cloud, new requirements from end-users and large-scale IoT deployment can also be met. These modern requirements span across communication, mobility, scalability, trust, and privacy, in addition to the traditional demands of latency, security, and load balancing. In particular for IoT with high demand to offload data processing and storage, edge and cloud can enable 5G to flexibly accommodate various IoT applications, ranging from urban sensing, smart farming, e-health, industrial control, and intelligent vehicles.

As a rising paradigm, edge computing, especially the MEC is currently standardized by the ETSI. The fundamental principle of edge computing is to complement cloud resources by bringing computing closer to devices and date sources. As part of 5G roadmap, MEC exploits a systematic integration of wireless access technologies, which is in line with the 5G evolution towards ultra-dense deployment of small-cells, such as micro, pico, and femtocells. This will directly enhance the access capacity and quality of the connections. As an example, MEC offers 5G with dual/multiple connectivity where smart devices are able to communicate simultaneously through both conventional macro and the new small cells. Furthermore, computational offloading schemes (Cuervo et al. 2010; Kosta et al. 2012; Cozzolino et al. 2017) on the edge will speed up computation and communication.

With a mature ecosystem, the cloud computing exploits the economies of scale through centralization and aggregation, which effectively press down the marginal cost of administration, operation, and maintenance. One clear advantage of 5G comes from the outsourcing of setting up data centers with significant capital cost. Instead, computing power can be obtained from large cloud service providers in a pay-as-you-go manner. In addition, cloud computing can support 5G with elasticity, which can adjust

5G MEG-cloud architecture

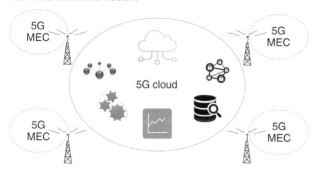

Figure 1 5G MEC-Cloud architecture.

the resource utilization to avoid under-provisioning and overprovisioning under very dynamic settings. Depending on the requirements, the service model of 5G cloud can be served via Infrastructure as a Service (IaaS), Platform as a Services (PaaS) or Software as a Service (SaaS). Figure 1 presents an architectural overview of 5G MEC-Cloud, where 5G cloud and 5G MEC form a two-tier structure to harness the benefits of both cloud and edge in terms of resource utilization, elasticity, and flexibility.

Given the recent progress in consolidating edge and cloud to better support 5G, numerous applications can benefit from such technology fusion, ranging from smart home to IIoT as illustrated in Table 1. The benefits of MEC-Cloud result from harnessing the power of both centralized and distributed resources for abstraction, programmability interoperability, and elasticity. In addition, there are four enabling technologies that support the vision of 5G MEC-Cloud, including network function virtualization (NFV), software-defined networking (SDN), information-centric networking (ICN) and network slicing (NS) (Liyanage et al. 2018a). By using those fast-evolving technologies, it is possible to optimize existing mobile infrastructure and implement novel ones for 5G. For instance, to realize the MEC-Cloud operational environment, it is necessary to run virtualized servers at various locations at the edge of 5G mobile networks. 5G base stations can be the physical hosts for such virtualized servers, which not only host edge services but also related services such as NFV and SDN. The setting can reduce the deployment costs and provide a common management infrastructure for all virtualized services. Furthermore, this can also enrich the mobile operators' existing business models by opening to versatile third party service providers.

Features and Synergies

The 5G MEC-Cloud is envisioned to be a federation of computing resources deployed across edge networks and provider data centers. The deployed MEC-Cloud servers can function independently and also collaborate with each other. The cloud computing part in MEC-Cloud can take the auxiliary role by offering reliable and powerful computing resources. In the 5G context, the general features of MEC-Cloud include virtualization, storage, networking, and multi-tenancy.

- Virtualization – an important feature in deploying services in heterogeneous settings. Virtualization allows the physical resources to be readily shared by multiple system

Table 1 MEC-Cloud can benefit 5G enabled services.

5G-enabled services	MEC-Cloud benefits
Smart home	Reduced communication latency, easy instantiation, and fast relocation. Moreover, MEC-Cloud can process sensitive data locally for privacy preservation
Smart urban sensing	Data can be processed at the edge of the network with location awareness and low latency, i.e. closer to sensor and hence removing the burden of sending raw data over a network with limited bandwidth
Intelligent vehicles	Improving the operational functions such as real-time traffic monitoring, continuous sensing in vehicles through infrastructure
Augmented reality	Migrating computationally expensive tasks to edge servers will increase the computational capacity of AR/VR devices and extend their battery life. MEC-Cloud also offers scalability by enabling high capacity and low latency wireless coverage for high populated places like smart cities or stadiums with a massive density of users to enjoy the AR/VR experience
Interactive gaming	Improving user experience for delay-sensitive game users by offloading the resource-intensive applications to the MEC-Cloud servers that are located in the proximity
Smart retail	MEC-Cloud servers can process local data generated by retailing systems such as intelligent payment solutions, facial recognition systems, smart vending machines
Smart farming	Reducing the overhead on data access, synchronization, and storage by using on-site MEC-Cloud servers to analyze collected farming data without real-time uploading to a remote cloud
Smart energy	Mitigating bandwidth bottlenecks and communication delays due to poor network connectivity and huge volume of data generation by allowing the computation to be performed closer to the data source. MEC-Cloud also reduces the attack propagation by enforcing security nearby the end power devices
Industrial IoT	MEC-Cloud can enable real-time edge analytics for future IIoT applications by addressing the challenges of predictive maintenance and M2M communication in terms of low power operation and reconfiguration

services/applications. By creating the virtual instance or devices such as operating systems, network interfaces, and storage devices, 5G can utilize existing infrastructure and hardware in more than one single execution environment.

- Storage – MEC-Cloud storage offers a hybrid of central and distributed storing possibilities for maintain, manage and backup data. The hybrid setting also allows flexible access from end-users in an on-demand fashion.
- Networking – to allow seamless connections among different servers of different purposes (e.g. data processing, storage) in a largely distributed manner. Secure networking is a mandatory feature of MEC-Cloud through both physical and virtual private networks, through which end-users can access their data and 5G services.
- Multi-tenancy – to support multiple 5G customers who do not share data but still share the same physical infrastructure and computing resources through a secure execution environment. This feature leads to the optimal utilization of 5G resources including radio access, data storage, and computing power.

Table 2 Comparing MEC-Cloud and conventional cloud on key properties.

Features	5G MEC-Cloud	Cloud/data centers
Ownership	5G providers	Private cloud providers
Deployment	Network edge and 5G core	Network core
Hardware types	Heterogeneous	Homogeneous
Service offering	Lightweight virtualization	Virtualization
Architecture	Decentralized and distributed	Centralized
Mobility	Yes	Limited
Latency	Low	Average
Location awareness	Yes	Limited

To understand the differences and similarities between MEC-Cloud and conventional data center cloud, Table 2 highlights the key properties including ownership, deployment, hardware types, service offering, architecture, mobility, latency, and location awareness. The consolidation of edge and cloud into the 5G MEC-Cloud has one goal: to bring cloud alike capability to the edge of 5G access networks. The technologies enabling MEC-Cloud support multi-tenant virtualization infrastructure. With NS and SDN, the MEC-Cloud infrastructure can dynamically adjust provision capacity to various demands in terms of location, speed, and/or privacy.

One important property for 5G is mobility, which is needed by mobile devices. MEC-Cloud supports mobility through several strategies such as hierarchical mobility management and live migration of lightweight virtual servers. Besides mobility, MEC-Cloud provides necessary scalability and availability for future 5G IoT deployment. This is crucial given the fact that devices and servers can be geographically widespread. MEC-Cloud can organize the distributed resources to assure that certain services can be efficiently set up and provided on-demand at the spot where needed. In practical setting, edge servers will provide redundancy at local level and function as proxy for the central cloud in case of temporary failure in core date centers. MEC-Cloud also offers third-party service providers to closely work with 5G providers to deploy edge specific services that can be integrated into telecommunication infrastructure.

As to the synergies of edge and cloud, the decentralization and proximity to date/devices bring obvious benefits but also deserve attention in terms of synchronization, interoperability, accountability, and usability. Given the cloud-edge-device architecture, both hard and soft states of services need to be timely synchronized across all the tier. Since edge servers can be managed by different infrastructure providers, it is necessary to form standards as to how the different parties in the MEC-Cloud architecture can collaborate with each other, how to discover services, and how to manage the life cycles of virtualized services in such distributed environment.

One important synergy MEC-Cloud is the management of virtual resources. The key concern is the optimization of resource utilization, e.g. to define when and where a virtual service instance needs to be set up, replicate, migrate, or merged. Another synergy concern is resource offloading where users can delegate the execution of tasks to external entities. As edge hardware in 5G is typically resource constrained, a fine-grained offloading design is needed to allow maximal usage of available resources.

MEC and Cloud Security in 5G

Threat Models

Given the core of MEC-Cloud consists of several enabling technologies such as NFV, SDN, and NS, security in 5G MEC-Cloud covers not only these fundamental building blocks but also to orchestrate different security schemes dedicated for each of these technologies. This requires unified management of available security mechanisms to achieve seamless integration. In particular for 5G inherent mobile features, security at networking and system level shall support mobility and can function in a decentralized manner without relying on centralized administration.

Besides potential threats emerged from edge computing, it is equally important to consider the security threats that are embedded in the enabling technologies and also the application domains. In this regard, IoT is a typical example of 5G that generates attack surface due to its scale and heterogeneity. Owing to the tight relation of IoT in our infrastructure and daily services, the implication for MEC-Cloud security is that the protection must consider all the layers of technologies not only about MEC-Cloud but also the threats from IoT.

In order to comprehend the security landscape of MEC-Cloud, security threats need to been investigated according to crucial aspects of edge and cloud (Roman et al. 2018). Table 3 summarizes the threats in MEC-Cloud under five categories, including network infrastructure, core servers, edge servers, virtualization, and end devices. The threats affect both edge and cloud computing which are the core of 5G MEC-Cloud.

Network Infrastructure

MEC-Cloud depends on the communication networks and protocols to connect various devices and servers, via both wired and wireless media. The infrastructure is a clear target for the adversaries and denial-of-service (DoS) is a common threat for all types of communication networks. The threat can take the form of distributed denial-of-service (DDoS) and wireless jamming.

As a typical threat to take control of the network, Man in the Middle is used by adversaries to launch further actions such as traffic injection, manipulation, and eavesdropping. In 5G mobile networks, such a threat is difficult to detect and affect several elements in MEC-Cloud including data and virtual images that are exchanged through the network.

Table 3 Threat models of MEC-Cloud.

Category	Threats
Network infrastructure	DoS, man-in-the-middle attacks, rogue mobile routers
Core servers	Privacy leak, service manipulation, rogue core server
Edge servers	Physical damage, privacy leak, privilege escalation, service manipulation, rogue edge server
Virtualization	DoS, resource misuse, privacy leak, privilege escalation
End devices	Data injection, service manipulation

Another threat of infrastructure is the rogue mobile routers. Since the deployment at the edge of network, adversaries can install their fake base stations. This threat is similar to the Man in the Middle threat where the rogue agents can impersonate, eavesdrop, and misguide the users in the network.

Core Servers

One major part of MEC-Cloud is the cloud support from the core. In practical setting, both edge and cloud can be managed by a single company or provider but can still share the infrastructure among several providers. In this context, privacy leakage is a common threat since it is hard to guarantee the information stored would be exposed to unauthorized adversaries.

Service manipulation in core servers is often from internal adversaries with privileges that can produce fake information to disturb the operation of MEC-Cloud. This threat has similar limitation as privacy leakage since the adversaries can only affect part of the system owing to the decentralized and distributed setting of MEC-Cloud.

Rogue core servers form another threat, which assumes that part of the core servers can be targeted by attackers. The rogue core servers are often compromised by the upper layer of software since the core infrastructure is typically protected. The rogue core servers mainly cause management disturbance and hence need fault tolerance scheme to tackle such threat.

Edge Servers

The edge servers in MEC-Cloud are mini data centers that host management services and virtual images. The attack surface of edge servers ranges from public APIs to physical tampering. Regarding physical damage, certain hardware elements of MEC-Cloud may not be guarded as compared with conventional data centers. The examples are edge servers managed by SME or small organizations. The physical damage threat requires the adversaries to approach the hardware to destroy it. Owing to this physical vicinity requirement, the impact is of local scope in that only the services in proximate to the attack will be affected.

Privacy leakage is one threat for edge servers but with limited scope. As edge servers often process and store only the information from users in the proximity of the service area. When there is migration of users or information, the leakage can be extended above the geographical coverage.

Privilege escalation is used by adversaries to take control services offered by edge servers. This type of threat is caused by the fact that edge servers could be managed by administrators with limited security knowledge nor training. Therefore, the edge infrastructure could be misconfigured and lack of maintenance. The privilege issue also arises from internal adversaries that result from social engineering, which is even harder to mitigate.

After gaining privileges, service manipulations can be another threat. Consequently, adversaries are able to launch complicated attacks such as DDoS using the manipulated resources in the existing infrastructure.

The rogue edge server is a threat that can be produced by either injection of servers into the infrastructure or manipulation of existing ones. This threat has a severe impact in that adversaries gain control over services in specific locations and hence having access to manipulate the information passing through those spots. If the rogue edge server is happening in a frequently used network section, the damage can be further escalated.

Virtualization

As MEC-Cloud is built upon virtualization, not only the virtualized hypervisors can be hacked, but also the virtual images with functionality can be the target as well. The DoS threat in this category is caused by malicious virtual functions that can deplete the computing, network, or storage resources. For edge setting, this threat is challenging to tackle since the hardware hosting edge servers are often of less resources.

Misuse of resources is another threat that is in the common form of botnet or bitcoin mining. This type is generated by adversaries that control the resources where they do not damage the MEC-Cloud infrastructure but use the resources for other purposes.

Privacy leakage is also a threat in a virtualization category due to the open APIs offered by virtualization can provide lots of contextual metadata about the status of the hardware and the network. Such information can be used by adversaries to derive other types of attacks.

Privilege escalation threat is from the vulnerabilities of hypervisors and tampered virtual images. Owing to potential isolation failure, compromised virtual service can manipulate other resources outside the regulated range. This threat can be escalated given that virtual images can migrate inside the MEC-Cloud infrastructure and hence quickly spread the threat.

End Devices

The end user devices are an important part of the MEC-Cloud since the devices also consume the resources offered by 5G. The mobile devices also participate in the distribution of the resources in the overall ecosystem. Injection of information is a user-driven threat where an end-device from adversary can be programmed to spread bogus data and even disturb the other devices sharing the same wireless communication link. Service manipulation is another threat from end devices where edge can be formed also by a cluster of end devices in device-to-device manner. In such case, adversaries that obtain control over one of the devices can gain access to other devices due to the fact that trust is often formed without strong verification.

Security Challenges and Considerations for Integrating MEC and Cloud to 5G

In 5G, one of the key challenges for deploying MEC-Cloud is security. In this regard, edge computing users in 5G face security challenges as being potentially vulnerable to security exploits since more and more IoT devices and applications are using the edge for data processing and storage. Such exposure of user data in MEC could introduce weak link where sensitive data can be breached. For example, IoT devices are typically programmed to trust other connected local devices and share data after simplified security check. In case such trust is natively enforced, it becomes difficult for MEC-Cloud to identify misbehaving ones. This can further create a disordered perimeter which prohibits security mechanism such as firewall to detect MEC security threats. In particular, it is challenging to balance the low-latency requirement and still being able to identify, authenticate and authorize data access in such distributed environment.

Besides general CIA requirements, MEC-Cloud needs to consider the following general factors: (i) privileged user access: offloading sensitive data to the MEC-Cloud can lead to the loss of direct physical and personal control over the data. (ii) Regulatory compliance: MEC-Cloud run by 5G providers should be willing to undergo external audits and security certifications. (iii) Data location: the exact physical location of the user's data is less transparent in MEC-Cloud, which may introduce confusion on specific jurisdictions and commitments on local privacy requirements. (iv) Data segregation: since data is usually stored in a shared space each user's data shall be separated from others with efficient encryption schemes. (v) Resilience: MEC-Cloud needs to offer proper recovery mechanisms for data and services in case of technical l failures or other disasters. (vi) Investigative support: since logging and data for multiple customers may be colocated, investigating illegal or malicious activities can be a time-consuming process. (vii) Long-term viability: to assure that users' data is safe and accessible even if the MEC-Cloud providers may become out of business. (viii) Device identity: as devices in MEC-Cloud typically authenticate themselves via a cryptographic key, the protection of device identity keys is of paramount importance. (ix) attestation: to prove that a device in MEC-Cloud is running up-to-date and patched code.

Given extensive discussions on edge and IoT security challenges (Hopkins et al. 2019; Hafeez et al. 2018; Singh et al. 2016), there are six major categories for 5G deployment ranging from hardware components to distributed logging that need to be tackled in order to establish a sustainable ecosystem of MEC-Cloud.

Hardware Component

As edge computing rests on edge hardware, there is a trend toward increasing computing power together with lower power consumption. This trend is enabling a rise in the capability of edge devices but not security yet. The cause is due to the fact there is little motivation at the current stage of edge computing since most of the companies prioritize the time-to-market instead of pushing security-first products. Even when hardware security features are available, the incorporation of such security support into the software is often missing. The trust of edge hardware is the very underpinning of secure edge computing but it only has value if integrated fully.

There are hardware-based methods to build trusted edge computing nodes such as trusted platform module (TPM) and hardware security module. Although such mechanisms for hardware trust can serve as a foundational building block for software security layers to depend on, there is a gap where the trust cannot be extended into the software infrastructure at the edge. This gap can affect the overall security of MEC-Cloud. Since deploying secure edge services to untrusted hardware can lead to fatal consequences, the hardware trust shall be the starting point on which all the other software components can be built.

In particular to distributed edge hardware, one key challenge is to the integrity of a reliable source of information about the condition of the device. Using insecure/untrusted information about edge hardware can result in potential vulnerabilities. For administration and orchestration, it is desirable to verify the information concerning CPU spec, RAM and disk storage of MEC-Cloud devices from a trusted entity. Such information can be valuable for detecting potential violations such as unexpected CPU consumption. Other attributes such as battery level or GPS coordinates can be combined as well. For example, a sudden change of device's GPS coordinates could be used to lock certain functionality of the software given the risk that such static device could have been compromised.

For 5G IoT deployment concern, physical security of MEC-Cloud cannot be taken for granted as to the data center environment. For devices out in the field, accessing the hardware is hard to control. It is hence useful to treat low-level components as viable targets. In this regard, even firmware and debugging interfaces, which are often considered as protected, can be vulnerable to leak sensitive hardware data. This calls for the usage of hardware security such as TPM to form the base of hardware trust and to validate the status of hardware components. Besides components that can be covered by TPM, other parts including USB ports and external buses not covered by TPM can be the next targets. For MEC-Cloud, it is equally important to protect those peripheral from unauthorized access, which is still an open challenge. As to intrusion detection mechanisms on edge hardware, it is typically related to the scenario of physical case opening. Given that certain intrusion can be linked to digital sphere such as disabling the boot of hardware or causing misleading notification to MEC-Cloud hardware owner. It is hence useful to restrict software functionality in case of detecting malicious access so that further damage can be alleviated.

On top of the hardware trust, the authenticity of the hardware shall be further ensured. Even when edge hardware contains a root of trust technology implementation, and even if that root of trust is integrated with software layers above it, there may still be a foundational breach of security if the authenticity of the hardware cannot be assured. For example, a MEC-Cloud device that masks espionage functionality as a trusted device may appear as a normal one. For 5G infrastructure, these threats are typically reserved for nation-state threat level and such security challenges can go beyond the technology or architecture scopes. For business operations and supply chain management, this is still a valid concern given that the cost of replacing deployed edge devices can be higher than centralized setting. In addition, the chance of detecting such rogue hardware can be limited since the owner may not be able to physically verify certain devices that are deployed under remote and challenging conditions.

Connected Devices

One advantage of MEC-Cloud is that computing resources can process data closer to the source. For 5G IoT use cases, the source of data is often physically apart from the MEC-Cloud hardware, e.g. sensing devices such as sensors and cameras deployed in the same network shared with MEC-Cloud. For mission and safety-critical scenarios, it is important to establish trust over the entire network of connected devices. Currently, this is one open challenge that is seeking scalable solutions.

Although it is necessary to establish trust among connected hardware, it is hard to verify the identity of large amount of sensing or actuators in 5G managed networks. In this regard, device identities shall be semantically defined to allow automatic verification. Currently, there is a lack of standardized approach, which is creating problem given the diversity of edge and IoT hardware devices. Even for devices that entail certain identity, spoofing such unverified identity can still compromise the security. One potential approach is to define a unified format for data exchange among connected hardware, but we still face the danger where malicious devices use proper format while the data is falsified. One example here is a smart home system that reacts to temperature reading for window opening. By generating false temperature readings, windows can be opened by attackers for stealing purposes.

In MEC-Cloud, once data enters the system, it can be stored, copied, forwarded, or used for analytics. When the data is passing through the network, it is hard to guarantee its integrity. This is similar to management commands between administration node and controlled devices. For instance, if door opening command can be spoofed or intercepted, it may result in intrusion and losses. For 5G industrial and mission-critical scenarios, commands such as "shut down" or "delete" can lead to catastrophic consequences. Even though we can add verification but it will introduce latency, which is unfavorable for many latency-sensitive services. In addition, protecting data and commands typically involve encryption but the problem of establishing a trusted communication channel in such distributed environment is yet to be solved. At the same time, managing connected devices includes several procedures. In practice, open APIs are used to manage and query device-related information. For services that involve production, one security challenge about how to prevent false information from generating wrong actions. In such cases, losses can be tangible in terms of money, time, materials.

System Software
System software security in MEC-Cloud is similar to the status of hardware security since existing methods are not integrated with service and infrastructure in the 5G context. Regarding secure boot at system level, hardware approach such as TPM can be used to verify drivers and boot loaders. However, such attestation also relies on BIOS software that executes the required procedures. It is hence necessary to protect the integrity of BIOS on MEC-Cloud devices to form a secure chain between hardware and operating system.

Given the tight linkage of system software and hardware, rogue software can falsify or cancel the process of monitoring before other actions are invoked. It is hence a challenge as to how can actions be enforced as soon as the detection of unauthorized software gets into the operating system. In addition, digital signatures for MEC-Cloud devices such as edge nodes and sensors can be used to verify device identifies but such methods rely on private keys. In cases where MEC-Cloud hardware can be physically accessed, attackers could copy such private keys and then impersonate with the stolen identity. One challenge here is how to ensure the trusted identity is of integrity. If blind trust is placed, it can motivate attackers to harness such false sense of trust.

Although hardware level security implemented by Intel and AMD provides a starting point, such security systems are often in the form of a black box including the add-on features and potential vulnerabilities that might be exploited. In 5G deployment, once a MEC-Cloud server is started, it is necessary to update the security mechanisms at the system software level. However, a secure delivery of updates also depends on the distribution servers which must be verified and trusted. This is yet not fully covered from communication perspective. For instance, using secure transport protocols such as TLS can effectively protect the data exchange in terms of integrity. Furthermore, even after the update package is pushed to the edge nodes, it is still crucial to verify the signature of any software updates before installing the binaries.

Networking and Communication

Networking issues in MEC-Cloud are due to the complexity of distributed architecture which is hard to protect in the same manner as to data centers with fixed cable connectivity. In this aspect, opening network ports with APIs is a common practice. Although such scheme is effective in a closed and centralized environment, it is challenging to copy the design to 5G scenarios where wireless communications are the dominant choice. For MEC-Cloud servers, open ports can expose vulnerability for local DoS attack. Therefore, port-based access shall be limited to the minimum. To avoid exposing public ports, one common approach is to use VPNs to interconnect different MEC-Cloud servers. However, similar to the case of using fixed private keys, such fixed VPNs in a distributed environment may cause security issues since a stolen VPN connection (e.g. by physical tampering) can give attackers direct access to the private networks.

In existing network design, credentials (e.g. in access control list) are used to prevent nonauthorized connections from utilizing the APIs via open ports, but those credentials are often not tied to any particular network identity in most cases. Under such circumstances, even without access credentials, being physically on the network may allow an attacker to harness the software with invalid requests to form DoS attacks. This challenge is critical for MEC-Cloud since physically tampering the network is distributed and wireless setting is of less effort in comparison with data center networks. Given the control role assigned to dedicated MEC-Cloud servers which entail credential stores, it is challenging to protect those servers from being physically hacked or attacked by rogue devices in the same network.

As MEC-Cloud heavily relies on wireless communications, wireless specific attacks such as jammers can result in similar damage as to DoS attacks in the wired networks. Preventing such attacks requires changes to the firmware but it is challenging and of high cost to low-cost IoT devices. Most edge use cases make use of devices from multiple vendors. To date, there has been no unified approach to solve this problem.

Another security issue in communication comes from the IoT devices. Sensing and actuation components in MEC-Cloud can be hacked to become part of the attack vector, and meanwhile, they are also the target of attacks. For example, battery-powered sensors can be maliciously turned off by requesting them to respond to invalid requests via wireless so that the power of sensors are quickly drained in using wireless interfaces. This is a special challenge for wireless-driven MEC-Cloud.

MEC-Cloud Service

MEC-Cloud runs virtualized and lightweight software, which is commonly referred as microservice. The integrity of such microservice needs to be secure in order to guarantee a reliable operating environment. To prevent MEC-Cloud servers from running arbitrary codes, microservice images must be verified in terms of integrity. Especially for multi-tenant environment, such verification also needs to be efficient so to avoid using extra hardware resources that can affect other users sharing the same hardware. On shared hardware, unauthorized microservices shall be efficiently detected and removed. In addition, reports must be generated if there is any attempt to launch unauthorized microservices.

Since microservices need configuration and credentials to function, it is important to avoid embed these sensitive data within the service images. One practical method is to obtain credentials via secure channel during runtime, and at the same time verify the integrity and identify. Furthermore, hardware components attached to MEC-Cloud servers such as serial port and wireless interfaces shall be accessible by microservices only when necessary. Since third-party software will be used by MEC-Cloud, access control must be enforced as well.

Regarding authorization boundary, microservices running on dedicated servers can still conduct activities that are outside the expected range. For example, even microservices can access the data in an authorized way, but if they transfer the data to undefined or malicious entities outside the control boundary, the protection of data simply fails. It is hence important to monitor unwanted connections to ensure the MEC-Cloud infrastructure has a clear boundary for data privacy.

Distributed Logging at Scale

Logging and audit are important to ensure MEC-Cloud services are properly functioning and in compliance with contracts, laws, and regulations. However, as the number of connected devices is increasing, the scale itself is becoming a challenge. In particular for MEC-Cloud environment, logging is often conducted in a distributed manner, i.e. over various deployed hardware, and can be diverse in terms of format. Even if the logging can be aggregated in a centralized location, it is challenging to meet the diverse goals from legal requirements by going through layers of hardware and software stacks. The multi-tenancy of MEC-Cloud further complicates the logging and audit since more volume and metadata need to be obtained to conduct fine-grained analysis.

Since logging in MEC-Cloud is partially decentralized, another consideration is the location awareness of logging. This is important for cases where many logging records are similar or duplicated. One approach to tackle this issue is by deploying logging analysis tools directly over such distributed logging data. In particular for storage overhead of logging, it is important to dynamically adjust the sampling ratio to capture threats while avoiding unnecessary logging.

Open Research Directions

Since the development of MEC-Cloud security is in its starting phase, the challenges and threats generate requirements for dedicated solutions. Specific for 5G, there are four open directions that deserve further investigations.

- *Trust Management* – for MEC-Cloud, trust goes beyond the basic authentication and shall handle the uncertainty as to what behaviors a device may take. Since the power of MEC-Cloud services comes from collaboration, which depends on trust, we need a reliable and deployable trust management solutions that can calculate trust metrics/reputation in an autonomous and distributed way. The interoperability is crucial for the trust management of MEC-Cloud given the data and devices are geographically distributed.
- *Machine Learning (AI)-based Security Enforcement* – given opportunities and data from the edge, machine learning is a potential domain to explore for security enforcement of MEC-Cloud. Owing to the increasing amount of security exploits on resource-constrained devices which will be the case for MEC-Cloud infrastructure, several security approaches for intrusion detection and classification can benefit

from the data-driven suggestions/prediction of machine learning (Hafeez et al. 2018). Furthermore, the combination of cloud and edge in MEC-Cloud provides nature support for the mode of centralized model training and edge-based predication. Given the opportunity to obtain high-quality data from the network edge at scale, machine learning-based security could help strike a balance between accuracy and latency.

- *Microservice Management* – fault tolerance and resilience are the key to success of 5G given the crucial role it will play in our society. The microservice is the core of MEC-Cloud that can promote the flexibility of service organization and diversity. Meanwhile, there is a lack of unified approach to securely delivery of secrets that are needed to manage a wide range of microservices in a decentralized manner. On top of migrating microservices, there is an urgent demand to allow guaranteed remote shutdown especially for safety or mission-critical services.
- *Hardware-Assisted Security* – protecting hardware and hypervisor is a core subject for MEC-Cloud. Given the active research on TPM, the virtualized environment of MEC-Cloud calls for more dedicated solutions (Tian et al. 2017). Meanwhile, it is equally important to investigate hardware acceleration for security by using dedicated components such as GPU and FPGA (Volos et al. 2018).

Conclusions

Securing MEC-Cloud is crucial for the success of 5G deployment on a large scale. The analysis of security requirements and the extensive discussions of existing solutions represent a clear interest and strong will from both academia and industry to address the open issues of the MEC-Cloud security in 5G. Although cloud security has been studied in the past, the combination of edge and cloud technologies has introduced new threats and risks. In particular for 5G IoT services, potential vulnerabilities are growing at an exponential rate as magnified by the rapid fusion of cyber and physical territory, the diversity of hardware, the heterogeneity of deployment scenarios, and lack of privacy awareness. All those add up the complexity to secure the inherently distributed 5G MEC-Cloud.

The discussions in this article serve as a stepping stone to expose key issues and reflect on the 5G specific security challenges for MEC-Cloud. By focusing on 5G domain, the presented overview and outlook shed light on the future development of security solutions that will tackle the threats and meet the dedicated requirements. As the field of MEC-Cloud matures, more robust and secure solutions are expected to be devised, and 5G customers will then embrace the benefits of MEC-Cloud. Nevertheless, the security of 5G MEC-Cloud is still in its initial stage. We need further investigations and research to address the open issues.

References

Cozzolino, V., Ding, A.Y., and Ott, J. (2017). FADES: fine-grained edge offloading with unikernels. Proceedings of ACM SIGCOMM Workshop on Hot Topics in Container Networking and Networked Systems (HotConNet '17), ACM, 36–41.

Cuervo, E., Balasubramanian, A., Cho, D.K. et al. (2010). MAUI: making smartphones last longer with code offload. Proceedings of the 8th International Conference on Mobile Systems, Applications, and Services (MobiSys'10), ACM, 49–62.

Ding, A.Y. and Janssen, M. (2018). Opportunities for applications using 5G networks: requirements, challenges, and outlook. International Conference on Telecommunications and Remote Sensing, ACM, 27–34.

Ding, A.Y., Liu, Y., Tarkoma, S. et al. (2015). Vision: augmenting WiFi offloading with an open-source collaborative platform. Proceedings of the 6th International Workshop on Mobile Cloud Computing and Services (MCS '15), ACM, 44–48.

Hafeez, I., Ding, A.Y., and Tarkoma, S. (2017). IOTURVA: securing device-to-device (D2D) communication in IoT networks. Proceedings of the 12th ACM MobiCom Workshop on Challenged Networks (CHANTS '17), ACM, 1–6.

Hafeez, I., Ding, A.Y., Antikainen, M., and Tarkoma, S. (2018). Real-time IoT device activity detection in edge networks. Proceedings of the 12th International Conference on Network and System Security (NSS '18), Springer, 221–236.

Haus, M., Waqas, M., Ding, A.Y. et al. (2017). Security and privacy in device-to-device (D2D) communication: a review. *IEEE Communications and Surveys* 19 (2): 1054–1079.

Hopkins, K., Bergquist, J., Ortner, B. et al. (2019). *Edge Security Challenges.* White Paper.

Kosta, S., Aucinas, A., Hui, P. et al. (2012). ThinkAir: dynamic resource allocation and parallel execution in the cloud for mobile code offloading. Proceedings of IEEE International Conference on Computer Communications (INFOCOM'12), 945–953.

Liyanage, M., Porambage, P., and Ding, A.Y. (2018a). Five driving forces of multi-access edge computing. *arXiv:1810.00827* [cs.NI], October 2018.

Liyanage, M., Ahmad, I., Abro, A.B. et al. (2018b). *A Comprehensive Guide to 5G Security,* 1ste. Wiley.

Morabito, R., Cozzolino, V., Ding, A.Y. et al. (2018). Consolidate IoT edge computing with lightweight virtualization. *IEEE Network* 32 (1): 102–111.

Roman, R., Lopez, J., and Mambo, M. (2018). Mobile edge computing, Fog et al.: a survey and analysis of security threats and challenges. *Future Generation Computer Systems* 78 (2): 680–698.

Singh, J., Pasquier, T., Bacon, J. et al. (2016). Twenty security considerations for cloud-supported internet of things. *IEEE Internet of Things Journal* 3 (3): 269–284.

Tian, H., Zhang, Y., Xing, C., and Yan, S. (2017). SGXKernel: a library operating system optimized for intel SGX. Proceedings of the Computing Frontiers Conference (CF'17), ACM, 35–44.

Volos, S., Vaswani, K., and Bruno, R. (2018). Graviton: trusted execution environments on GPUs. 13th USENIX Symposium on Operating Systems Design and Implementation (OSDI 18).

Further Reading/Resources

Ardagna, C.A., Asal, R., Damiani, E., and Vu, Q.H. (2015). From Security to Assurance in the Cloud: A Survey. *ACM Computing Surveys* 48 (1): 50.

Byma, S., Steffan, J.G., Bannazadeh, H. et al. (2014). FPGAs in the cloud: booting virtualized hardware accelerators with OpenStack. Proceedings of IEEE 22nd Annual International Symposium on Field-Programmable Custom Computing Machines, IEEE, 109–116.

9

Security in Network Slicing

Pawani Porambage[1] and Madhusanka Liyanage[1,2]

[1] *Centre for Wireless Communication, University of Oulu, Oulu, Finland*
[2] *School of Computer Science, University College Dublin, Dublin, Ireland*

Introduction

The fifth generation (5G) mobile networks will drive the concept of network of entities toward the network of virtual functions. It makes a very complex network architecture that accommodates heterogeneous networking and communication technologies (Liyanage et al. 2018). Furthermore, 5G brings an evolution in terms of capacity, performance, and spectrum access in radio network segments along with native flexibility and programmability conversion in all non-radio network segments (Wang et al. 2014). The most compelling 5G use cases are identified in three main areas such as mobile broadband, machine-to-machine or massive-Internet of Things (IoT), and critical communications where each category has different levels of requirements in terms of latency, reliability, capacity, and mobility (Foukas et al. 2017). For instance, a mission-critical IoT use case like autonomous driving requires very high reliability with low latency whereas augmented/virtual reality (AR/VR) or smart city types of applications need high capacity and massive connections.

Under such a circumstance, "one-fits-all" kind of architecture is not sufficient and efficient enough to support the different needs of 5G services, which have very diversified requirements. Therefore, the realization of 5G foresees a new mobile network architecture, network slicing (NS), mainly based on Software-Defined Networking (SDN), Network Function Virtualization (NFV), and Cloud Computing (CC). The term "slice" coined with GENI project that introduced a shared testbed to run multiple isolated experiments (Global Environment for Network Innovations (GENI) 2018). By definition, a slice is a logical network that provides specific network capabilities and characteristics to provide flexible solutions for different market scenarios, which have diverse requirements, with respect to functionalities, performance, and resource allocation (International Telecommunication Union 2017). Basically, NS provides specialized services for different traffic classes in the network, which are concurrently operated on the same infrastructure (Figure 1). The concept of using NS in 5G was officially announced by the Next-Generation Mobile Network (NGMN) Alliance in 2015 with some high-level examples (Wang et al. 2014). Third generation partnership project (3GPP) also considers

The Wiley 5G REF: Security. Edited by Rahim Tafazolli, Chin-Liang Wang, Periklis Chatzimisios and Madhusanka Liyanage.
© 2021 John Wiley & Sons Ltd. Published 2021 by John Wiley & Sons Ltd.

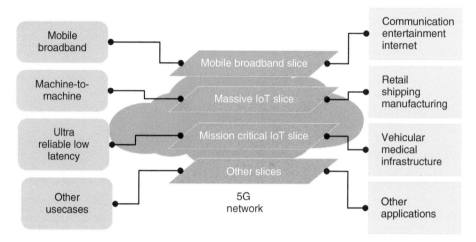

Figure 1 Network slicing for 5G.

NS as a key feature of 5G where they present the 3GPP architecture for NS in 3GPP (2018).

In any shared infrastructure, security is a paramount property to guarantee safe and accurate operations which ensure a fair share of preserved error-free resources to each user. Similarly, security is a carefully considered feature throughout the design, run-time, and termination phases of NS architecture. Not only the slice customers but also rest of the key roles of NS architecture should be protected against security and privacy attacks. They must be appropriately authenticated, their rights should be enforced by authorization mechanisms, and the operations should be accounted in such way that further auditing can be applied in case of any problem is detected. Beyond this, measures have to be in place to proactively detect and address active security attacks, avoiding a security breach in one slice propagates into the infrastructure and/or other slices.

Not only the novelty of NS concepts but also NS security is a barely investigated research area. To the best of our knowledge, this is the very first effort of surveying the security implications of NS. In this article, we discuss the security requirements of NS architecture and possible security threats, and the potential solutions that can be adopted to mitigate those vulnerabilities. Of course, NS will be affected from the security issues related to SDN and NFV since they are the foundation for realizing NS into practice. Apart from those NS will face unique security challenges during the process of strong slice isolation. The importance of security in a slice is also dependent on its use case. Moreover, we identify some interesting research areas of NS security beyond the state of the art.

Network Slicing Architecture

A network slice can be vertical, horizontal, static, or dynamic (Samdanis et al. 2016).

- *Vertical slicing* focuses on a mass scale market (e.g. public safety or autonomous car (Samdanis et al. 2016)).

- *Horizontal slicing* serves for one particular use case from that pool.
- *Static slicing* may follow a vertical or horizontal framework for serving fixed devices or IoT type applications for perpetual use.
- *Dynamic slicing* remains the most prominent approach for next-generation networks that encourages the emergence of slice-as-a-service paradigm (Zhou et al. 2016).

A fully functioning network slice should be able to control its own packet forwarding from user equipment (UE) to cloud servers in the core network (CN) without affecting the other slices. Furthermore, it should be defined to meet different service, application, and communication requirements. Basically, the end-to-end (E2E) NS architecture is defined using three segments as illustrated in Figure 1. First is the access slice that supports either radio or fixed access and close to UE. Second is CN slice which provides one or multiple virtualized network functions (VNFs). Third is the slice pairing function that connects access and CN slices to formulate a complete network slice.

The key design aspects for RAN slicing include the management of radio resources, the configuration rules for control plane (CP) and user plane (UP) functions, slice specific admission control policies, and UE awareness on the RAN configuration for the different services. Moreover, CN slices are designed with the logical separation between CP and UP functions and the corresponding NFs implemented as VNFs. Certain NFs can be common to multiple slices whereas some are customized for specific slices.

Security Threats of NS High-Level Functional Roles

The entire life-cycle of a NSI (Network Slice Instances) is achieved with the contribution of multiple logical and physical entities. Figure 2 illustrates the high-level functional roles of a NSI life-cycle. Based on the actual deployments, these roles can be played by single or multiple organizations. Among them, customer, service provider (i.e. CSP or

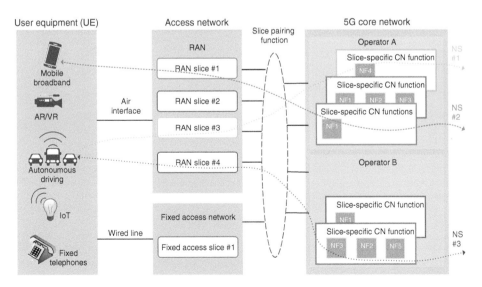

Figure 2 High-level illustration of network slicing architecture.

Table 1 Security threats on the key roles involved in NS life-cycle and potential solutions.

Key role	Description and contribution of NSI life-cycle	Possible security threats	Defense techniques							
			1	2	3	4	5	6	7	8
Communication service customer (CSC)	End users who experience the communication services may create requests for Network Slice-as-a-Service (NSaaS)	Impersonate attacks	X							
		Privacy attacks				X				
		Attacks on secrecy			X					
		Fraud attacks	X	X		X				
Communication service provider (CSP) and network operator (NOP)	NOP offers communication services and responsible for designing, building, and providing NSaaS based on customer requirements. This can also be the CSP	DoS and DDoS attacks	X		X				X	
		Man-in-the-middle attack	X	X				X		
		Unauthorized access	X							
		Traffic and data modification	X	X						
		Eavesdropping			X			X		
Virtualization infrastructure service provider (VISP)	They provide virtualized infrastructure (VI) services based on the NOP's specifications. They are responsible for designing, building, and operating VIs	Man-in-the-middle attack	X	X				X		
		Flooding attack					X			
		Unauthorized access	X							
		Data leakages			X					
		Logging and reporting issues				X				
		Hypervisor attacks				X			X	
		DoS attacks on VM	X			X			X	
		VM escape attacks				X			X	
Data center service provider (DCSP)	Those who provide data center services based on NOP's requirements to design and operate NSs	Identity thefts	X	X		X				
		Physical attacks								X
		Unauthorized access	X		X					
		Data "leakages," alterations, and thefts			X					
		DoS attacks	X		X				X	

1. Authentication; 2. Encryption; 3. Access control; 4. Privacy preservation; 5. Attack prevention by puzzle mechanism; 6. Key derivation; 7. Slice isolation; 8. Physical protection.

NOP), virtual infrastructure service provider, and data center service provider have the direct involvement with the NSI life-cycle. Therefore, in Table 1, we describe the contribution of these roles in NSI life-cycle, their common security threats, and the potential solutions.

Information related to customer privacy can be compromised during the phase of NS selection. An attacker can exploit the security vulnerabilities of a VNF to manipulate the records of user data from the database and violate user privacy. Therefore, privacy-preserving and data protection mechanisms need to be followed since the time when the customer initiates the request for NS creation. In addition to that couple of other attacks are identified in terms of the NS customer such as impersonate attacks, fraud attacks, and attacks on user secrecy. Unauthorized access to a third-party service provider may create security vulnerabilities on the services provided by a NS belong to a NOP. Moreover, the attackers may illegally manage slices or on-going services and launch attacks to slices such as terminating a slice or compromise a critical network function. The third party may need to authenticate the users those who access the slice via suitable application program interfaces (APIs).

When there are malicious parties that continuously ask for new NSIs there can be a congestion at the NS manager they can create denial-of-service (DoS) and distributed denial-of-service (DDoS) attacks at all the entities including CSP, NOP, and DCSP. Moreover, when a NSI life-cycle is not terminating at the correct point (i.e. when it is not needed), the resources can be exhausted and poorly managed. Strong authentication and access control mechanisms are the commonly used defense techniques for DoS and DDoS type of attacks. The impact of DoS attacks that can occur on VNFs offered by VISPs can be also eliminated by proper slice isolation.

Many security threats are identified due to the lack of logical and physical isolation between distinct VNF hosted by the same hypervisor. VISPs may encounter possible attack types that can occur in virtualized environments. Unlike in monolithic network architectures, the entrance of an attack cannot limit to single entry point and gain access to network resources. Instead in virtualized networking environments could provide entry points into multiple network domains and/or network slices. In VM escape attacks, the attackers will break out from a compromised VM and take the control of the hypervisor by invoking VNFs.

Data centers may encounter physical attacks and inbuilt security threats due to erroneous network topology validations and implementation failures at DCSPs. These need to be addressed by following strong physical security techniques and providing softwarized security solutions such as firewalls. Secure and efficient key management algorithms are important to mitigate eavesdropping attacks. Key derivation also explicitly supports encryption and authentication functions in many ways.

Security Considerations in Network Slicing and Related Work

Security Considerations in Network Slicing

Security requirements and related issues become complex when NS is implemented on the multi-domain infrastructure which require excellent coordination mechanisms. In NGMN Alliance (2016), NGMN has identified several security considerations in NS and associated technologies. In addition to the key requirement of strong isolation between slices, there are multiple other security requirements, which may affect the safety and reliability of slicing technology. These security considerations incur on managing the different phases of NSI life-cycle which includes preparation, commissioning, operation, and decommissioning (3GPP 2018).

Preparation is the very first stage of planning the composition of a slice, where security-by-design principles can be incorporated. Commissioning is the creation of a NSI. While allocating resources in a NSI the service level agreements should be fine-tuned and well defined. This should also ensure access control and proper authentication mechanisms are integrated along with NSI policy management. Operation or run-time phase include activation, supervision, performance reporting, resource capacity planning, modification, and deactivation of NSI. Slice isolation is the most critical security consideration in run-time phase. Finally in decommissioning phase, the network slice manager (NSM) demolishes the NSI specific configuration from the shared constituents. This is very important in security point of view as the complete the termination of NSI needs to be ensured. As illustrated in Figure 3, the key security considerations can be related to different phases of the NSI life-cycle. Furthermore, it shows the distribution of contributing roles in each phase.

1) *Isolation between slices*: Isolation is the most critical and the obvious security requirement of NS. Security issues related to slice isolation arise while resource isolation and allocation policies are defined at different levels of network service management for multiple tenants. If there is a malfunction or a security attack occurs on one slice, it will not affect the other slices due to the strong isolation (NGMN Alliance 2016). Therefore, adding several levels of isolation will protect the slices when intruders or attackers are deliberately trying to attack through the gap between slices. Side-channel attacks can occur across slices. When an attacker can learn about how the cryptographic codes are running on underlying hardware platforms, they may use that knowledge to extract keys. In particular, when two slices are sharing common hardware platforms, the attacker can observe the crypto code running on one slice and affect the security functions running on other slice. Therefore, strong slice isolation is highly recommended. To prevent such attacks, isolation of virtual machines is important. It is advisable to eliminate the possibility of avoiding the influence of code running on one virtual machine/slice hardware to reveal the status of another similar entity running on the same hardware.
2) *Reliability of a network slice*: A network slice can be subjected to node or link failures in the physical infrastructure. Such issues may directly affect the network capabilities and the performance level which a network slice can provide. Thus, it is required to identify and quantify the reliability of a network slice when a disruption occurs in a network slice.
3) *Secure inter-network slice communications*: This implies the secure communication between slices and between the VNFs. The network slices and their components are required to securely manage signaling and management plane communications. Therefore, security mechanisms should aware of the operations within the expected parameters and the security needs of the operator. Having the assumption that the NSIs can be dynamically offered on demand basis by multiple network operators (NOs), multi-domain security solutions are needed for orchestrating security policies.
4) *Security at NSM*: A NSM is responsible for dynamically creating and destroying instances of a network slice and map, and loading them to available physical host platforms such as routers, switches, or servers. Strong authentication is required between the NSM and the host platforms before initiating the slice instances. When

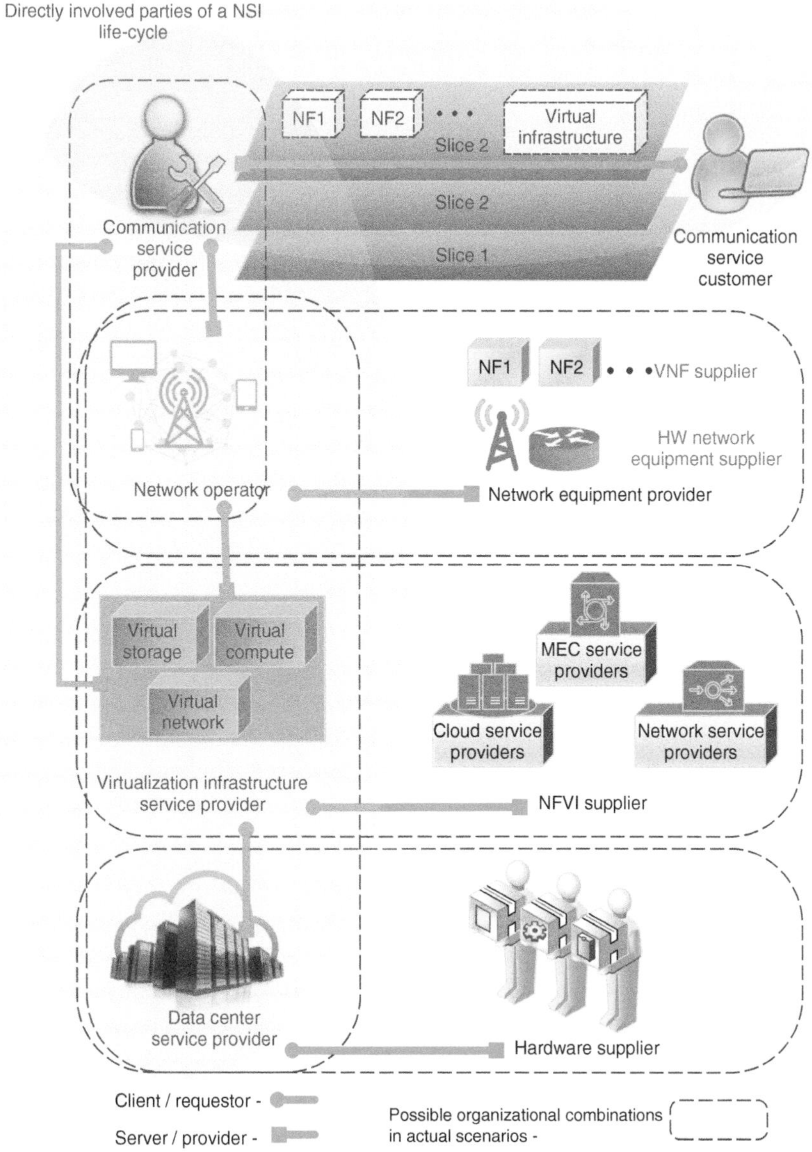

Directly involved parties of a NSI
life-cycle

Figure 3 Illustration of high-level functional roles of a NSI life-cycle.

multiple operators or host platforms are contributing for a particular slice, NSM
needs to achieve mutual authentication among each other before any negotiation
in order to overcome impersonate attacks. Network slicing management function
(NSMF) is aware of the composition of NSI and NSSI and the corresponding network
functions and resources. Whenever there is a security threat is identified related to
either one of it, NSMF needs cease the associated NSI and NSSIs. This is also relevant
to automated healing of a NSSI. When there is a malfunctioning NF, it may affect all

the related NSSI. In order to identify the malfunctioning of the NF, there should be a continuous monitoring system in the NSSMF. AI-trained systems can be exploited for identifying such anomalies in the system and then automated healing actions are also required.

5) *Secure NSI*: In order to achieve 5G seamless enhanced mobile broadband (eMBB) service with fixed mobile coverage (FMC), both wireless and fiber access technologies should support NSI. While optimizing resource efficiency in 5G systems, NS may allow multiple access technologies to be used simultaneously for single or multiple services active on an UE. This may create serious security threats while switching between multiple access technologies. It is important to investigate the security considerations while creating a communication service spanning multiple NSI hosted across multiple operators. Moreover, all virtual network functions available in a network slice instance should be authenticated and integrity should be verified. The CN components related to NSI security architecture include authentication of UE, storing security credentials of UEs, and maintaining security control policies.

6) *Slice heterogeneity*: When different slices offer distinctive services, their constraints and security requirements will also vary. For instance, services with low latency will require very fast key derivation techniques and fine-grained key management mechanisms, whereas services that need extremely long battery life need to determine how often reauthentication should be performed. Similarly, services with high privacy risks may require frequent reallocation of temporal identities. Security protocols should not only acceptable for one slice, if it is weak for 5G as a whole. With the slice heterogeneity, it is important to differentiate user end (UE) accessing of multiple slices at one instance with different security levels. UE can sometimes be attached to different slices. Moreover, the operators' preferences should be taken into account while defining the security strength of NFV in NS for 5G deployments.

7) *Protection against DoS attacks on slices*: An attacker can exhaust resources in one or multiple slices and cause service degradation of that particular slices. Sometimes resources can be common to multiple slices, which can be also exhausted by deliberate DoS attacks. While allocating the network resources for security to individual slices, they should be guaranteed with minimum level of resources. An attack on one slice may cause the exhaustion of resources in another slice by explicitly imposing a DoS attack. These pitfalls can be avoided by running ring-fencing resource allocation policies in the security protocols among slices.

State of the Art for NS Security

Apart from identifying security challenges related to NS security, not many technical research findings have not yet been published providing the security solutions for NS architecture. Majority of the work considers authentication protocols from the user end or for the inter-slice communication. Slice isolation, privacy, and managing trust among different stakeholders and slices also matter the most. Customizing NS security by SDN using micro-segmentation is another approach to isolate traffic flows related to different applications or users. The concept of micro-segmentation is originally introduced for securing data centers (VMWare 2014). Typically, a data center has one firewall at the perimeter, which filters all the incoming traffic. When an attacker is somehow able to gain access to the data center through the firewall, he is free to move and perform attacks. Micro-segmentation aims to get rid of single point of failure in security by allowing security monitoring both inside the data center and at the perimeter. Recently, micro-segmentation is adopted to secure 5G network slices to provide fine-grained isolation, specific access control, and security policies in Mammela et al. (2016); Suomalainen et al. (2018).

Future Research Directions

Artificial Intelligence

On one hand, existing security mechanisms for NS are either human or machine-centric. In addition, automated techniques such as anomaly detection are still required human intervention to address false negatives. However, the NS-based future mobile network will face more and more automated and advanced attacks due the advancement of communication technologies and machine learning techniques (You et al. 2019). Attackers will use AI to obtain the control of future mobile networks. On the other hand, it is not possible to run human or machine-centric security mechanism, when the network services and users are exponentially increasing in NS systems. To prevent much massive scale network, sophisticated and intelligent security solutions are required. AI can be used as a tool to design such intelligent security solutions for NS. Moreover, AI algorithms and models (e.g. Markov models, neural networks, genetic algorithms, and machine learning techniques) can be used to find configuration errors, security vulnerabilities, and threats to reduce the human intervene. New AI-based algorithms and security models can be designed to mitigate both identified and predicted attacks and threats on NS-based systems.

One Slice for Security

Several security mechanisms must be implemented to achieve a secure NS system. However, these security mechanisms must be coordinated and securely implemented with minimum security overhead and impact to other security mechanisms. To achieve this goal, allocation of an independent network slice for security is beneficial. For instance, security-related communication such as authentication messages, firewall updates, security policy updates can be transported over this slice. In addition, network monitoring and security incident handling systems can be run on top of this security slice to make sure the proper operation of the network.

A dedicated security slice can ensure the end to end supply chain security of the systems. For instance, security-related services such as security service management, security incident and event management (SIEM), security monitoring, security service change management, cryptographic service, authentication and access control, security auditing and security service life-cycle management can be implemented on top of this security slice. When a security slice is available, resources allocated for security services can be dynamically changed. More importantly, it can ensure the availability of network resources for security.

Context-Aware Security

With the development of ubiquitous computing and rapid increment of IoT ecosystems, it is expected that context-aware communication and networking will dominate in 5G and beyond 5G era. Many of the future 5G and beyond 5G applications will require reliable access to various sources of context information, for example, detailed localization information both indoors and outdoors will be required, for data and multimedia delivery every time and everywhere, rapid file sharing in the form of cellular broadcasting and wireless car video. Future mobile communication networks (beyond 5G) are also very frequently integrated with IoT/IoE (Internet of Everything) networks to provide wide range of novel services. Thus, at one end, these heterogeneous kind of networks will be considered crucial for improving context awareness but on the other end, security risks will also emerge. Therefore, context awareness-based security mechanism requires intelligent and controlled solutions by the NO and other involved stockholders. On one hand, NS has identified as one of the prominent solutions to provide such context-aware security. By using NS, security services can be dynamically scaled in and out according to the context of the network. On the other hand, NS will play a crucial role to enable above context-aware services. Thus, it is also important to consider the security of slicing system to achieve secure context-aware services.

Security Orchestration and Automation

The use of security orchestration is mandatory softwarized networks where the operator needs to control both virtual and physical network segments. The primary goal of security orchestration is removing the need for manually on figure with human interaction. Human central security management is no longer feasible due to high dynamicity of the future mobile network. The security orchestrator will be responsible for deployment, configuration, maintenance, monitoring and life-cycle management all security functions in a softwarized mobile network. It should be able ensure the E2E security by automatically aligning the security policies inside the both virtual and physical network segments. European Telecommunications Standards Institute (ETSI)-Industry Specification Groups (ISGs) group has already defined the security orchestrator for NFV systems. The group has also defined different tasks of security orchestrator in NFV systems and the required interfaces to interact with the existing ETSI NFV components such as NFV orchestrator, the VNF managers, the element managers, and the virtual infrastructure managers. Since, NS will also be a part of softwarized network and the functions of security orchestration should be extended to manage the security of NS system as well. The role of security orchestrator in slicing systems should be defined along with the new interfaces to communicate with 5G slice control elements.

Security-By-Design

Secure-by-design (SbD) is an approach that will consider the security concerns already at the beginning of a design a product, service or software. SbD can secure the foundation of the product or service by minimizing impact anticipated security vulnerabilities. Many software systems i.e. AWS (Amazon Web Services) is using SbD to automates security controls and streamlines auditing. In current software systems, the current SbD approaches can offer benefits such as establishment reliable operation of controls and enabling continuous and real-time auditing. However, the core concept of SbD is not limited to software systems. It can be extended to any system including mobile networks. For instance, SbD approach can be used in two occasions in NS systems. First, SbD approach can be used to design slicing systems (including relevant VNFs) with a secure foundation. Second, SbD approach can be used during the slice creation process for different network services. In both occasions, SbD approach can reduce the impact of know attacks on the system.

Security-as-a-Service

5G networks provide services for a large variety of verticals including smart grids, transportation, health care, smart city, and future factories. However, most of these vertical operators will not have up-to-date security expertise to manage all security aspects of their network (Sriram et al. 2019). Therefore, they must obtain a wide range of security services by security service providers. In this content, security-as-a-service (SaaS) is an approach where service providers can offer security services for cooperate customers. Typically, these security services are ranging from authentication, security monitoring intrusion detection, penetration testing, and security event management, among others. SaaS concept can be further extended to provide NS as well. As typical vertical operators do not have expertise on both network security and network softwarization, SaaS concept can offer easy integration route for them. It is an interesting research domain that to look it to possibility to provide slicing security as a SaaS solution.

Conclusions

In summary, the ultra-low latency attribute of 5G will encourage the attackers to impose the security threats in a similar high speed. Therefore, as a key driving force of 5G, NS will also require preventive security mechanisms since the reaction will be too late. The security vulnerabilities that are inherent to different logical roles of NS architecture may also impose security threats to the NSI life-cycle. Considering the consequences and applicability of security attacks in NS, strong slice isolation is the best practice to mitigate the spreading of most of the threats. Undoubtedly, security in NS is still in its infancy and there are many open research areas and security challenges to tackle.

References

3GPP (2018). Study on management and orchestration of network slicing for next generation network. Technical specification. https://portal.3gpp.org/desktopmodules/Specifications/SpecificationDetails.aspx?specificationId=3091 (accessed June 2018).

Foukas, X., Patounas, G., Elmokashfi, A., and Marina, M.K. (2017). Network slicing in 5G: survey and challenges. *IEEE Communications Magazine* 55 (5): 94–100.

Global Environment for Network Innovations (GENI) (2018). www.geni.net (accessed 19 November 2018).

International Telecommunication Union (2017). Terms and definitions for IMT-2020 network: ITU-T Y.3100 (September 2017).

Liyanage, M., Ahmed, I., Abro, A.B. et al. (2018). *A Comprehensive Guide to 5G Security*. Wiley.

Mammela, O., Hiltunen, J., Suomalainen, J. et al. (2016). Towards micro-segmentation in 5G network security. European Conference on Networks and Communications (EuCNC) Workshop on Network Management, Quality.

NGMN Alliance (2016). 5G security recommendations package 2: Network slicing (April 2016).

Samdanis, K., Costa-Perez, X., and Sciancalepore, V. (2016). From network sharing to multi-tenancy: the 5G network slice broker. *IEEE Communications Magazine* 54 (7): 32–39.

Sriram, P.P., Wang, H.-C., Jami, H.G., and Srinivasan, K. (2019). 5G security: concepts and challenges. In: *5G Enabled Secure Wireless Networks*, 1–43. Cham: Springer.

Suomalainen, J., Ahola, K., Majanen, M. et al. (2018). Security awareness in software-defined multi-domain 5G networks. *Future Internet* 10 (3): 27.

VMWare (2014). Data Center Micro-Segmentation: A Software Defined Data Center Approach for a Zero Trust Security Strategy, White Paper.

Wang, C.X., Haider, F., Gao, X. et al. (2014). Cellular architecture and key technologies for 5g wireless communication networks. *IEEE Communications Magazine* 52 (2): 122–130.

You, X., Zhang, C., Tan, X. et al. (2019). AI for 5G: research directions and paradigms. *Science China Information Sciences* 62 (2): 21301.

Zhou, X., Li, R., Chen, T., and Zhang, H. (2016). Network slicing as a service: enabling enterprises' own software-defined cellular networks. *IEEE Communications Magazine* 54 (7): 146–153.

Further Reading

Porambage, P., Miche, Y., Kalliola, A. et al. (2019). Secure keying scheme for network slicing in 5G architecture. 2019 IEEE Conference on Standards for Communications and Networking (CSCN). IEEE, 1–6.

10

VNF Placement and Sharing in NFV-Based Cellular Networks

Francesco Malandrino[1] *and Carla Fabiana Chiasserini*[2]

[1] *CNR-IEIIT, Torino, Italy*
[2] *Politecnico di Torino, CNR-IEIIT, Torino, Italy*

Introduction

Network function virtualization (NFV) has revolutionized the concept of service, conceiving it as a set of software, namely, virtual, interconnected functions, referred to as virtual network functions (VNFs): the service data traffic then flows through the sequence of VNFs composing the service, and is processed by each them. In parallel, new generation cellular networks have become capable not only to efficiently transfer traffic but also to process and store data, thus they can effectively implement a large variety of services that may be requested by the so-called verticals, i.e. content providers, automotive, or e-health industries. Deploying services within the cellular network instead of the cloud can indeed bring significant advantages, among which, lower service latencies, local data processing (hence, lower bandwidth consumption due to data transfer through the network infrastructure), and lower energy consumption.

Network providers and vertical industries have therefore built a new relation, regulated by Service Level Agreements (SLA), which define the level of quality of service that a network provider has to ensure in order to match the fee paid by a vertical. As a consequence, upon receiving a service request, a network provider has to put into place service deployment strategies that allow the fulfillment of the target key performance indicators (KPIs), e.g. throughput, delay, or reliability, while minimizing the cost, i.e. the amount of resources necessary to run the service. In the following paragraphs, we describe in more detail these two important issues and briefly discuss some relevant works that have dealt with them.

Target KPI. To effectively address the target KPIs, network providers can resort to network slices (NGMN Alliance 2016), i.e. reserve a suitable amount of computing, network, and memory resources for a service, or a set of services with no isolation requirements and similar target KPIs. As an example, such safety services as forward collision warning or intersection collision avoidance (Malinverno et al. 2019) by the same automotive industry, can be implemented using virtual machines (VMs) in servers, along with network links for data transfer, that are sufficiently capable to match millisecond-order latencies. Creating a network slice thus implies that the network provider needs to identify the VMs where to place the VNFs composing the service,

The Wiley 5G REF: Security. Edited by Rahim Tafazolli, Chin-Liang Wang, Periklis Chatzimisios and Madhusanka Liyanage.

how much CPU and memory to assign to each of such VMs, and which links to use to transfer data traffic from one VNF to the next one, if they are located at different servers. This problem, often referred to as VNF placement, is typically formulated as a Generalized Assignment Problem (GAP) (Cohen et al. 2015), which minimizes the cost assignment of VNFs to VMs subject to capacity and KPIs constraints. Optimization is indeed an especially popular approach, aiming at minimizing load imbalance (Hirwe and Kataoka 2016), network utilization (Kuo et al. 2016), or energy consumption (Pham et al. 2017). Other works have built more complex cost functions, accounting for several network-related aspects (Mechtri et al. 2016; Gu et al. 2016) and/or energy consumption (Marotta and Kassler 2016; Khoury et al. 2016).

Optimization approaches usually result in a mixed-integer linear programming (MILP) formulation; since MILP problems are impractical to solve in real-world scenarios, the aforementioned works have envisioned finding near-optimal solutions through heuristic approaches. An alternative approach is represented by works tying VNF placement to more general problems, e.g. shortest path with limited resources (Martini et al. 2015), and set covering (Tomassilli et al. 2018). These works prove competitive ratio properties for the algorithms they propose, and such algorithms are also valid for problems other than VNF placement. The recent work (Bega et al. 2019) aims at simplifying the problem of VNF placement by foreseeing the evolution of the load to be served; specifically, the authors observe that the needed capacity is easier to foresee than the actual traffic demand, and employ deep neural networks to that end.

Cost minimization. To minimize deployment costs, instead, one may exploit the fact that several services may have one or more VNFs in common, i.e. services may be composed of a same VNF sub-graph, corresponding to a child service (Rost et al. 2017). This implies that several services, as well as several slices, may contain common sub-slices (5G PPP Architecture Working Group 2017; IETF 2017), and that such sub-slices can be shared instead of being replicated for each single service. For instance, the aforementioned safety services may share the same LTE eNB or 5G gNB when their service coverage overlap; similarly, they can share the same database where the vehicles' information, like position, speed, or heading, can be stored before data are processed (Rost et al. 2017). Importantly, VNF sharing is a multi-faced problem, which requires that the network operator not only identifies which VNFs (sub-slices) can be shared and among which services, but also how the amount of CPU and memory assigned to the VMs implementing them should be set so as to still fulfil the target KPIs of all involved services. VNF sharing is a new aspect that has been scarcely addressed so far. In 2017, the brief contribution in (Yi et al. 2017) proposed a scheduling scheme for VNF-based networks where the same VNF instance may be used for multiple services. In the same year, Soualah et al. (2017) mentioned VNF sharing as one of several viable strategies to improve the energy efficiency of networks. More recently, Malandrino et al. (2019) has identified VNF sharing as a distinct problem from VNF placement, arising within datacenters (points of presence, PoPs) and thus calling for different decisions on (i) whether a given VNF instance should be shared, (ii) how much capacity it shall be assigned, and (iii) the priority to give to different services sharing the same VNF instance.

In this article, we first summarize in the section titled "Related Issues" some important issues related to VNF placement and sharing. Then in the section titled "System Model and Decisions to Make" presents the 5G-PPP 5G reference architecture, highlighting the role of the orchestrator, as well as possible multi-access edge computing (MEC)-based architectures, where, due to resource scarcity, effective service deployment is utmost important. In the section titled "System Model and Decisions to Make" also introduces the main quantities that need to be taken into account, the decision to be made in terms of resource scaling and service/traffic priority setting, and the objective one seeks to optimize. In the section titled "The FlexShare Algorithm" describes the state-of-the-art solution strategy FlexShare (Malandrino et al. 2019), which, as mentioned, can make effective decisions in polynomial time, thus permitting a swift and efficient network and system management. In the section titled "Numerical Results" discusses some numerical results, obtained through FlexShare, considering real-world services and realistic network scenarios. Finally, we draw our conclusions and highlight possible directions for future work in the section titled "Conclusions".

Related Issues

It is worth mentioning that both VNF placement and VNF sharing are related to the network (NGMN Alliance 2016) concept and are actually part of a network slice creation process. Indeed, network (NGMN Alliance 2016) is a network paradigm whereby the same physical infrastructure – including networking and computing equipment – is concurrently used by multiple services, each of which is guaranteed isolation from the others. Earlier works like Zhang et al. (2017) and Rost et al. (2017) focus on architectural aspects, including the entities in charge of making decisions on infrastructure usage.

Specifically, Zhang et al. (2017) identifies mobility management as one of the main challenges arising in NGMN Alliance (2016)-based 5G networks, due to their multi-RAT (radio-access technology) nature. Indeed, 5G networks embed several different radio access technologies, from Wi-Fi to mmWave, with significantly different characteristics, e.g. availability and reliability. This poses two challenges when users move from a network to another: first, ensuring that the new network is consistent with the user's quality-of-service (QoS) needs; second, ensuring that the end-to-end, service performance experienced by the user is not jeopardized under the new network. The latter becomes especially complex when slices have not only data transfer capabilities but also processing ones, e.g. VNFs deployed on edge servers: the edge servers may be too far away from the new RAT the user is connected to, and a VNF migration may be triggered. The result is that handover procedures in 5G are much more complex than their counterparts in 4G/LTE, and require deeper coordination among the involved actors.

Focusing on the same multi-RAT scenario, Rost et al. (2017) focuses on scheduling issues arising from the integration of different network technologies. Among the challenges to tackle, the authors identify the need to decide at which level RATs shall be shared, e.g. whether MAC-level decisions for different RATs shall be made jointly by centralized entities or locally at the individual RATs. Furthermore, the authors raise the issue of network (NGMN Alliance 2016) request brokerage: since spectrum is a finite resource and overprovisioning is impossible, an admittance strategy for new slice requests must be defined. Upon denying a slice request, the authors envision that the network may reply with a counter-offer, suggesting to offer a different RAT that is more plentiful and offer comparable performance.

Other works, including Samdanis et al. (2017) and Vassilaras et al. (2017), tackle instead the problem of how those decisions should be made, including algorithmic and complexity challenges. Indeed, Samdanis et al. (2017) identifies software-defined networking (SDN) and NFV as two of the main enabling technologies network (NGMN Alliance 2016), which itself one of the main innovations of 5G networks. The authors also point out the potential challenges due to the increased complexity of the network control plane, and identify two main ways to tackle such a complexity: on the one hand, using network slice templates to reduce the number of decisions to make; on the other, crafting efficient and effective decision-making algorithms. Relatedly, Vassilaras et al. (2017) identifies two main decisions to be made concerning network (NGMN Alliance 2016), namely (i) which resources should be used to build the slice and (ii) how to connect them. The authors map both decisions to solving a virtual network embedding (VNE) problem, whereby the physical infrastructure at the operator's disposal is mapped to the virtualized slice to build. Such a problem is NP-hard, as proven via a reduction from the multi-way separator problem; indeed, even deciding the connection between resources is NP-hard, as proved via a reduction from the unsplittable multicommodity flow problem.

The process of creating, updating, and deleting network slices is known as orchestration. The work (Li et al. 2018), supported by the European project 5G-TRANSFORMER, identifies service orchestration as one of the main tasks 5G networks have to perform. Orchestration decisions need to account for a variety of different factors, including service requirements, the available infrastructure, and inter-operator agreements concerning resource sharing (multi-domain federation). To make all these decisions, the authors envision a three-layer architecture, whereby a vertical slicer (VS) translates business-related service goals into technical slice requirements, a service orchestrator (SO) selects the resources to use, possibly from different domains, and a mobile transport platform (MTP) manages the virtual and physical hardware. The entity in charge of most orchestration decisions is the SO, implements and extend the functionality standardized by ETSI in standard NFV-MAN 001 ("Management and Orchestration").

The authors of Santos et al. (2017) consider multi-domain federation scenarios, and address the unique security challenges therein. A first one concern information: on the one hand, operators need to exchange information in order to make effective orchestration decisions; on the other, they do not want to disclose critical information on the (mis)configuration of their own infrastructure, and the global resources at their disposal. A second one is represented by isolation: while all services should be transparent to each other, some may require stronger isolation, e.g. by avoiding to share VNF instances with other services. Such stronger isolation increases security – in the example, a malformed or malicious input at one service cannot compromise the other – but increase the resource consumption and therefore the cost.

As discussed earlier, services can be decomposed into virtual network functions (VNFs), which can then be placed at different locations throughout the infrastructure. In many relevant scenarios, services can be described as sequences or chains of VNFs that incoming flows have to traverse; in this case, the problem of placing VNFs across the available hosts takes the name of VNF chaining. Works talking VNF chaining usually aim at optimizing a cost function, subject to constraints concerning the available resources and the target delays (Pham et al. 2017). Liu et al. (2017) follows a similar approach but also accounts for the coexistence between already-deployed and newly requested services, which may share some of their VNFs; the resulting optimization problem is solved via a column generation-based approximation. Other works aim at bringing into the picture additional real-world limitations, including memory access issues in multi-core servers, e.g. Zheng et al. (2019). The authors formulate the problem as a non-linear, integer optimization problem; then, in view of its complexity, follow a heuristic approach whereby a performance drop index is defined, measured, and used to efficiently make near-optimal decisions.

More recent works have accounted for additional metrics beyond sheer performance, the most relevant of which is reliability. As an example, Zhang et al. (2019a,b) pursues the goal of resilience to server failures, and tackles the problems of (i) how many additional VNF instances shall be created, (ii) where they shall be placed, and (iii) how many resources they should be assigned to. The resulting problem is NP-hard and is solved through a greedy, iterative algorithm. Chemodanov et al. (2019) considers instead the physical location of servers (especially those at the network edge), and makes VNF placement decisions able to guarantee the geographical availability of services. Making such decisions requires to solve a multi-commodity-chain flow (MCCF) problem, which is tackled through a metapath composite variable approach, reaching near-optimal performance in the average case.

Several recent works tie network (NGMN Alliance 2016) and resource allocation with other problems, hitherto considered separate or orthogonal to it. As an example, Mandelli et al. (2019) MAC-level scheduling, with the objective of ensuring that network slices are able to deliver the data they process to the users needing them. In a similar spirit, Zhang et al. (2019a,b) jointly addresses the problems of choosing (i) the best location within the network to process the data at, and (ii) the best radio technology to deliver the results to the users, in order to reduce interference and increase the throughput.

Jošilo and Dán (2019) addresses a hybrid scenario, where individual devices can choose between performing their computation tasks themselves or leveraging the resources available at the network edge. The authors model the resulting interaction between devices and network operators a Stackelberg game, and propose several decentralized algorithms converging to the equilibrium. Liu and Han (2019) addresses cross-domain (also known as federation) scenarios, whereby different mobile operators decide to pool their resources in order to provide their services with a lower cost. In this context, the authors propose a distributed decision-making algorithm based (i) decomposing the original problem into subproblems via the alternating direction method of multipliers (ADMM) method, and (ii) solving the individual subproblems via learning-assisted optimization (LAO).

System Model and Decisions to Make

Several NFV-architecture for 5G networks have been proposed; examples include the one envisioned by the EU 5GPPP (and further interpreted by the EU 5G-PPP 5G-TRANSFORMER project), those proposed by the NGMN alliance and the IETF, and those considered in the works addressing VNF placement such as Cao et al. (2016), Cohen et al. (2015), Agarwal et al. (2018), Einziger et al. (2019). These architectures foresee that all decisions on VNF placement and resource allocation are made by a centralized entity, namely the Network Function Virtualization Orchestrator (NFVO) (see the ETSI Management and Orchestration (MANO) framework (ETSI 2014; IETF 2017). Thus, it is the NFVO that makes fine-grained decisions on the allocation and usage of individual hosts and links.

With reference to VNF placement in real-world implementations, it is important to underline that ETSI (2016) specifies four hierarchical levels, namely,

- individual host;
- zone, defined as a set of hosts with similar features;
- zone group, defined as a set of zones;
- point of presence (PoP), e.g. a datacenter.

Real-world 5G networks, also demonstrated through testbeds (Antevski et al. 2018; Sayadi et al. 2016; De la Oliva et al. 2018), consider that the NFVO makes VNF placement decisions at the PoP level, while different entities are in charge of placement and sharing decisions within each PoP (such entities are referred to using different names depending on the standardization body or association, e.g. ETF (IETF 2017), NGMN alliance (NGMN Alliance 2017), or 5G-PPP (5G PPP Architecture Working Group 2017).

In the following, we focus on the architecture foreseen by ETSI and reported in the 5GPPP document entitled "View on 5G Architecture" (2019). With reference to such proposal, the pictorial representation of the architecture in Figure 1 underlines the relationship of Software Defined Networking (SDN) and MANO controllers and their relation with the network, storage, and computing resources, abstracted through the Virtual Infrastructure Manager (VIM). The VIM summarizes the details of such deployed units as containers and VMs. Importantly, the VIM also abstracts the edge computing or multi-access edge computing (MEC) resources. The SDN controller's task is to correctly configure the network, while the MANO's task is to control storage and computing resources, both acting on behalf of the NFVO. Slice management can be part of the

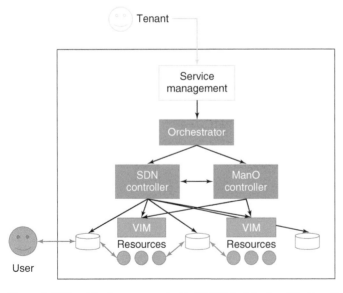

Figure 1 ETSI architecture as reported in 5G PPP Architecture Working Group (2017).

NFVO's tasks, or another entity can be included with this specific purpose. Finally, the service management block should interact with tenants, and the same or similar entities with vertical industries requiring a service to be deployed.

In this scenario, it is worth emphasizing how MEC resources can be managed and allocated. We again refer to the ETSI specifications (ETSI 2018). Possible scenarios all foresee one or more edge PoPs, each hosting a server and leveraging radio access functionalities. Such servers can be directly connected to points of access of the Radio Access Network (RAN), or to entities such as the Evolved Packet Core (EPC) in LTE, i.e. dealing with IP packets. Importantly, the ETSI NFV MANO considers applications at the mobile edge as regular VNFs. It follows that the MEC computing, storage, and network resources have to be orchestrated as well as the deployment of VNFs composing services that require to be in the MEC must be carefully instantiated so as to meet the service KPIs (e.g. ultra-low latency) and maximize the resource efficiency utilization. The latter is indeed particularly critical at the MEC, due to the limited available resources therein.

Turning our attention back to the problem of VNF placement and sharing, given the above architecture and upon receiving a new service request, the NFVO should make decisions about:

- whether any of the VNFs composing the newly requested service shall be provided through existing instances of such VNFs;
- if existing instances can be reused, how to set the priorities for the traffic flows belonging to the services sharing the same VNF instance;
- else, which VM to exploit to deploy the VNF instance;
- how to scale up or scale down the computing resources assigned to the VMs within the PoP.

Below, we will focus on a single PoP and consider that the VNFs composing a service should be instantiated or be co-located within the PoP. This implies that network latencies due to traffic flows transiting from one VNF to another can be neglected, thanks to the high-speed switching that is possible between VMs within the same datacenter (Xia et al. 2017). As a consequence, the only contribution that is relevant to the latency KPI is the data processing time within the VNFs composing the service.

Relevant Quantities for VNF Placement and Sharing

Without loss of generality, we consider that VNFs instances are deployed within VMs and that each VM can host exactly one VNF instance; also, VNF are assumed not to require isolation.

As considered in many recent works (Cohen et al. 2015; Agarwal et al. 2018; Bhamare et al. 2017), each VNF instance running within a VM is represented as an M/M/1 queue with FIFO queueing and preemption. For simplicity, we focus on adapting computing capabilities of VMs and neglect instead memory and storage. In order to reflect real-world conditions, we account for the fact that the computing resources assigned to a VNF instance can be varied (i.e. scaled up/down). For instance, a VNF requiring 1 computational unit and running on a VM with capability equal to 1 unit, it takes 1 time unit to process the traffic associated to the service that includes that VNF. Using instead that VM for a VNF requiring 2 computational units leads to a processing time of 2 time units. Importantly, the requirement values do not depend on the service that includes that VNF, but only on the VNF itself. Crucially, varying the amount of computing resources per VM influences the processing time at flows at the VNF hosted by that VM. As in real-world implementation, the amount of computational resources used by a VM cannot exceed a given maximum value.

As mentioned, services may be composed of a number of VNF; in the following we consider as target KPI the maximum average delay of a service, although the model and discussion can be extended to other KPIs as well.

Objective and Constraints

Deployment cost is one of the main concerns for the mobile network providers as well as for the vertical industries. Such a cost typically consists of two components: the cost for a VM instantiation, which is a fixed contribution, and a variable cost, which depends on the computing resources consumed by the VM hosting a given VNF instance (a proportional dependency is commonly assumed).

The main constraints instead to account for when making placement and scaling decisions are:

- at most one instance of at most one VNF can run at each VM;
- VMs cannot be scaled beyond their maximum capability;
- all traffic of all services must be served;
- per-service processing time targets must be honored.

The last constraint is especially important to honor and complex to formulate: indeed, the service time experienced by flows of a given service at a VNF depend upon (i) the computational capability available to the VNF; (ii) the arrival rate of flows (of any service) to the VNF; and (iii) the priority of the service, relative to other services sharing the same VNF.

Due to the many factors to account for, as well as the sheer number of available options, making optimal decisions about VM scaling and VNF placement is exceedingly complex, and impractical in most real-world scenarios. To this end, in the section titled "System Model and Decisions to Make" we present a simpler, efficient and effective solution strategy called FlexShare, able to perform near-optimal decisions in a short – namely, polynomial – time.

The FlexShare Algorithm

The problem formulated above is too complex to be directly solved with off-the-shelf solvers like CPLEX or Gurobi. Therefore, Malandrino et al. (2019) proposes an efficient and effective solution methodology named FlexShare, whose high-level approach is summarized in Figure 2. FlexShare considers service requests one by one, and previously made priority decisions – though not placement ones – can be adjusted as new services are deployed.

The first step of FlexShare is described in the section titled "Relevant Quantities for VNF Placement and Sharing", and deals with placement decisions, i.e. which VNF deploy at which VM. To this end, FlexShare builds a bipartite graph, whose nodes correspond to VNFs to deploy and VMs the latter can be deployed to. Edges of the

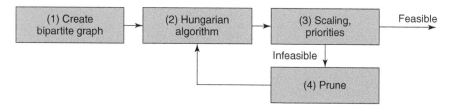

Figure 2 The FlexShare strategy. Step 1 builds a bipartite graph showing which VMs *could* run each VNF. Step 2 runs the Hungarian algorithm on such a graph to obtain the placement decisions. Step 3 solves a convex optimization problem to make scaling and priority assignment decisions. If such a problem is infeasible, the bipartite graph is *pruned* (step 4) and the procedure restarts from step 2.

bipartite graph connect (VNF, VM) pairs such that the newly-requested service can use the VNF instance deployed at the VM. The weights of edges correspond to the cost of providing the VNF with that VM, including the proportional component (which is influenced by the service traffic) and the fixed component (which is not incurred if the VM is already active, that is, if the VNF instance can be shared).

Given the bipartite graph, step 2 leverages the Hungarian algorithm, Kuhn (1955) to obtain the minimum-weight (hence, cheapest) assignment to VNFs to VM. Such an assignment is solely based on the weights on the bipartite graph and, critically, is not guaranteed to be feasible.

In step 3, the decisions made in step 2 are used as constraints of a convex (hence, simple to solve efficiently) optimization problem, detailed in the section titled "Priority, Scaling, and Pruning". The purpose of the problem is to (i) setting the priorities of each service within every VNF, (ii) set the capabilities of each VM, and, crucially, (iii) verify if the placement decisions made in step 2 are feasible. If the optimization succeeds, then FlexShare terminates successfully.

If step 3 fails, i.e. the problem solved therein is infeasible, then FlexShare moves to step 4 and tries to *prune* the bipartite graph. Indeed, an infeasible problem in step 3 can be due to too much sharing in step 2, i.e. too many VNF-to-VM edges in step 1. To correct this, FlexShare removes one of the edges from the bipartite graph, and restarts from step 2. This implies that placement decisions will foresee less VNF sharing and thus have a higher cost, but also a higher likelihood to result in a feasible problem in step 3.

Placement Decisions

On the left-hand side of the bipartite graph created in step 1 of FlexShare, we find the VNFs to place; on the right-hand one, the VMs they can be placed at. Edges are drawn between (VM, VNF) pairs such that the VNF can be provided by the VM. This happens in two cases:

- the VM is currently unused, therefore, a new instance of the VNF can be deployed therein;
- the VM already hosts an instance of the VNF, and such an instance can be shared between already-deployed services and the newly requested one.

The weight of the edge represents the cost of providing the VNF through a given VM; if the VM is currently inactive, the edge weight also includes the fixed activation cost.

Importantly, edges are not drawn if the VM cannot be scaled up to a capacity sufficient to serve the newly requested service while keeping stability. This, however, does not imply that service time requirements are met; indeed, such a condition is checked in step 3 as discussed in the section titled "Priority, Scaling, and Pruning."

Once the bipartite graph is ready, the Hungarian algorithm (Kuhn 1955) is employed to find a minimum-cost matching between VNFs and VMs. Specifically, the Hungarian algorithms selects a set of edges such that (i) each VNF is connected (hence, is deployed) in exactly one VM, and (ii) the total weight of the selected edges is as low as possible. Importantly, the Hungarian algorithm has polynomial (namely, cubic) complexity in the size of the graph. The edges selected by the Hungarian algorithm correspond to placement decisions, including:

- activation of currently inactive VMs, if need be;
- sharing of already-deployed VNF instances, if warranted.

These decisions are fed to the optimization problem in step 3, as described in the section titled "Priority, Scaling, and Pruning."

Priority, Scaling, and Pruning

Step 3 of FlexShare takes as an input the deployment decisions made by the Hungarian algorithm in step 2, and then solves a convex optimization problem where:

- the objective is to minimize the total cost;
- the constraints concern VM capability and per-service, end-to-end target delays;
- the decision variables are the capability to assign to each VM and the priorities to give to each service sharing every VM.

Importantly, all decision variables are real, hence, the problem can be solved in polynomial (cubic) time through commercial solvers (Boyd and Vandenberghe 2004); indeed, embedded convex optimization is routinely used in real-time applications. If the optimization succeeds, i.e. if the problem is feasible, then FlexShare terminates and the placement decisions made in step 2, along with the scaling and priority decisions made in step 3, can be applied.

If the optimization fails, then FlexShare proceeds to step 4, i.e. pruning the bipartite graph. The intuition behind the pruning procedure is that one cause for infeasibility is too aggressive sharing of VNF instances. To fix that, one edge of the bipartite graph is removed; to select such an edge, FlexShare resorts to the irreducible infeasible set (IIS) (Chinneck 2007). The IIS contains all constraints that, if removed, would render the problem feasible; intuitively, it provides an explanation as to why the optimization failed. Among constraints in the IIS, step 4 of FlexShare identifies the one that:

- concerns VM capability;
- involves a VNF used by the newly deployed service;
- involves the VM closest to its maximum capability,

The intuition behind the latter item is that VMs close to their maximum capability are more likely to introduce long delays, hence, lead to infeasible problem instances. By removing the corresponding edge from the bipartite graph, we ensure that such a placement decision is not made in subsequent iterations of FlexShare.

Note that it is possible to prove that the IIS contains at least one VM-capability constraint, hence, it is always possible to perform the procedure in step 4.

Reference Scenarios and Benchmark Strategies

We evaluate the performance of FlexShare using two reference scenarios, namely, a synthetic, small-scale scenario allowing us to perform a comparison against the optimum, and a large-scale scenario including real-world services.

Synthetic Scenario

In order to understand how FlexShare operates and to compare its performance with alternative approaches, we first leverage a simple, synthetic scenario. The scenario includes three services and five VNFs, with a many-to-many relationship among them. Services have different request rates and target delays, as reported in Table 1. The available infrastructure is composed of 10 VMs, with capability varying between 5 and 10 units; all VMs have an activation cost of 8 units and a proportional cost of 0.5 units.

Thanks to its small size, in the synthetic scenario it is possible to compare the performance of FlexShare against the optimum; specifically, optimal decisions are found through brute force.

Real-World Scenario

The FlexShare performance is also studied in a real-world, large-scale scenario, including five services belonging to the domains of smart city and smart factory. The VNFs composing each service, as well as the traffic each of them has to process, are based on Casetti et al. (2018) and Taleb et al. (2014, 2019), as reported in Table 2. The reference topology is the Luxembourg City center (Codeca et al. 2015).

Three services, namely, Intersection Collision Avoidance (ICA), vehicular see-through (CT), and entertainment (CDN) concern vehicles and their drivers/passengers. In ICA, the cooperative awareness messages (CAMs) broadcasted by vehicles are processed by a collision detector in order to check whether some vehicles are set on a collision course and, if so, alert them. All vehicles within 50 m of an intersection send a CAM message every 100 ms.

In the CT service, vehicles can display on their on-board screen a video feed coming from preceding vehicles (e.g. a truck blocking the view), so that their driver can be aware

Table 1 Services in the synthetic scenario.

Service	Arrival rate (flows/ms)	Max. delay (ms)
s_1	2	10
s_2	1.5	7.5
s_3	1	5

Table 2 Services and arrival rate in the realistic scenario.

VNF	Flow arrival rate
Intersection collision avoidance (ICA)	
eNB	117.69
EPC PGW	117.69
EPC SGW	117.69
EPC HSS	11.77
EPC MME	11.77
Car information management (CIM)	117.69
Collision detector	117.69
Car manufacturer database	117.69
Alarm generator	11.77
See through (CT)	
eNB	179.82
EPC PGW	179.82
EPC SGW	179.82
EPC HSS	17.98
EPC MME	17.98
Car information management (CIM)	179.82
CT server	179.82
CT database	17.98
Sensing (IoT)	
eNB	50
EPC PGW	50
EPC SGW	50
EPC HSS	5
EPC MME	5
IoT authentication	20
IoT application server	20
Smart factory (SF)	
eNB	50
EPC PGW	50
EPC SGW	50
EPC HSS	5
EPC MME	5
Robotics control	50
Video feed from robots	5

Table 2 (Continued)

VNF	Flow arrival rate
Entertainment (EN)	
eNB	179.82
EPC PGW	179.82
EPC SGW	179.82
EPC HSS	17.98
EPC MME	17.98
Video origin server	17.9
Video CDN	179.82

of impeding obstacles and/or hazards. Messages are sent every 200 ms by any vehicle within 100 m of an intersection.

The CDN service is used by 10% of all vehicles on the topology, randomly chosen with uniform probability; such vehicles consume a 25-fps video.

For the smart-city domain, an Internet-of-Things (IoT) service is considered, including a total of 200 sensors deployed throughout the topology; as per the 3GPP standard (Taleb et al. 2014), each sensor transmits 10 packets per second.

Finally, smart-factory applications are represented by a smart-robot service, controlling 50 robots in real time (which requires one packet per millisecond per robot); furthermore, 10% of all robots also need to transmit a 25-fps video feed.

As for the operator infrastructure, we assume it contains a total of 10 VMs, whose capability can be scaled up to 1000 units, and whose fixed and proportional cost are (respectively) 1000 units and 1 unit.

Benchmark Strategies

The performance evaluation includes several priority assignment strategies, with different levels of flexibility.

The lowest-flexibility option is represented by service-level priorities (**service** in plots), whereby priority levels are associated to whole services (the lower the delay target, the higher the priority); all requests of a given service have the same priority.

An intermediate option is represented by VNF-level priorities, where different services can have different priority levels at different VNFs (but all requests of the same service in the same VNF have the same priority). Priority levels can be decided through FlexShare (*VNF/FS* in plots) or through brute-force (*VNF/brute*).

Finally, at the highest level of flexibility, there are per-request priorities, assigned via FlexShare (*req./FS* in plots).

All solutions are implemented in Python, and a Xeon E5-2640 server with 16 GByte of RAM is used to run all tests.

Numerical Results

We begin from the synthetic scenario described in the section titled "Synthetic Scenario." Figure 3a shows the cost sustained by the MNO as the traffic changes; such costs become, consistently with our intuition, higher as the traffic grows. It is more interesting to remark that, for a given quantity of traffic, more flexibility always means smaller costs.

In Figure 3b, we turn our attention to sharing, and display the number of services using, on average, a given VNF instance. Again, more flexibility results in more sharing; intuitively, operators are able to more fully utilize their VNF instances, hence, need to deploy fewer of them.

Accordingly, Figure 3c shows that the used VM capability, as well as the maximum capability to which used VMs could be scaled to, decrease with more flexible priority assignments. This is consistent with Figure 3b: how higher flexibility is associated with more sharing, hence a better usage of the deployed VMs, hence the need to deploy fewer VMs.

Moving to the realistic scenario – where, due to its size, comparison with the optimum is impossible – it is possible to observe the same trends. Figure 4a highlights how more flexibility consistently results in smaller costs. Also notice how, when n becomes very high, the costs associated with all strategy overlap; this is because, in very high traffic conditions, no VNF can be shared for any priority assignment.

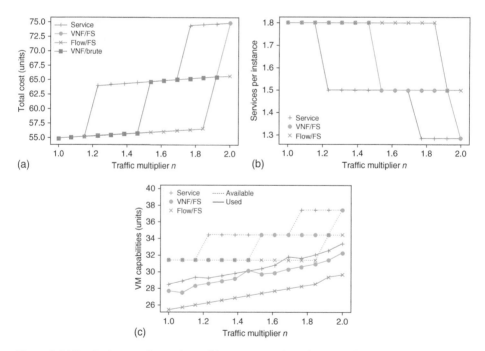

Figure 3 (a) Synthetic scenario: total cost; (b) average number of services sharing a VNF instance; (c) used and maximum VM capability. Per-VNF and per-flow priorities are assigned via FlexShare; per-service priorities are assigned by giving higher priorities to lower-delay services.

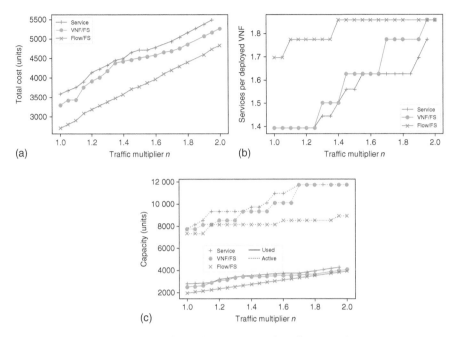

Figure 4 Realistic scenario: (a) total cost; (b) average number of services sharing a VNF instance; (c) used and maximum VM capability. Per-VNF and per-flow priorities are assigned via FlexShare; per-service priorities are assigned by giving higher priorities to lower-delay services.

Consistently with Figure 3b, Figure 4b confirms that more flexibility results in more sharing. Furthermore, the intermediate approach of per-VNF priorities tends to perform better than it does in Figure 3b, which suggests that such an approach can be a viable option in those cases where per-request priorities are too complex to implement.

Moving to VM capabilities, Figure 4c shows a very significant difference between used and potential VM capabilities; in other words, VMs are used *less* efficiently than in the synthetic scenario, for all priority assignments. This may seem surprising; however, recall that, as seen in Table 2, the real-world scenario has fewer common VNFs between services, hence, it presents fewer sharing opportunities in the first place.

We now ask a different question, namely, how efficient FlexShare is, that is, how long it takes to make its decisions. As summarized in Table 3, FlexShare takes at most a few minutes to run, even for the complex, real-world scenario summarized in Table 2. Consistently with our intuitive expectations, FlexShare takes longer to run in the realistic scenario; the main reason is that such a scenario has more alternatives to explore and compare.

It is also interesting to observe how, in general, more traffic tends to correspond to longer run times, but such a trend is by no means monotonic. The explanation for this effect lies in Figure 2. Indeed, the runtime of FlexShare is associated with the number of times the cycle in Figure 2 is repeated, and such a cycle is repeated every time a deployment is found to be infeasible. How often this happens does not depend upon the traffic *per se*, but rather upon how close VMs operate to their maximum capacity.

Table 3 Running time (in minutes) of our FlexShare implementation, in the synthetic and realistic scenarios.

Traffic multiplier	Synthetic scenario	Realistic scenario
1	4	6
1.2	5	7
1.4	6	6
1.6	5	10
1.8	7	9
2	7	12

Finally, it is important to stress that all runtimes can be substantially reduced if need be, by replacing the current Python implementation of FlexShare, which leverages the venerable but old optimization routines of the SciPy library, with more optimized code leveraging more modern, state-of-the-art solvers like CPLEX and Gurobi.

Conclusions

The problem of VNF placement and sharing in NFV-based networks is one of the main issues to overcome in 5G-and-beyond network systems. The support of the services target KPIs, along with the efficiency of resource utilization, are main concerns of both network providers and vertical industries wishing to make their services available to mobile users.

In this article, we first introduced the most widely accepted system architecture and highlighted the main challenges that it poses, along with some solutions that have been proposed to address them. Among the most relevant existing approaches, the FlexShare scheme promises to make swift decisions on resource allocation and sharing among different services, achieving a performance close to the optimum.

Such an approach, however, considered a single PoP, and neglected the network latency due to traffic transfer from one PoP to another. Interesting directions for future research thus include the definition of strategies for making the same VNF instances be reused by several services even when such instances are not deployed within the same datacenter. Furthermore, different target KPIs can be considered, beside the maximum average latency. To tackle this last point and introduce stricter delay guarantees, new modeling strategies are needed. Finally, the envisioned algorithmic solutions should be fully implemented in real-world networks to actually verify their ability to cope with practical issues and requirements.

Acknowledgment

This work was partially supported by the EU 5GROWTH project (Grant No. 856709).

References

5G PPP Architecture Working Group (2017). View on 5G Architecture.

Agarwal, S., Malandrino, F., Chiasserini, C.-F., and De, S. (2018). Joint VNF placement and CPU allocation in 5G. IEEE INFOCOM.

Antevski, K., Jorge, M.-P., Nuria, M. et al. (2018). Resource orchestration of 5G transport networks for vertical industries. IEEE PIMRC.

Bega, D., Gramaglia, M., Fiore, M. et al. (2019). DeepCog: cognitive network management in sliced 5G networks with deep learning. IEEE INFOCOM.

Bhamare, D., Samaka, M., Erbad, A. et al. (2017). Optimal virtual network function placement in multi-cloud service function chaining architecture. *Computer Communications* 102: 1–16.

Boyd, S. and Vandenberghe, L. (2004). *Convex Optimization*. Cambridge University Press.

Cao, J., Zhang, Y., An, W. et al. (2016). VNF placement in hybrid NFV environment: modeling and genetic algorithms. IEEE ICPADS.

Casetti, C., Chiasserini, C.F., Molner, N. et al. (2018). Arbitration among vertical services. IEEE PIMRC.

Chemodanov, D., Calyam, P., and Esposito, F. (2019). A near optimal reliable composition approach for geo-distributed latency-sensitive service chains. IEEE INFOCOM.

Chinneck, J.W. (2007). *Feasibility and Infeasibility in Optimization: Algorithms and Computational Methods*. Springer.

Codeca, L., Frank, R., and Engel, T. (2015). Luxembourg SUMO Traffic (LuST) Scenario: 24 hours of mobility for vehicular networking research. IEEE VNC.

Cohen, R., Lewin-Eytan, L., Naor, J.S., and Raz, D. (2015). Near optimal placement of virtual network functions. IEEE INFOCOM.

De la Oliva, A., Li, X., Costa-Perez, X. et al. (2018). 5G-transformer: slicing and orchestrating transport networks for industry verticals. *IEEE Communications Magazine* 56 (8): 78–84.

ETSI (2014). Network functions virtualisation (NFV); management and orchestration.

ETSI (2016). Network functions virtualisation (NFV); management and orchestration; Or-VNFM reference point – interface and information model specification.

ETSI (2018). MEC deployments in 4G and evolution towards 5G.

Einziger, G., Goldstein, M., and Sa'ar, Y. (2019). Faster placement of virtual machines through adaptive caching. IEEE INFOCOM.

Gu, L., Tao, S., Zeng, D., and Jin, H. (2016). Communication cost efficient virtualized network function placement for big data processing. IEEE INFOCOM Workshops.

Hirwe, A., and Kataoka, K. (2016). LightChain: a lightweight optimization of VNF placement for service chaining in NFV. IEEE NetSoft.

IETF (2017). Network slicing management and orchestration.

Jošilo, S., and Dán, G. (2019). Wireless and computing resource allocation for selfish computation offloading in edge computing. IEEE INFOCOM.

Khoury, N.E., Ayoubi, S., and Assi, C. (2016). Energy-aware placement and scheduling of network traffic flows with deadlines on virtual network functions. IEEE Cloudnet.

Kuhn, H.W. (1955). The Hungarian method for the assignment problem. Wiley Naval Research Logistics.

Kuo, T.W., Liou, B.H., Lin, K.C.J., and Tsai, M.J. (2016). Deploying chains of virtual network functions: on the relation between link and server usage. IEEE INFOCOM.

Li, X., Mangues-Bafalluy, J., Pascual, I. et al. (2018). Service orchestration and federation for verticals. IEEE WCNC Workshops.

Liu, Q., and Han, T. (2019). DIRECT: distributed cross-domain resource orchestration in cellular edge computing. ACM Mobihoc.

Liu, J., Lu, W., Zhou, F. et al. (2017). On dynamic service function chain deployment and readjustment. *IEEE Transactions on Network and Service Management* 14 (3): 543–553.

Malandrino, F., Chiasserini, C.F., Einziger, G., and Scalosub, G. (2019). Reducing service deployment cost through VNF sharing. *IEEE/ACM Transactions on Networking* 27 (6): 2363–2376.

Mandelli, S., Andrews, M., Borst, S., and Klein, S. (2019). Satisfying network slicing constraints via 5G mac scheduling. IEEE INFOCOM.

Marotta, A., and Kassler, A. (2016). A power efficient and robust virtual network functions placement problem. IEEE ITC.

Martini, B., Paganelli, F., Cappanera, P. et al. (2015). Latency-aware composition of virtual functions in 5G. IEEE NetSoft.

Mechtri, M., Ghribi, C., and Zeghlache, D. (2016). A scalable algorithm for the placement of service function chains. *IEEE Transactions on Network and Service Management* 13 (3): 533–546.

Malinverno, M., Avino, G., Casetti, C. et al. (2019). MEC-based collision avoidance for vehicles and vulnerable users. *IEEE Vehicular Technology Magazine*, in press.

NGMN Alliance (2016). Description of network slicing concept.

NGMN Alliance (2017). 5G network and service management including orchestration.

Pham, C., Tran, N.H., Ren, S. et al. (2017). Traffic-aware and energy-efficient VNF placement for service chaining: joint sampling and matching approach. *IEEE Transactions on Services Computing* 13 (1): 172–185.

Rost, P., Mannweiler, C., Michalopoulos, D.S. et al. (2017). Network slicing to enable scalability and flexibility in 5G mobile networks. *IEEE Communications Magazine* 55 (5): 72–79.

Samdanis, K., Wright, S., Banchs, A. et al. (2017). 5G network slicing – part 2: algorithms and practice. *IEEE Communications Magazine* 55 (8): 110–111.

Santos, M.A.S., Ranjbar, A., Biczók, G. et al. (2017). Security requirements for multi-operator virtualized network and service orchestration for 5G. In: *Guide to Security in SDN and NFV*. Springer.

Sayadi, B., Gramaglia, M., Friderikos, V. et al. (2016). SDN for 5G mobile networks: NORMA perspective. Springer Crowncom.

Soualah, O., Mechtri, M., Ghribi, C., and Zeghlache, D. (2017). Energy efficient algorithm for VNF placement and chaining. IEEE/ACM CCGRID.

Taleb, T., Ksentini, A., and Kobbane, A. (2014). Lightweight mobile core networks for machine type communications. *IEEE Access* 2, 1128–1137.

Taleb, T., Afolabi, I., and Bagaa, M. (2019). Orchestrating 5G network slices to support industrial internet and to shape next-generation smart factories. *IEEE Network* 33 (4): 146–154.

Tomassilli, A., Giroire, F., Huin, N., and Pérennes, S. (2018). Provably efficient algorithms for placement of service function chains with ordering constraints. IEEE INFOCOM.

Vassilaras, S., Gkatzikis, L., Liakopoulos, N. et al. (2017). The algorithmic aspects of network slicing. *IEEE Communications Magazine* 55 (8): 112–119.

Xia, W., Zhao, P., Wen, Y., and Xie, H. (2017). A survey on data center networking (DCN): infrastructure and operations. *IEEE Communications Surveys and Tutorials* 19 (1): 640–656.

Yi, B., Wang, X., and Huang, M. (2017). A generalized VNF sharing approach for service scheduling. *IEEE Communications Letters* 22 (1): 73–76.

Zhang, H., Liu, N., Chu, X. et al. (2017). Network slicing based 5G and future mobile networks: mobility, resource management, and challenges. *IEEE Communications Magazine* 55 (8): 138–145.

Zhang, J., Wang, Z., Peng, C. et al. (2019a). RABA: resource-aware backup allocation for a chain of virtual network functions. IEEE INFOCOM.

Zhang, Q., Liu, F., and Zeng, C. (2019b). Adaptive interference-aware VNF placement for service-customized 5G network slices. IEEE INFOCOM.

Zheng, Z., Bi, J., Yu, H. et al. (2019). Octans: optimal placement of service function chains in many-core systems. IEEE INFOCOM.

Further Reading

Nguyen, Q.-H., Morold, M., David, K., and Dressler, F. (2019). Adaptive safety context information for vulnerable road users with MEC support. IEEE/IFIP WONS.

11

Security Monitoring and Management in 5G

Edgardo Montes de Oca

Montimage, Paris, France

Introduction

Monitoring is an essential functionality that is required to provide the awareness and automation necessary to manage both performance and security of 5G (and beyond) networks. 5G has introduced new concepts that both facilitate the monitoring and security functions but have also introduced new vulnerabilities and challenges that need to be addressed (Liyanage et al. 2018). The concepts introduced include: network programmability, e.g. software-defined networks (SDN); virtualization, e.g. network function virtualization (NFV), network slicing, cloud computing, mobile edge computing (MEC); massive distribution, e.g. Massive Internet of Things (mIoT), massive machine-type communications (mMTC), vehicle-to-everything communications (V2X), fog computing; and more intelligence, e.g. cognitive and intent-based networks, artificial intelligence (AI), machine learning (ML). Through these, 5G enables agile and dynamic creation, reallocation and suppression of processes and services in response to changing customer demands and information flows, increased interactions between machines, between humans and machines and between humans through new communication modes (e.g. gestures, facial expressions, sentiments, sound, *haptics*).

The new applications that are being deployed have very different performance and security requirements. For instance, the following application types have different strong requirements:

- mMTC: high availability, high energy efficiency, and high device density
- Enhanced mobile broadband (eMBB) services: improved mobility and high availability;
- Virtual reality: low latency and high throughput;
- Dense urban information society: high user and device density;
- Connected vehicles: low latency, high reliability, high availability, and improved mobility;
- Industry 4.0: in many cases, ultra reliable low latency communications (URLLC).

5G is the first mobile architecture designed to support multiple, specific use cases, each with their own unique cybersecurity requirements. This introduces new requirements on security monitoring and management. Furthermore, the omnipresence of

The Wiley 5G REF: Security. Edited by Rahim Tafazolli, Chin-Liang Wang, Periklis Chatzimisios and Madhusanka Liyanage.
© 2021 John Wiley & Sons Ltd. Published 2021 by John Wiley & Sons Ltd.

these new applications in all layers of society and infrastructure makes them extremely critical with respect to privacy, security, and safety. Unfortunately, the use of monolithic legacy solutions, e.g. firewalls, intrusion detection and prevention systems (IDPS), security information event management systems (SIEMs), to manage the networks and their security are no longer valid and have to be re-thought and re-engineered. These functions now have to adapt to these new complex and critical environments.

However, there is no such thing as perfect security or safety so security will always depend on an optimal balance between resource costs and the effectiveness of the protection, i.e. obtaining efficient security mechanisms. In many cases it could also mean incrementing attack tolerance or robustness of the network and application services so that at least some essential level of service is maintained at all times.

5G and beyond mobile networks are aiming at converting the networks into a set of energy-efficient distributed programmable computers. This enables the full automation of network and service management and operation, but introduces the need to protect the network and system against potential cybersecurity risks inherent in the Internet and the software. Full automation is needed to manage complex and massive communications and devices but also introduces the ability to replicate a small isolated error or attack that can put the entire critical system or system of systems in danger.

New vulnerabilities and attack vectors that are introduced by network *softwarization* include the following:

- 5G networks are now more easily vulnerable to attacks on both the control and data planes. Monitoring has to be performed at both of these levels and the collected information needs to be correlated.
- The hijack of Internet of Things (IoT) devices (facilitated by lack of standards and best practices) or the compromise of services can serve to carry out distributed denial-of-service (DDoS) attacks. Monitoring needs to be able to detect abnormal activity using behavior analysis and ML techniques.
- Cloud and edge computing, and the convergence of mobile and traditional IT networks create new attack vectors. The monitoring framework needs to deal with highly distributed and dynamic systems, as well as legacy networks.
- Certain centralized network elements, e.g. controllers and orchestrators, become single points of failures. Different 5G entities and segments, such as user equipment (UE), the radio access network (RAN), the core network, virtualized network functions (VNFs), and operator or third-party-hosted applications and services, could be targets for attackers. The open, flexible, and programmable nature of SDN and NFV architectures introduces new threat vectors. Monitoring agents (MA) need to be deployed and collocated with critical elements and analyze not only network traffic but also the system and application traces to be able to detect any anomaly. The Microsoft STRIDE model divides threats into six categories, namely spoofing, tampering, repudiation, information disclosure, denial of service (DoS), and privilege escalation (Open Networking Foundation 2016). In the other side, European Telecommunications Standards Institute (ETSI) has identified the threat surface of NFV as the union of generic virtualization threats (e.g. memory leakage, interrupt isolation), generic networking threats (e.g. flooding attacks, routing security), and the threats due to combining virtualization technology with networking (ETSI GS NFV-SEC 001 2014). All software components are vulnerable to software

vulnerabilities including both implementation and design flaws. The authors of (Ahmad et al. 2018) provide a list of many of the different vulnerabilities.

On the other hand, the flexibility introduced by 5G facilitates the introduction of new security-by-design and security functions. For instance:

- Network segmentation (network slices) is a common, proven way to mitigate security risks. Each network slice customer can obtain customized services for fault and performance management, security monitoring, service life-cycle management and network optimization.
- New highly distributed and configurable monitoring and management techniques bring the awareness, visibility and control necessary to help protect the networks and applications.
- Security controllers and orchestrators analyze and determine, based on policy, the mitigation and controls to be applied. Automation, orchestration, and NFV can be made to work with security functions to prevent and contain both known and new attacks.
- Secure path computation is used so that the VNFs can exchange sensitive data, which must be performed according to some predefined security constraints, e.g. defined by security policies or security service level agreements (SSLAs). These security constraints may have to be enforced in the virtual network as well as in the underlying physical network. Such path is called a "trusted path." Monitoring of the sensitive traffic along this path is carried out to verify the planned trusted path is followed as expected.

A concrete attack example and how monitoring can address it (5G Americas Whitepaper 2018):

- First, an attacker finds a zero-day vulnerability and exploits it to create a botnet by infecting a very high number of IoT devices with a malware that allows remotely rebooting the device. Currently, this is rather easy since IoT devices have default passwords that in general are not modified by the users, as shown by the Mirai botnet (Antonakakis et al. 2017).
- Next, the attacker instructs the malware to reboot all the devices of a targeted 5G coverage area at the same time. These cause excessive *attach requests*, creating a signaling storm that overloads the 5G RAN resources making the RAN unavailable.
- Several solutions have been proposed, for instance in Antonakakis et al. (2017), Kumar and Lim (2019), Meidan et al. (2018). Most require detecting malicious scanning performed to prepare the attack and detecting malicious behavior, i.e. the signaling storm. These detection mechanisms can be carried out at different levels: devices, gNodeB, depending on the possibilities: in the devices themselves, which would be added to the gNodeB's central unit–control plane (CU-CP), and access and mobility management function/session management function (AMF/SMF) component functions (5G Americas Whitepaper 2018).

To resume, the main security enhancements in 5G as defined by 3GPP include the following:

- Network slicing enabling isolation, customization of security functions, and preventing propagation of attacks from one slice to another.

- Control plane – Data plane separation enabling flexible and dynamic placement (or reconfiguration) of security functions where and when they are needed. NFV orchestration includes deploying, instantiating, and managing a composition of VNFs that form a service function chain (SFC) and, thus, can be used to connect to different security functions.
- State-of-the-art encryption and integrity protection are adopted at all levels including the data and control planes.
- Unified authentication framework and identity management provide seamless mobility across different access technologies and tenants.
- User privacy protection provides protection of the different identifiers (e.g. subscription permanent identifier (SUPI), International Mobile Subscriber Identity (IMSI), and International Mobile Equipment Identity (IMEI)).

5G security platform relies on key emerging trends and technologies. All depend on monitoring techniques to provide the awareness and analytics required for efficient security management. These include:

- Architecture for supporting zero-touch end-to-end smart network and service security management. The architecture introduces flexibility of *softwarization* technologies (e.g. SDN/NFV) and *cognitivity* (i.e. AI/ML techniques).
- Software-defined security (SDS) orchestration and management that enforces and controls security policies based on real-time analysis of network and application events. In this way security can adapt more easily to dynamic changes in topology, services, security requirements and threat landscape.
- AI-driven security, including moving target defense (MTD) mechanisms and cyber threat intelligence (CTI), to empower smart security management with proactive defensive strategies. AI/ML is used to improve the prediction and detection accuracy, but the distribution of the computation also is required for the monitoring and security analysis using, for instance, microservices or advanced programming techniques (programming protocol-independent packet processors (P4), field-programmable gate array (FPGA), data plane development kit (DPDK), etc.).
- Mechanisms to foster trustworthiness in multi-tenant/multi-domain environments based on real-time monitoring and analysis, trusted execution environments (TEEs), digital rights management (DRM), SSLA, AI-powered validation, and distributed ledger techniques.
- Mechanisms to enforce liability of involved parties when security breaches occur or systems fail, including smart contracts, VNF certification, trust level agreements (TLA), AI-based liability, and root cause analysis (RCA) techniques.

Network Security Monitoring

Network monitoring is a fundamental enabler for both managing the performance and the security of the networks. It is necessary for understanding the state of the network, detecting malfunctions and security breaches, and triggering remediations either manually by the human operators or automatically for self-organizing networks (SON). In the past, security has depended on firewalls or IDPS placed at the perimeter of the networks where all incoming traffic is considered un-trusted and all inside traffic is considered trusted. This model is no longer valid in highly dynamic, multi-domain, multi-tenant, multi-provider, virtualized, and programmable environments introduced by SDN/NFV, and integrating massive number of devices. Attackers can leverage the openness, flexibility, and vulnerabilities introduced by open application programming interface (APIs), software, and the Internet to circumvent any firewalls at the perimeter or perform insider attacks. 5G considers issues such as unauthorized use and access, slice isolation, traffic hijacking, Security, Trust and service level agreement (SLA) compliance, forensics, etc.

Monitoring and analysis based on existing techniques such as deep packet inspection, complex event processing, SIEM, and intrusion detection and prevention are still essential but have to be adapted to these new environments. The monitoring function also needs to consider critical network elements and applications, behavior analysis, and business activity monitoring, and detect different attacks and evasion techniques even when the traffic is encrypted, all of this without contravening privacy and general data protection regulation (GDPR) issues. Real-time packet and flow analysis, behavior analysis, and ML techniques are used to detect DDoS attacks, zero-day attacks, ransomware, etc. In many cases, for instance IoT networks, it is not possible to log, query, or analyze all exchanges; so, to detect compromises one needs to analyze network behavior and identify changes that indicate security breaches.

To deal with the highly distributed and dynamic SDN/NFV architecture, network- or application-level probes can be dynamically and automatically generated and placed where needed. This security monitoring framework is composed of MA and a centralized monitoring operator. For instance, Figure 1 represents this framework mapped to

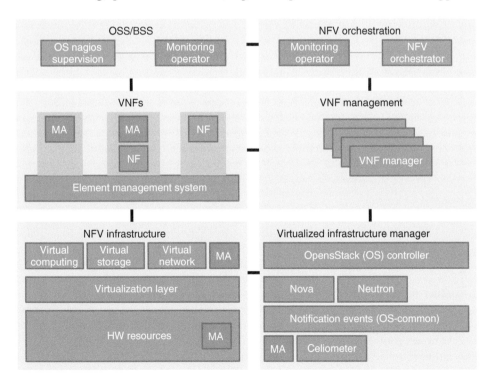

Figure 1 Deployment of a security monitoring framework in an NFV-based OpenStack platform.

the Open Stack cloud computing platform (https://www.openstack.org/). The MA capture the metadata from the traffic, system, and applications, and do a local analysis. They can be deployed as standalone VNF, collocated with other VNFs but also analyze events occurring within the NFV infrastructure, HW (hardware), and VIM (virtualization infrastructure manager) levels. The monitoring operator provides a global view and correlates heterogeneous information from different agents; visualizes statistics, historical views, and trends in network traffic; presents and manages alarms and notifications; provides details on each server or network element in the network; represents the results of the analysis of the behavior or usage of each network link; interacts with other services to trigger prevention and mitigation actions; and, deploys, configures, updates, maintains, and controls the agents.

Solutions exist or are being investigated that deal with orchestrating the monitoring function but also interacting with the orchestrators for reacting or countering attacks. Examples of finished or ongoing projects are:

1) SIGMONA project (SIGMONA 2016): proposed SDM solutions (Liyanage et al. 2017), as for instance a monitoring framework based on MMT agents (highly configurable monitoring probes that can be deployed as VNF or collocated with another VNF) was used to analyze mirrored traffic in OpenStack via Neutron API to analyze the traffic going through the ports that are mirrored by the tap as a service and accessed via its API (Liyanage et al. 2018). This framework was also used in the case of named data networks, where the content is accessed by users using names instead of IP addresses, to show how DDoS-type attacks (i.e. content poisoning and interest flooding attacks) can be detected (Nguyen et al. 2018a, 2019). It is also being exploited to monitor and secure 4G/5G networks created on the fly using an eNodeB based on a software-defined radio device, the OpenAirInterface (https://www.openairinterface.org/), and a commercial Evolved Packet Core (https://www.montimage.com/products/EPC_in_a_box.html).

2) CHARISMA (Parker et al. 2016) project proposes a real-time, automated Security Management Framework for 5G telecommunications networking, by implementing a continuous and closed loop real-time environment inspection regime, based on analytics, policy-based decisions, and actuation/enforcement via cloud and SDN orchestration procedures. The monitoring framework used (Angelopoulos et al. 2017) is based on an open-source solution called Prometheus (Prometheus 2012) that uses a time-series database LevelDB and a query language PromQL. It focuses on performance analysis more than security.

3) COGNET (http://www.cognet.5g-ppp.eu/) project applies ML techniques, using a double closed-loop data streaming architecture, for threat detection, attack analysis, and security incident mitigation and response. It focuses on TCP SYN flooding, DNS DoS, and ICMP flooding attacks. The following testbeds have been provided in the context of this project:
   a) Distributed security enablement testbed uses the SFlow tool (https://sflow.org/) that monitors the network using packet sampling and is embedded within the switches/routers.
   b) Honeynet testbed is proposed for generating realistic network traffic containing brute-force and web application attacks. The monitoring tool used is based on ELK (Elasticsearch, Logstash, and Kibana, https://www.elastic.co/products/kibana).

c) NFV security anomaly detection testbed is proposed as proof of concept to show how dynamically distributed security mechanisms can be embedded directly into the NFVI. A time series–based monitoring system is used to obtain number of flows and data per flow.

d) Dense urban area testbed is used to detect performance degradation. It uses a monitoring server Zabbix (https://www.zabbix.com) to obtain the following system parameters: incoming/outgoing traffic, CPU, memory, and disk usage metrics.

4) SELFNET (Jiang et al. 2017b) project provides a new definition of perimeter security in virtualized 5G infrastructures, implications in workloads associated with multi-tenancy infrastructures, and how this perimeter security in virtual infrastructures can be protected against cyberattacks by providing mechanisms to allow the inclusion of security control points along the 5G architecture. These control points allow the deployment of security monitoring components and the deployment of security enforcement components in key architectural locations of the 5G infrastructure. It proposes a proof-of-concept testbed (Strufe et al. 2018; Jiang et al. 2017) that shows self-protection mechanisms against DDoS attacks. ML algorithms explored in SELFNET include decision tree, support vector machine, and nearest neighbor.

5) 5G-ENSURE (ENSURE 5G 2017) project proposed a 5G trust model and risk analysis that allows modeling the system, highlighting potential risks, and demonstrating the effect of adding controls or changing the design. Security monitoring provides the evidence needed to obtain trust, but also needs to be trusted with respect to the extent of coverage and the precision of the detections and diagnosis.

6) SLICENET (Wang et al. 2018) aims to achieve an end-to-end cognitive network slicing management and orchestration framework in multi-domain 5G networks. The proposed SLICENET's framework architecture envisages a "slice security manager" component to enable security of slice instances, considering end-to-end encryption, management of virtual security functions (VSFs) (e.g. DPI (deep packet inspection), IDS (intrusion detection system)), and AI-driven analytics.

7) 5Genesis (Koumaras et al. 2018) proposes an ML-enabled security framework to reinforce the security of a 5G experimentation platform. The cybersecurity framework provides the capability of performing near-real-time traffic analytics to detect and classify security incidents, leveraging on Big Data and ML technologies (e.g. Apache Spot platform (Apache Spot Platform 2016)).

8) SENDATE (http://www.sendate.eu/) proposes solutions to several key challenges related to 5G security. These include:

a) SSLA verification and vertification: enabler to formalize the commitment of each infrastructure or slice provider to fulfill the main static security requirements of a tenant;

b) Identity and access management as a service: enabler to allow slice and service providers to support mutual authentication;

c) Software protection as a service: enabler to achieve trusted software execution; and more specific to security monitoring;

d) Distributed monitoring framework of Figure 1;

e) SSLA monitoring: enabler to implement a complete SLA process model by allowing capturing the raw information needed from different sources (e.g. network service traces, network elements, and network traffic), translating it into events, correlating this information, and generating notifications or alerts that will help identifying violations and diagnosing the problems detected.

Software-Defined Security

Software-defined security (SDS or SDSec) is a security model in which the information and communication security is controlled and managed by software. The network security hardware devices, such as firewalling, intrusion detection, access control, and network slicing are abstracted and controlled through a software layer. SDS exploits SDN to enhance network security and makes the network control plane programmable through protocols such as OpenFlow (Open Networking Foundation 2015).

SDS also enables SECaaS (Security as a Service) delivery models that provide cost-effective and agile security enhancements in a fully virtualized infrastructure. In Khettab et al. (2018) the authors propose an architecture that takes advantage of SDN/NFV capabilities to empower SECaaS in inter-domain environments. The proposed architecture focuses on applying security in 5G slices by enabling predictive automatic scaling of VSFs from predefined policies and performance metrics. The authors of (Blanc et al. 2018) have designed a security architecture that supports the management of network slices with built-in security features using the SDS and SECaaS mechanisms. The architecture allows the application of tenant security in a multi-tenant infrastructure with dynamic placement and chaining of network security functions. In addition, network programmability allows end-to-end orchestration of the network and its resources, namely microservice chaining, VNFs, and so on, in accordance with the security policies defined for the protection of the architecture and its tenants.

AI also plays an important role by enabling self-managing of the security functions, to achieve better robustness and lower operational costs. Recent academic research (Maimo et al. 2018; Ali-Tolppa et al. 2018) and standardization initiatives (Experiential Networked Intelligence (ENI) 2018) have been working on the development of AI-driven SDS solutions for improving the security management of the prevention, detection, and mitigation capabilities.

Software entities involved in the security management operations (SDN controller, NFV Orchestrator, VNFs, microservices, etc.) have become critical elements that are subject to software vulnerabilities. These elements can be compromised causing performance degradation or even network outage (ETSI GS NFV-SEC 001 2014; ETSI GS NFV-SEC 003 2016; Collberg and Thomborson 2002). Compromised virtualized components could provide incorrect monitoring data that can mislead the security analysis and reaction functions. Outsourcing of SECaaS to external cloud providers introduces new risks, such as introspection, that allows monitoring of the virtual machines by a hypervisor or virtual machine monitor. To address these issues several techniques are proposed: hardware security module (HSM), trusted platform module (TPM), and virtual trusted platform module (vTPM) to provide trusted protection for VNFs (ETSI GS NFV-SEC 003 2016); software obfuscation (Collberg and Thomborson 2002) to complicate the analysis and tampering of code; tamper proofing (Wurster et al. 2005) to cause altered software to fail; fully homomorphic encryption (Gentry 2009) for assuring data integrity by performing computations directly on encrypted data; and TEEs (Lefebvre et al. 2018) to assure integrity of code execution (e.g. Intel's Software Guard Extensions (SGX) and Advanced Micro Devices (AMD's) secure encrypted virtualization (SEV) that are compared in Mofrad et al. (2018)).

Network Slicing

Network slicing can be defined as a collection of functions and resources upon a common infrastructure, specialized to offer a dedicated logical network (i.e. network as a service) that offers network services adapted to the requirements coming from specific application domains and vertical industries. A slice can cover multiple technical domains that include: terminals, access/transport/core networks, and datacenters hosting applications.

The introduction of network slices in 5G enables new envisioned use cases or verticals by providing improved management, resource usage, and isolation of concerns. But it also introduces more complex configurations and management, including the need to guarantee SLAs and SSLAs, and automating the decision-making process. These challenges are difficult to solve since they involve a great number of variables, stakeholders, and use cases. AI/ML techniques are used to help solve the security issues in network slicing by providing the ability to learn from the environment, plan response actions, and perform the proper configuration to solve the issues that may arise (Li et al. 2017). The authors of (Suárez et al. 2018) provide a thorough study of the different architectural approaches to slicing and identify how slice challenges and threats can be addressed by different AI techniques. For each challenge they suggest a possible approach, as for instance: to control network slice behavior, the CPU and bandwidth usage trends can be analyzed to determine the normal behavior model of a given network slice, e.g. an IoT slice would show lower CPU usage and short bursts of traffic, and if this behavior changes an alert could be issued and trigger RCA or remediation actions.

Cognitive Network Security Management

Security management specifically aims at protecting the network data and performance through precise detection of intrusions, privacy breaches, and denial of services, but also managing the prevention, mitigation, and reaction to these breaches. For achieving effective and efficient security in a highly dynamic network environment it is essential to introduce autonomous security functions. As previously presented, SDN/NFV and AI/ML techniques are the enablers that make improving the security management functions possible and support the advent of what is called cognitive network management to obtain self-aware, self-configuring, self-optimization, self-healing, and self-protecting systems.

Not much work has been published on cognitive network management specifically addressing security. The 5G Whitepaper (Cognitive Network Management for 5G 2016) by the 5GPPP Working Group on Network Management and quality of service (QoS) identifies some of the high-level requirements and dangers. The authors of (Ayoubi et al. 2018) extend IBM's MAPE (Monitor, Analyze, Plan and Execute) control loop consisting of four functions monitor, analyze, plan, and execute. While MAPE applies ML only to the analysis part, the authors propose C-MAPE that applies it to all the functions. In this way, for instance, the monitor function is able to determine the what, when, and where to monitor, and achieve better efficiency. A cognitive security manager is presented as a use case of C-MAPE. It relies on the resource orchestrator (e.g. OpenStack) and the SDN controller (e.g. OpenDaylight). The resource orchestrator administers the physical and virtual resources, while the SDN controller facilitates automated and flexible configuration of the network resources. C-monitor recuperates the needed packet and flow statistics and metadata. The C-analysis relies on an outlier detection model using an ML algorithm (e.g. k-Means, k-NN) for anomaly inference. The C-plan function employs reinforcement learning to select the optimal change plan based on the criticality of the anomaly. The C-execute function directs the resource orchestrator and SDN controller to perform further investigations or remediate any faults.

It must be noted that AI/ML models can be biased and result in erroneous prediction and decision making, potentially causing performance degradation and financial losses (Barreno et al. 2006; Boutaba et al. 2018). Furthermore, the establishment of trust in AI/ML and SDN/NFV-driven systems does not provide complete prevention of breaches and failures due to unforeseen vulnerabilities and attacks (e.g. zero-day and advanced persistent threats). Consequently, the question of liability and responsibility for failures needs to be addressed. This need is further enhanced by concerns related to the liability in AI/ML and the obligation to fulfill regulations (AI HLG 2017; Expert Group on liability and new technologies 2018; European Commission 2018).

Optimization Techniques

Programmability and microservices have shown their ability to improve the performance and security of network functions compared to their monolithic equivalent. The end-to-end global orchestration of network microservices can be used to optimize security features to reduce their impact on network performance. The introduction of a comprehensive end-to-end monitoring and orchestration framework providing a multi-level and multi-technology abstract view for the optimization of network services finely cut into microservices is essential for adaptations and optimizations of the security functions required by 5G networks based on highly virtualized and dynamic environments.

P4, a programming language for packet forwarding planes, extends the concept of network programmability (da Costa Cordeiro et al. 2017). Standard traffic engineering functions, such as security analysis, routing/switching, filtering, field translation, flow classification, etc., can be implemented through different means and at different locations, thus opening the way to fully end-to-end programmable networks. However, network programmability exhibits advantages and limitations related to execution time, resource consumption, protocol stack layer, ease of deployment, configuration, migration, etc. Also, a given network programmability technology can only orchestrate a limited number of services (Kuklinski et al. 2016). The intrinsic feature of each network function also impacts its deployment and orchestration; stateful functions being challenging to migrate, unlike stateless ones. To date, several orchestration solutions exist (de Saraiva et al. 2018) (e.g. ONOS1, OpenDaylight 2, Open Source MANO3, and ONAP4). Nevertheless, although existing solutions now provide the technological means required to enable fully programmable network solutions, advanced network service orchestration algorithms are missing (Katsalis et al. 2016). Deployment in large-scale networks, dependability and efficiency to monitor and secure services are now the crucial questions raised by network programmability (da Costa Cordeiro et al. 2017). New self-management techniques based on AI are being used, as well as TEEs for microservices.

The expected benefits of dynamic deployment of microservices (Dragoni et al. 2017) rely on an easier development and maintenance, better quality, scalability, and responsiveness to new scenarios than monolithic approaches (Alshuqayran et al. 2016), while offering more possibilities for operators and management facilities through orchestration as demonstrated in Nguyen et al. (2018b) when testing a new protocol stack. Initiatives such as ClickOS (Martins et al. 2014) and Unikernels (Madhavapeddy et al. 2013) aim at virtualizing network processing to offer more flexibility to operators while reducing their investment and operating costs. However, even if few initial research works propose automatic methods to help monolithic application to be split into microservices by identifying possible decoupling points by static analysis of the code (Levcovitz et al. 2016) and the creation of a dependency graph to manage microservices (Toffetti et al. 2015), selecting essential network functions to convert to microservices (and their level of division) as well as their placement in the network is still a complex problem to study (Toffetti et al. 2015).

Conclusions

The new paradigms introduced by 5G (e.g. SDN, NFV, massive distribution, multi-tenancy, and network slicing among others) both facilitate security management and introduce new vulnerabilities. Legacy security solutions based on fixed perimeter-based security solutions are no longer effective. No given technology (e.g. encryption) can solve all security problems. Monitoring is required to bring the awareness necessary for understanding the dynamicity and complexity introduced in 5G mobile networks and by new emerging vertical applications. More than ever, self-management of networks becomes a must that needs to be protected from all kinds of threats (known, unknown, DDoS, etc.).

These new challenges need to be addressed by research, industry, and other stakeholders including regulation/standardization bodies and governments. The main challenges are the need to (i) consider the new vulnerabilities due to the software and openness introduced by cloud and virtualization technologies, SDN, and NFV; (ii) profit from the flexibility and modularity of SDN/NFV-based architecture to adapt to highly dynamic network environments; (iii) consider multi-tenancy and multi-domain, as well as equipment and technology heterogeneity; (iv) adapt security monitoring to the needs to obtain cost-effective efficient and effective security that integrates risk and reputation; (v) provide new business models based on SDS and SECaaS; (vi) adapt to new massive distribution (e.g. mIoT, mMTC, V2X); (vii) integrate emerging technologies to obtain optimized and distributed security analysis computations and remediations (e.g. P4, microservices, Fog, and MEC); (viii) profit also from the programmability introduced at all levels (hardware, control plane, data plane); (ix) introduce AI and ML to support security self-management and automation; (x) consider the limits of ML, e.g. adversarial ML; (xi) bring awareness to improve trust (e.g. SSLAs) and liability management; (xii) integrate CTI and information sharing to make security detection and remediation or self-repair more effective; and (xiii) detect security evasion techniques. But this list is certainly not complete and other aspects need to be considered when defining security monitoring and management, such as security by design architectures, correlation of cyber and physical events, achieving attack tolerant systems, etc.

Related Article

5G Security – Complex Challenges

References

5G Americas Whitepaper (2018). The evolution of security in 5G. https://www.5gamericas
.org/wp-content/uploads/2019/07/5G_Americas_5G_Security_White_Paper_Final.pdf
(18 October 2019).

Ahmad, I., Kumar, T., Liyanage, M. et al. (2018). Overview of 5G security challenges and
solutions. *IEEE Communications Standards Magazine* 2 (1): 36–43.

AI HLG (2017). High-Level Expert Group on Artificial Intelligence. https://ec.europa.eu/
digital-single-market/en/high-level-expert-group-artificial-intelligence (accessed 18
October 2019).

Ali-Tolppa, J., Kocsis, S., Schultz, B. et al. (2018). Self-healing and resilience in future 5G
cognitive autonomous networks. Proceedings of the 10th ITU Academic Conference,
Machine Learning for a 5G Future, 35–42 (November 2018).

Alshuqayran, N., Ali, N., and Evans, R. (2016). A systematic mapping study in microservice
architecture. Proceedings of the IEEE 9th International Conference on Service-Oriented
Computing and Applications (SOCA), 44–51, IEEE.

Angelopoulos, I., Trouva, E., and Xilouris, G. (2017). A monitoring framework for 5G
service deployments. CAMAD 2017: 1–6. https://www.researchgate.net/publication/
318563341_A_monitoring_framework_for_5G_service_deployments

Antonakakis, M., April, T., Bailey, M. et al. (2017). Understanding the Mirai Botnet.
USENIX Security Symposium 2017, 1093–1110. https://www.usenix.org/system/files/
conference/usenixsecurity17/sec17-antonakakis.pdf

Apache Spot Platform 2016. A Community Approach to Fighting Cyber Threats. http://
spot.incubator.apache.org (18 October 2019).

Ayoubi, S., Limam, N., Salahuddin, M.A. et al. (2018). Machine learning for cognitive
network management. *IEEE Communications Magazine* 56 (1): 158–165. doi: 10.1109/
MCOM.2018.1700560.

Barreno, M., Nelson, B., Sears, R. et al. (2006). Can machine learning be secure?
Proceedings of the ASIACCS'06, 16–25.

Blanc, G., Kheir, N., Ayed, D. et al. (2018). Towards a 5G security architecture: articulating
software-defined security and security as a service. Proceedings of the 13th International
Conference on Availability, Reliability and Security (ARES 2018), Article No. 47 (August
2018).

Boutaba, R., Salahuddin, M., Limam, N. et al. (2018). A comprehensive survey on machine
learning for networking: evolution, applications and research opportunities. *Journal of
Internet Services and Applications* 9: 16.

Cognitive Network Management for 5G (2016). Whitepaper by the 5GPPP working group on network management and QoS. https://5g-ppp.eu/wp-content/uploads/2016/11/NetworkManagement_WhitePaper_1.0.pdf (18 October 2019).

Collberg, C.S. and Thomborson, C.D. (2002). Watermarking, tamper-proofing, and obfuscation: tools for software protection. *IEEE Transactions on Software Engineering* 28 (8): 735–746.

da Costa Cordeiro, W.L., Marques, J.A., and Gaspary, L.P. (2017). Data plane programmability beyond openflow: opportunities and challenges for network and service operations and management. *Journal of Network and Systems Management* 25 (4): 784–818.

Dragoni, N., Giallorenzo, S., Lafuente, A.L. et al. (2017). Microservices: yesterday, today, and tomorrow. In: *Present and Ulterior Software Engineering* (ed. M. Mazzara and B. Meyer). Springer.

ENSURE 5G (2017). 5G Enablers for Network and System Security and Resilience. http://www.5gensure.eu/ (accessed 18 October 2019)

ETSI GS NFV-SEC 001 (2014). NFV Security; Problem Statement. V1.1.1 (October 2014).

ETSI GS NFV-SEC 003 (2016). NFV Security; Security & Trust Guidance. V1.2.1 (August 2016).

European Commission (2018). Report from the Commission to the European Parliament, the Council and the European Economic and Social Committee on the Application of the Council Directive on the approximation of the laws, regulations, and administrative provisions of the Member States concerning liability for defective products (85/374/EEC). Brussels (May 2018).

Experiential Networked Intelligence (ENI) (2018). ENI use cases: ETSI GR ENI 001 V1.1.1. https://www.etsi.org/deliver/etsi_gr/ENI/001_099/001/01.01.01_60/gr_ENI001v010101p.pdf (accessed 18 October 2019).

Expert Group on liability and new technologies (E03592) (2018). http://ec.europa.eu/transparency/regexpert/index.cfm?do=groupDetail.groupDetail&groupID=3592 (accessed 18 October 2019).

Gentry, C. (2009). Fully homomorphic encryption using ideal lattices. (STOC'09), 169–178.

Jiang, W., Strufe, M. and Schotten, H.D. (2017a). Experimental results for artificial intelligence-based self-organized 5G networks. PIMRC 2017, 1–6.

Jiang, W., Strufe, M. and Schotten, H.D. (2017b). Intelligent network management for 5G systems: The SELFNET approach. EuCNC. 1–5.

Katsalis, K., Nikaein, N., and Edmonds, A. (2016). Multi-domain orchestration for NFV: challenges and research directions. Proceedings of the 15th International Conference on Ubiquitous Computing and Communications-Symposium on Cyberspace and Security (IUCC-CSS), 189–195, IEEE.

Khettab, Y., Bagaa, M., Dutra, D.L.C. et al. (2018). Virtual security as a service for 5G verticals. Proceedings of the IEEE Wireless Communications and Networking Conference (WCNC) (April 2018).

Koumaras, H., Tsolkas, D., Gardikis, G. et al. (2018). 5GENESIS: The Genesis of a Flexible 5G Facility. CAMAD.

Kuklinski, S., Dinh, K.T., Destre, C. et al. (2016). Design principles of generalized network orchestrators. International Conference on Communications (ICC), 430–435, IEEE.

Kumar, A. and Lim, T.J. (2019). Early detection of Mirai-Like IoT bots in large-scale networks through sub-sampled packet traffic analysis. CoRR abs/1901.04805.

Lefebvre, V., Santinelli, G., Muller, T. et al. (2018). Universal trusted execution environments for securing SDN/NFV operations. ARES 2018 (August 2018).

Levcovitz, A., Terra, R., and Tulio Valente, M. (2016). Towards a technique for extracting microservices from monolithic enterprise systems. preprint arXiv:1605.03175.

Li, R., Zhao, Z., Zhou, X. et al. (2017). Intelligent 5G: when cellular networks meet artificial intelligence. *IEEE Wireless Communications* 24 (5): 175–183.

Liyanage, M., Okwuibe, J., Ahmed, I. et al. (2017). Software defined monitoring (SDM) for 5G mobile backhual networks. Proceedings of the 23th IEEE International Symposium on Local and Metropolitan Area Networks (LANMAN), Osaka, Japan (June 2017).

Liyanage, M., Ahmad, I., Abro, A.B. et al. (ed.) (2018). *A Comprehensive Guide to 5G Security*, 231–243. Wiley https://www.researchgate.net/publication/322466640_ Software_Defined_Security_Monitoring_in_5G_Networks.

Madhavapeddy, A., Mortier, R., Rotsos, C. et al. (2013). Unikernels: library operating systems for the cloud. SIGPLAN Not. 48, 4, 461–472, ACM Press.

Maimo, L.F., Gomez, A.L.P., Clemente, F.J.G. et al. (2018). A self-adaptive deep learning-based system for anomaly detection in 5G networks. *IEEE Access* 6: 7700–7712.

Martins, J., Ahmed, M., Raiciu, C. et al. (2014). ClickOS and the art of network function virtualization. Proceedings of the 11th Conference on Networked Systems Design and Implementation, 459–473. USENIX Association.

Meidan, Y., Bohadana, M., Mathov, Y. et al. (2018). N-BaIoT – network-based detection of IoT botnet attacks using deep autoencoders. *IEEE Pervasive Computing* 17 (3): 12–22.

Mofrad, S., Zhang, F., Lu, S. and Shi, W. (2018). A comparison study of intel SGX and AMD memory encryption technology. HASP@ISCA 2018, 9:1–9:8.

Nguyen, T.N., Mai, H.L., Doyen, G. et al. (2018a). A security monitoring plane for named data networking deployment. *IEEE Communications Magazine* 56 (11): 88–94.

Nguyen, T., Mai, H.-L., Doyen, G. et al. (2018b). A security monitoring plane for named data networking deployment. *Communications Magazine* Special Issue on ICN Security, IEEE.

Nguyen, T., Mai, H.-L., Cogranne, R. et al. (2019). Reliable detection of interest flooding attack in real deployment of named data networking. *IEEE Transactions on Information Forensics and Security* 14 (9): 2470–2485. doi: 10.1109/TIFS.2019.2899247.

Open Networking Foundation (2015). OpenFlow switch specification. https://www. opennetworking.org/wp-content/uploads/2014/10/openflow-switch-v1.5.1.pdf

Open Networking Foundation (2016). Threat analysis for the SDN architecture. TR-530, Version 1.0 (July 2016) (18 October 2019).

Parker, M.C., Koczian, G., Adeyemi-Ejeye, F. et al. (2016). CHARISMA: converged heterogeneous advanced 5G cloud-RAN architecture for intelligent and secure media access. EuCNC.

Prometheus (2012). Open source monitoring platform. https://prometheus.io/docs/ introduction/overview/ (18 October 2019).

Saraiva de Sousa, N.F., Lachos Perez, D.A., Rosa, R.V. et al. (2018). Network service orchestration: a survey. Submitted to IEEE Communications Surveys and Tutorials (March 2018).

SIGMONA. (2016). D5.3-SIGMONA (SDN Concept in Generalized Mobile Network Architectures) White Paper. https://www.sigmona.org/?attachment_id=150 (accessed 18 October 2019).

Strufe, M., Jiang, W. and Schotten, H.D. (2018). A 5G NFV test bed for the evaluation of AI based network management and security (Concepts). EuCNC 2018, https://www. researchgate.net/publication/325934178_A_5G_NFV_Test_Bed_for_the_Evaluation_of_AI_Based_Network_Management_and_Security_Concepts

Suárez, L., Espes, D., Le Parc, P. et al. (2018). Enhancing network slice security via artificial intelligence: challenges and solutions. Conférence C&ESAR 2018 (November 2018), Rennes, France.

Toffetti, G., Brunner, S., Blöchlinger, M. et al. (2015). An architecture for self-managing microservices. Proceedings of the First International Workshop on Automated Incident Management in Cloud (AIMC'15), ACM, 19–24.

Wang, Q., Alcaraz Calero, J.M., Weiss, M.B. et al. (2018). SliceNet: End-to-End Cognitive Network Slicing and Slice Management Framework in Virtualised Multi-Domain, Multi-Tenant 5G Networks. BMSB. 1–5.

Wurster, G., van Oorschot, P.C. and Somayaji, A. (2005). A generic attack on checksumming-based software tamper resistance. Proceedings of the 2005 IEEE Symposium on Security and Privacy (S&P'05) (8–11 May 2005).

Further Reading

Penttinen, J.T.J. (2019). 5G Explained: Security and Deployment of Advanced Mobile Communications. Wiley. ISBN: 9781119275732.

Ylianttila, M., Gurtov, A., Bux Abro, A. et al. (2018). A Comprehensive Guide to 5G Security. Wiley. ISBN: 9781119293040.

12

Security for Vertical Industries

Mehrnoosh Monshizadeh[1,2], Vikramajeet Khatri[3], and Iris Adam[4]

[1] *Nokia Bell Labs France, Nozay, France*
[2] *Aalto University, Helsinki, Finland*
[3] *Nokia Bell Labs Finland, Helsinki, Finland*
[4] *Nokia Bell Labs Germany, Stuttgart, Germany*

Introduction

Due to the vital role of mobile operators in providing Internet services and the fast growth of cloud computing technology, mobile operators have considered reforming themselves as one of the cloud providers for networking services. A physical mobile network can host several network operators called Mobile Virtual Network Operators (MVNOs). On the other hand, Telecommunication network as a Service (TaaS) is a platform for sharing physical and virtual resources of cloud infrastructure among multiple MVNOs. TaaS is composed of software, hardware, and application functions, which are also called virtualized Network Functions (vNFs). In TaaS, each mobile operator has interconnection with cloud layers, such as Infrastructure as a Service (IaaS), Platform as a Service (PaaS), and Software as a Service (SaaS) depending on the type of services it provides to the customers (Monshizadeh and Khatri 2017).

However, cloudification of mobile operators introduces several advantages, but security is still one of the biggest challenges. Although there are numerous studies to address security threats on cloud environments and introduce their mitigation mechanisms, still few cover security solutions at each layer of cloud environment for MVNOs. Some of these threats and their mitigation mechanisms are briefly discussed as follows.

Due to resource sharing, various internal or external cyberattacks (data leakage, data corruption, etc.) can target MVNOs. Therefore, security challenges of TaaS and the new threats that are introduced by TaaS should be investigated and prevention mechanisms to resist them must be introduced. On the other hand, combination of cloud service providers, mobile network operators, and various customers from the vertical industries in 5G requires multilayer and multitenancy capable security management architectures. In addition, telecom service providers face new problems regarding the management of Service Level Agreements (SLAs). 5G communication technology will bring applications with very high bandwidth and very low latency requirements and SLA violations may lead to legal and financial issues. In the parallel, security is a big concern and additional resources are needed to fulfil the strict security requirements from cloud consumers. Security management and orchestration functions should

The Wiley 5G REF: Security. Edited by Rahim Tafazolli, Chin-Liang Wang, Periklis Chatzimisios and Madhusanka Liyanage.
© 2021 John Wiley & Sons Ltd. Published 2021 by John Wiley & Sons Ltd.

enable the enforcement of performance and security-related SLAs in multiple layers of a distributed cloud environment and should automatically mitigate SLA violations considering the mutual impact from performance and security requirements. Therefore, in 5G networks, an automated security event management is necessary to offer security monitoring and correlation capabilities to mobile network operators, infrastructure service providers, and tenants like verticals.

Despite the fact that virtualization has many attractive features, it makes the networks more vulnerable to Denial of Service (DoS) attacks. DoS attacks are attempts by a nonlegitimate user to degrade network resources. Attacker can cause too many resources to be used by a Virtual Machine (VM) and therefore denying service to other VMs. With the same concept, an attacker can send small requests to a Domain Name System (DNS) server and ask it to send the victim a large reply and overwhelm the target. To solve this problem, virtual load balancers and virtual DNS servers can be used along with Hardware Security Models (HSM) to prevent unauthorized access via third-party vNFs.

In a cloud environment for Network Functions Virtualization (NFV)-based security services, regardless where they are located and which operator they belong to, common interfaces can be used. In mobile network environment, security applications include security configuration, security function negotiation, and security request from a user device. For access networks, security applications include traffic inspection, traffic manipulation, and traffic impersonation. Required security functions for these applications are Deep Packet Inspection (DPI), Intrusion Prevention System (IPS), firewall, Virtual Private Network (VPN), and honeypots.

MVNOs can apply SLAs and security management to overcome virtualization security challenges such as VM hopping, VM DoS, VM mobility, and VM diversity. With VM hopping, an attacker can access from one VM to other VMs that can be mitigated by proper policy enforcement on resource usage. VM mobility emphasizes spread of vulnerable configuration.

The European Telecommunication Standard Institute (ETSI) introduces a distributed deployment for security management in NFV environments, which compromises of NFV Security Manager (NSM) and the NFV Infrastructure Security Manager (ISM) as logical functional blocks as shown in Figure 1 (Dutta et al. 2017). However, for adaption to support multiprovider and multitenant cloud environments, this framework needs to be extended with security management capabilities provided directly within multiple layers of the distributed cloud environment and within the tenants' network slices.

In addition, ETSI defines a logical architecture for security monitoring and management in cloud environments considering that the security monitoring requirements may vary within the same NFV deployment for each tenant and service. Network Functions Virtualization Security Service Providers (NFV SSPs) are comprised as logical functions within the Virtualized Infrastructure Manager (VIM) and the virtual network function management (VNFM) to receive, and optionally analyze, monitoring data from heterogeneous security functions. A separate NFV Security Monitoring Analytics System collects telemetry data from various NFV SSPs in the same NFV deployment to detect threats and anomalies and to initiate activities for remediation.

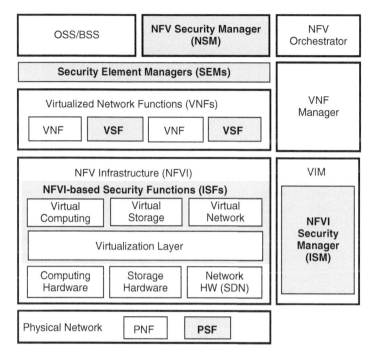

Figure 1 ETSI high-level NFV security management framework. Source: Dutta et al. (2017). Reproduced with permission of ETSI.

However, in distributed multiprovider and multitenant infrastructures, such a centralized monitoring entity has too complicated analytic mechanisms and troubleshooting activities.

Today, a cloud management system enforces SLAs by dynamically allocating available resources to cloud services and monitoring the software and physical cloud resources to maintain the conditions defined in the SLAs. Those SLAs are static and refer to resource requirements without considering security. Though, there is a lot of research on network management in distributed cloud environments, even especially on telecom SaaS, SLA management is mentioned only with limited scope. In general, the focus is on performance requirements and specific functionality dealing with SLAs, e.g., the challenges from cloud computing on telecom SaaS are analyzed and an SLA resolution process is defined to determine infrastructure allocations for optimized resource management (Vajda et al. 2012). Some research refers to the management of dynamic SLAs to adapt to changing demands from performance-critical applications (Lee and Sill 2014), but do not consider the fact that security functions may decrease the performance.

Both automated security management to provide security monitoring, analytics, and response and SLA management are seen as gap in the currently ongoing 5G standardization efforts.

This article reviews the concept of mobile operators' cloudification and discusses how TaaS can help MVNOs optimizing their networks for specific use cases or verticals to provide a wide range of services to the customers. Based on the concept of TaaS platform, this article investigates MVNOs' security challenges and mitigation mechanisms for cloud layers and cloud deployments. Furthermore, the challenges for MVNOs to correlate security events in 5G scenarios are analyzed and a framework for automated security event management is proposed to offer security monitoring and correlation capabilities to mobile network operators, infrastructure service providers, and tenants like verticals.

Cloudification of the Network Operators

The shift to cloud computing technology introduces diverse delivery models to telecom operators. In this transition, mobile operators could act as cloud network providers and based on common characteristics, such as geographical zone and availability offer services either to end users or other operators. For this purpose, TaaS is applied to create functionalities and services for commercial Mobile Network Operator (MNO) business. TaaS is combination of software, hardware, and application functions (also known as vNFs), which could be sold as a service product to emerging MNOs or MVNOs and a single TaaS platform may host various mobile operators. Each mobile operator has interconnection with cloud layers (IaaS, PaaS, and SaaS) depending on the type of services it provides to the customers. Figure 2 shows the TaaS stacks in the cloud layers (Monshizadeh and Khatri 2017).

Infrastructure as a Service (IaaS) provides virtualized infrastructures. Mobile operators can rent out their network elements, storage resources, computing system, and licenses to other operators.

Platform as a Service (PaaS) is an interface between applications in SaaS and VMs in IaaS. PaaS controls VMs. This virtual platform is provided to developers for programming and web management. This programming could be related to network optimization, adding new features, and so on. The main added value for PaaS comes from providing easy to use mechanism to deploy customer's software applications to the cloud service and providing scaling for server capacity.

Software as a Service (SaaS) is an application layer that provides different kinds of application software services to mobile operators, when they are relying on cloud base services. The applications can be used for bandwidth control, Quality of Service (QoS) management, network configuration, system backup, and so on.

The proposed software stack of TaaS can be implemented in a combination of cloud deployment model: public cloud, private cloud, community cloud, and hybrid cloud. The combination is based on security consideration and will be discussed in the following section.

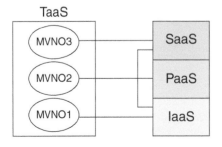

Figure 2 *TaaS stack*. Source: Monshizadeh and Khatri (2017). Reproduced with permission of John Wiley & Sons.

MVNO Security in 5G Networks

In general, it is possible to build 5G security based on 4G security mechanisms considering their similarities and significant robustness in 4G security mechanisms. However, there are new elements and technologies introduced by 5G that requires further security investigation. Programmable network architectures such as Software Defined Network (SDN) carries most of the three-layer threats, such as configuration, authorization, and access control, as well as software and image vulnerabilities. Similarly, NFV carries all attacks due to virtualization. Virtualization mechanisms, such as multitenancy, application sharing, network slicing, and open-source software, lead to security threats like information leakage, misconfiguration, and data corruption (Monshizadeh and Khatri 2017).

On the other hand, due to the nature of 5G networks to support extremely fast communication, these networks are expected to massively expand IoT (Internet of Things) devices that sometimes carry critical missions. Therefore, it is important how and which authentication mechanisms are chosen since even the lowest additional latency can have considerable effect on communication. Moreover, in 5G networks, delay in setting up control plane (CP) security will impact the delay of sensitive applications such as machine to machine (M2M). Common Non-Access Stratum (NAS) signaling (state transition) is an example of CP signaling, which may introduce delay in 5G security. User-enabled security features are another aspect to be considered in 5G networks. In current mobile networks, it is the MVNO who decide which and when security mechanisms are to be applied, while in 5G, it would be more feasible to offer corresponding security services to customer (Schneider and Horn 2015). Therefore, in a 5G network, MVNOs must consider (Nokia 2017):

- Security mechanisms with highest robustness

  Increased robustness against cyberattacks, e.g. in protocol design, consider more the possibility of misbehavior or attacks of user devices or network nodes in roaming networks.

  Enhanced privacy, an example could be protection against International Mobile Subscriber Identity (IMSI) catching.

  Security Assurance: Methods to ensure security properties of entities (or even systems, e.g. network slices).
- Security mechanisms with highest flexibility.

  Alternative identification and authentication procedures: In Long-Term Evolution (LTE), user identification and authentication are based on credentials on a removable Universal Integrated Circuit Card (UICC). Another option is to use an embedded Universal Integrated Circuit Card (eUICC). For simple devices like sensors, one may consider cheaper solutions with no dedicated hardware to hold the credentials. The security implications must be balanced against factors such as manufacturing and provisioning costs.

  User plane (UP) encryption and integrity protection optional to use: In LTE, there is only confidentiality protection (encryption) for the UP, and the network decides whether to use it. For 5G, we consider both encryption and integrity protection, where the user decides what to apply. The idea is not to forego security, but to

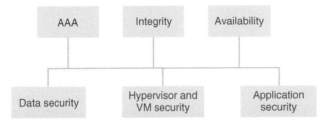

Figure 3 *TaaS security classification*. Source: Monshizadeh and Khatri (2017). Reproduced with permission of John Wiley & Sons.

avoid, for example double encryption if the application uses end-to-end security mechanisms anyway.

Adjust security mechanisms per network slice according to the requirements from the tenants like verticals.

- Security mechanisms with highest automation.

Holistic security management and orchestration: Highly automate the management of security functions, virtual functions as well as physical boxes, with a holistic view over the complete network and all deployed security functions.

Self-adaptive, intelligent security controls: More futuristic measures to protect the network against yet unknown attacks. This involves anomaly detection, self-learning, artificial intelligence, big data processing, global intelligence gathering, etc. Our proposed architecture will be a contribution toward automated security mechanism.

As it is shown in Figure 3, TaaS security is divided into three main aspects: Data Security, Hypervisor and VM security, and finally Application Security. To protect TaaS platform against threats and vulnerabilities, security requirements, such as availability; integrity; Authentication Authorization, and Accounting (AAA) will be reviewed for each of the mentioned aspects.

AAA refers to mechanism such as certificate-based authentication that should be utilized to avoid unauthorized access. Keys and signatures must be stored in a secure storage such as HSM (Alcatel-Lucent 2013) to make it invisible to third parties. Hardening should be applied to infrastructure layer and wherever needed to block any backdoor access. Virtual firewalls must be used inside VMs and proxy and traditional firewalls must be used where needed to prevent unauthorized traffic. Backup must be maintained for all VMs so that data can be restored in case of failure. AAA should be maintained by logging actions from each VM and modules; and logs should be stored in a safe storage so that in case of attack or failure, logs should not be affected and would help revealing the root cause. Encryption must be used so that data is not readable to unintended party even if it is accessed without authorization. Security policies should be enforced to make sure that all users in the cloud have similar security policy and are in line with SLA (Petcu 2014). Integrity refers to protection against threats on virtualized network. These threats could vary from attacking different VMs such as virtual Mobility Management Entity (vMME); virtual Home Subscriber Server (vHSS); and their virtual functions, misconfigurations, or abuse of resources, corrupting operating systems, switches, and management software (in SDN) and inducing any kind of malicious applications. Availability can be improved by applying techniques, such as load balancing, redundancy, and data backup as discussed earlier.

Data Security in TaaS. Mobile operators who act as cloud providers are responsible for data protection of their customer who are either end users or other mobile operators. In addition, location of the cloud provider where the parent company is registered should be considered since different countries have diverse laws. Regardless of the location of data centers, in special circumstances, authorities could have access to customer data. The legal aspects such as security warranties and compensation agreements among operators belonging to TaaS are other aspects of TaaS data security.

Hypervisor and VM Security in TaaS. Concept of virtualized threats refers to every kind of attack against availability, integrity, and confidentiality of the hardware and software in a virtualized mobile network. There are three elements in a virtualized network: hypervisor, VMs (virtual hardware and images), and applications; all these elements should be adequately secured against unauthorized access, change, and destruction. In a virtualized mobile network, hypervisor itself is not directly connected to any end user, and threats arise through malicious VMs mostly, therefore having a reliable hypervisor requires secure VMs. While traditional security techniques such as Intrusion Detection System (IDS), antivirus, and firewalls are still applicable for virtualized networks, isolation could be an important approach toward security of VMs. Isolation will ensure that if one VM would be attacked, other VMs would not get infected (Ali et al. 2015; Doelitzscher et al. 2012). There are different methods such as security zones and traffic separation for VM isolation. VMs with similar functionality and security requirements could be grouped in same hardware. Each zone could be controlled by different access list defined in firewall or dedicated IDS, and so on.

In a hypervisor, SDN is a new approach to separate UP and CP in mobile networks. SDN can be considered in IaaS (SDN switch) or in PaaS (SDN controller). SDN controller in mobile networks only carries CP including Mobility Management Entity (MME), Serving/Public Data Network Gateway (S/P GW) VMs, and is located in PaaS. In this case, SDN is an interface between infrastructure and application layer. SDN in its switching functionality (S/P GW VM) only carries UP and considers part of infrastructure layer. From the security point of view, however SDN bring several advantages, there are also disadvantages for MVNOs and such for TaaS as listed below.

- Potential of single attack on centralized controller
- Vulnerable southbound interface between controller and data forwarding that may result to network degradation and unavailability via DoS attack
- Vulnerable northbound interface between controller and applications
- Applications access to controller and programming the network
- Reduced isolation of network functions
- Expensive and vulnerable cryptographic keys

Though SDN security introduces new threats, yet their mitigation mechanisms are almost similar to traditional networks. However, traditional network implementations rely on dedicated hardware and private connections between network elements. Therefore, their CP connections are not exposed to public, unless there is somewhere a configuration error. In addition to commonly used mitigation techniques such as IDS, DPI, secure protocols, security zones, and virus scanners, there are also Software Defined

Network Monitoring (SDNM) and Remote Monitoring (RMON) techniques that are SDN-specific security mechanisms.

NFV in hypervisor refers to any network function that runs on mobile network equipment over a hypervisor. Three attack profiles are introduced in NFV: Intra-MVNO attacks include attacks on an MVNO by its own employee to occupy and degrade network services. Inter-MVNO attack refers to any type of attack from one MVNO toward other MVNO(s) to extract competitor's information, corrupt, or misuse their services. Attacks performed by end users within the same MVNO or other MVNO fall into third attack profile known as attacks by end users.

In a cloud environment with NFV, network functions will be deployed as vNFs that bring security challenges. Different solutions such as security zone and grouping, isolating applications by VMs, and licensing are recommended for NFV security. NFV acts in hypervisor and other parties could see the encryption keys, therefore providing signature beside the keys looks necessary (Alcatel-Lucent 2013). Both firewall and security orchestration are recommended for NFV and platform security. Some of the major threats on vNFs and their mitigation mechanisms are explained here.

- Malicious loops that are caused by routing loops and unavailability of management network due to network failure: To prevent these threats, network should be logically validated to be sure that management interfaces are accessible even if vNFs are down.
- Improper data removal due to VM crash, execution of malicious vNF, and therefore unauthorized changes to Basic Input/Output System (BIOS) or Unified Extensible Firmware Interface (UEFI), hypervisor, and Operating System (OS): For mitigation secure boot, Trusted Platform Module (TPM) and crash protection can be used.
- Abuse of hypervisor resources by malicious VM (impacting other VMs) and QoS degradation: Performance isolation by segregating resources to each VM is recommended as prevention mechanism (Lukyanenko et al. 2014).
- Insufficient vertical and horizontal VM AAA mechanism: to prevent this threat, AAA mechanisms among vNFs, between vNFs and application layer, and between vNFs and management stations should be revised.
- Software unreliability, such as:
  Coding flaws that affect all MVNOs using same software: Correction and security patches should be applied on all VMs using same software.
  Configuration changes or correction patches that need reboot and cause service outage on MVNOs: Backup and load balancing are the mitigation mechanisms for such threat.
  Test and monitoring backdoors: Closing test and monitoring and debug interfaces is recommended.
  Stored password and private keys in VM images: Using unique private key for each image could prevent these threats.

Application Security in TaaS. Another aspect of virtualized network security refers to protection against threats that are related to an application server or a web server connected to the Internet. Based on the concept of SaaS, software applications should be accessible over the Internet that makes security a very critical challenge for mobile operators. Beside the mechanisms, such as data encryption, access control and authentication, backup, and redundancy, mobile operator could implement sensitive applications that do not require end user intervention (such as billing application) using

PaaS that is accessible only to limited professional users among mobile operators (Yrjo and Rushil 2011).

Security Management and Orchestration

Security management in 5G networks should meet the needs of heterogeneous business models in the 5G ecosystem and address the motivations from the different stakeholders. SaaS providers utilize resources like compute, storage, and inter-datacenter connectivity from other IaaS and PaaS providers based on negotiated SLA contracts. Even for telecom service providers (as in 4G and 5G), there may be single infrastructure providers that hosts several telecom SaaS as tenants on a shared infrastructure, whereas the infrastructure is partitioned for the isolation of tenants (cloud slices). A "Broker" may orchestrate the services from heterogeneous clouds to enable the collaboration between tenants and service providers, and during SLA negotiation the requirements from tenants are mapped to the constraints from service providers aimed to create slices for resource allocation. Tenants have the possibility to evaluate whether a certain cloud can meet their demands and to choose the most appropriate one.

Standard SLAs lack specific cybersecurity measures, and security-related SLAs need to be developed, as part of the SLA management process. According to standard SLAs, security-related SLAs should be negotiated and executed between service providers and tenants. The security SLAs are reflected as security requirements on tenants' side and security capabilities and constraints on providers' side. Security-related SLAs might be defined for, e.g.:

- Measures against Distributed DoS (DDoS), malware, etc.
- Monitoring and reporting anomalies
- Security log and forensics review capability
- Security protocols and algorithm types
- Response time to incidents
- Monitor and secure optical networks
- Regulation compliance level
- Vulnerability score

Monitoring, correlation of events, and mitigation strategies are essential to have consistent security management in 5G networks. IaaS providers manage the underlying shared infrastructure including computing, storage, networking resources, and the like to ensure that the infrastructure is secure and robust for all tenants. A tenant can only control its own virtual network and make sure that its security requirements for the infrastructure are met (Dutta et al. 2017; Adam and Jing 2018). In addition, the transport of false and duplicated alarms as well as raw data may lead to performance issues. Further, security functions like anomaly detection systems may suffer from false positive alarms and need further investigations to reduce false alarms. Therefore, security management hierarchy should be considered across multiple layers including the vertical business, telecom services, and infrastructure services.

Each cloud service provider has its own Security Management and Orchestration entities with monitoring, correlation, and remediation capabilities, as shown in Figure 4. While considering private and sensitive information, data are transferred between cloud

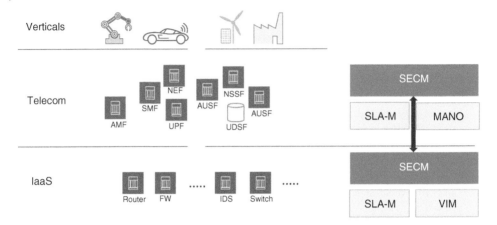

Figure 4 Security management and orchestration in multiple layers.

service providers and tenants to implement security mechanisms against cyberattacks like DDoS attacks. Events, logs, and correlated data may be exchanged in both directions based on corresponding security SLAs (Adam and Jing 2018).

Verticals do not have any knowledge of the infrastructure nor of other tenants. For control and visibility, security management function should provide Security Functions (SFs) as a service to tenants like verticals aimed to monitor VMs and virtual Network Functions as well as to correlate all relevant event and log data for detecting attacks and anomalies. The cloud service providers may enable a self-service interface to create and configure virtual security functions and appliances enhanced with a usage-based billing model, or the virtual security functions and appliances may be fully managed by the cloud service provider offered as managed service.

Furthermore, in 5G, the cloud service providers will be responsible to ensure the compliance with SLA requirements. In parallel, "self-healing" of the cloud service providers is critical to support various devices and applications in future networks (e.g. real-time IoT in 5G) (Mullins and Barros 2017). In this sense, self-healing means that SLA violations are mitigated before the tenants are affected. Security management should monitor security-related SLAs from all tenants like security-related Key Performance Indicators (KPIs), e.g. security management is capable to predict that incident response time could be delayed. To resolve the possible violation, more security functions like analytic entities need to be deployed and configured. The security management automatically triggers the cloud management function for additional resources.

Today, security functions are separately managed without interference with performance management. The security management should enable cloud service providers to deploy and configure security functions like firewalls and IDS under strong performance requirements during set-up of a service. In addition, security management should automatically adjust security controls for services during run-time without violating the performance requirements from IoT applications in the 5G ecosystem. It should be possible to predict the performance required from the deployment and configuration of security functions to secure the services as well as tenants' slices while not allowing lowered QoS of business applications. The performance prediction and guarantee may be based on simulation tools or algorithms. Furthermore, conflicts between performance and security-related SLAs may arise during deployment of cloud services (SLA specification) as well as during run-time (e.g. while mitigating incidents, scaling up vNFs, or congestion in peak time due to a big social event). The SLA management function should enable the cloud service providers to decide whether to trigger the cloud management or the security management to take appropriate actions aimed to mitigate different kinds of SLA violations (e.g. performance, security, or costs). It might be decided automatically, and the kind of mitigation depends on the capabilities and capacities of the cloud provider as well as the defined methods to resolve conflicts between performance and security-related requirements.

TaaS Deployment Security

With a combination of cloud layers and deployments, the proposed Cloud Security Framework for Operators (CSFO) not only recommends for each layer a proper deployment but also emphasizes on specific detection–prevention mechanism for different layers (Brunette and Mogull 2009). Figure 5 shows our proposed security framework for TaaS (Monshizadeh and Khatri 2017).

IaaS Security based on the layer. Since infrastructures are fully managed by a mobile cloud provider, the security mechanism is also the responsibility of the provider. The tenants usually have minimum control and interaction on the network elements. They do not have access to the CP VMs, even though they still could reach some of the network elements such as Home Location Register (HLR) or Policy Control and Charging Function (PCRF) server to pull their subscribers' information (such as subscriber profile, billing information). However, IaaS is less accessible by customers (end users or tenants); still insider attackers need to be highly considered. For this layer, techniques such as data isolation through VMs, ciphering to protect data against unauthorized access, backup and recovery for data reliability, firewalls, and IDS for preventing malicious attacks should be considered by the cloud provider.

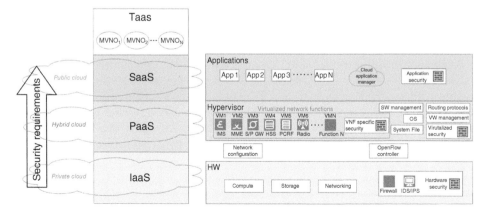

Figure 5 Cloud security framework for operators (CSFO). Source: Monshizadeh and Khatri (2017). Reproduced with permission of John Wiley & Sons.

IaaS Security based on the deployment. Considering high-security requirements for infrastructures, limited accessibility, geographic location, and high cost of network elements, mobile operators are recommended to use private cloud for this layer.

PaaS Security based on the layer. Layer point of view: This layer is normally used by developers to program and run their applications. Generating software bug or file system corruption, unauthorized access or privilege upgrade, and DoS are the security threats that should be considered at this layer. Strong authentication and access right control is required for this layer to limit user base that can make critical modification to configuration. Logging of all management actions are important to trace misbehaving users and to learn from mistakes. Therefore, a policy control mechanism could evaluate the requested access and decide whether to grant the access to developer or not.

PaaS Security based on the deployment. This layer will be used by limited group of professionals and does not need to be accessible by all end users; therefore, community or hybrid deployment for this layer is recommended. Mobile operators could take the advantages of public cloud, while for sensitive part of system software they just provide private cloud.

SaaS Security based on the layer. Since application layer is the closest layer to the end users, they could easily install different kinds of malware or spyware and steal the information or cause the data corruption at this layer. Secure protocols and malware detection methods are some of the prevention mechanisms that should be considered in this layer.

SaaS Security based on the deployment. To gain the initial cloud computing benefits, such as elasticity and economies of scale, SaaS should be available to all customers (end users and other tenants), therefore public cloud is recommended for SaaS.

Conclusions

Emerging traditional mobile operators who follow similar interests introduce potential demand for a new service model in cloud computing called TaaS.

Firstly, TaaS helps MVNOs optimizing their networks for specific use cases or verticals in order to provide a wide range of services to the customers. According to a location-based, customer-based, or service-based agreement, mobile operators could be grouped to considerably improve their cost structure, time, and quality efficiency and therefore their speed to market.

Secondly, TaaS gives the possibility to understand mobile operator's threats in a wide range and based on their provided cloud layers and various use cases since each vertical requires different level of security. In a 5G network, network that provides connectivity to IoT, devices may carry critical communications and services such as public safety and remote surgery, so these networks will have an increased attack surface, and new threats may arise from these devices. Therefore, in 5G networks, new use cases are needed for verticals; security must be built by considering security mechanisms with highest robustness, flexibility, and automation; so that security mechanisms are self-learning, self-adaptive, detect and react to attacks, and a high degree of reliability for services is maintained.

Thirdly, design criteria in 5G may provide ultra-high reliable and low-latency networks, and the correlation processes must be able to cope with these requirements.

For minimizing latency associated with the detection and remediation of attacks, monitoring data should be analyzed and correlated at different layers in 5G network architecture. Furthermore, 5G Network Slicing requires automated security and orchestration functionality, especially in complex network environments comprising multiple providers and multiple tenants. MNVOs must not only define security-related SLAs for new business models but also they should offer automated security management capabilities to tenants like verticals.

Finally, having a changing ecosystem (new stakeholders, new business models, MVNOs mode of operation e.g. an enterprise customer may rent a network slice) in 5G, there is a need to coordinate the cybersecurity capabilities between the different stakeholders of such an ecosystem, and it is necessary to adapt to new security-related prevention mechanisms. Cloud service providers should offer corresponding security services to MNVOs, services regarding deployment and configuration of security functions.

Therefore, this chapter proposed a cloud security model to help mobile operators to understand how to trade-off and merge their services based on the deployment and importance of the use cases (for various verticals) and provided services. In Figure 5, private deployment is assigned to IaaS that requires highest security consideration. For SaaS which is the closest layer to the end users, public cloud is recommended. Some of the reasons for this recommendation come from application availability to a wide range of end users, scattered end users (geographic location), and roaming condition. Finally, hybrid cloud is the proposed deployment for PaaS layer; that means private and public deployment could be considered for provided platforms and based on their sensitivity and security concerns. Discussed CFSO model is a combined security model that considers different threats and vulnerabilities for each layer, their modules, services, and protocols and helps TaaS to find the best combination of deployment solution.

Related Articles

5G-Core Network Security
SDMN Security
Security in Network Slicing

References

Adam, I. and Jing, J. (2018). Framework for security event management in 5G. Workshop on 5G Networks Security, 13th International Conference on Availability, Reliability and Security (ARES), 51:1–51:7.

Alcatel-Lucent (2013). Why service providers need an NFV platform (white paper). Tech. Rep. C401-01087-WP-201409-1-EN.

Ali, M., Khan, S.U., and Vasilakos, A.V. (2015). Security in cloud computing: opportunities and challenges. *Information Sciences* 305: 357–383.

Brunette, G. and Mogull, R. (2009). Security guidance for critical areas of focus in cloud computing v2. 1. Cloud Security Alliance. 1–76. https://cloudsecurityalliance.org/guidance/csaguide.v2.1.pdf

Doelitzscher, F., Reich, C., Knahl, M., and Clarke, N. (2012). Understanding cloud audits. In: *Privacy and Security for Cloud Computing* (ed. S. Pearson and G. Yee), 125–163. Springer.

Dutta, A., Sood, K., and Lu, W. (2017). Network Functions Virtualisation (NFV) Release 3; Security; Security Management and Monitoring specification. ETSI GS NFV-SEC 013 V3.1.1.

Lee, C.A. and Sill, A.F.F. (2014). A design space for dynamic service level agreements in OpenStack. *Journal of Cloud Computing* 3.

Lukyanenko, A., Nikolaevskiy, I., Kuptsov, D. et al. (2014). STEM+: Allocating Bandwidth Fairly To Tasks. Technical Report TR-14-001, ICSI, April 2014. https://www.icsi.berkeley.edu/icsi/publication_details?n=3651

Monshizadeh, M. and Khatri, V. (2017). Mobile virtual network operators (MVNO) security. In: *Comprehensive Guide to 5G Security*, 323–346. Wiley. ISBN: 978-1-119-29304-0.

Mullins, R. and Barros, M. (2017). Cognitive network management for 5G. 5GPPP Working Group on Network Management and QoS version 1.2.

Nokia (2017). Security challenges and opportunities for 5G mobile networks (white paper). SR1612004464EN.

Petcu, D. (2014). SLA-based cloud security monitoring: challenges, barriers, models and methods. In: *Euro-Par 2014: Parallel Processing Workshops* (ed. L. Lopes, J. Žilinskas, A. Costan, et al.), 359–370. Springer.

Schneider, P. and Horn, G. (2015). Towards 5G Security. 2015 IEEE Trustcom/BigDataSE/ISPA, Helsinki. 1165–1170.

Vajda, A., Baucke, S., Catrein, D. et al. (2012). Cloud computing and telecommunications: business opportunities, technologies and experimental setup. World Telecommunications Congress (WTC).

Yrjo, R. and Rushil, D. (2011). Cloud computing in mobile networks—case MVNO. 2011 15th International Conference On Intelligence in Next Generation Networks (ICIN). 253–258.

Further Reading

Ali, M., Khan, S.U., and Vasilakos, A.V. (2015). Security in cloud computing: Opportunities and challenges. *Information Sciences* 305: 357–383.

Khan, R., Kumar, P., Jayakody, D.N.K., and Liyanage, M. (2018). A Survey on Security and Privacy of 5G Technologies: Potential Solutions, Recent Advancements and Future Directions. *IEEE Communications Surveys & Tutorials* doi: 10.1109/COMST.2019.2933899.

13

Introduction to IoT Security

Anca D. Jurcut, Pasika Ranaweera, and Lina Xu

School of Computer Science, University College Dublin, Dublin, Ireland

Introduction

The rapid proliferation of the Internet of Things (IoT) into diverse application areas such as building and home automation, smart transportation systems, wearable technologies for healthcare, industrial process control and infrastructure monitoring and control is changing the fundamental way in which the physical world is perceived and managed. It is estimated that there will be approximately 30 billion IoT devices by 2020. Most of these IoT devices are expected to be of low-cost and wireless communication technology based, with limited capabilities in terms of computation and storage. As IoT systems are increasingly being entrusted with sensing and managing highly complex eco-systems, questions about the security and reliability of the data being transmitted to and from these IoT devices are rapidly becoming a major concern.

It has been reported in several studies that IoT networks are facing several security challenges (Alaba et al. 2017; Kim and Lee 2017; Jurcut et al. n.d., 2008, 2018; Deogirikar and Vidhate 2017; Pasca et al. 2008) including authentication, authorization, information leakage, privacy, verification, tampering, jamming, eavesdropping, etc. IoT provides a network infrastructure with interoperable communication protocols and software tools to enable the connectivity to the internet for handheld smart devices (smart phones, personal digital assistants [PDAs] and tablets), smart household apparatus (smart TV, AC, intelligent lighting systems, smart fridge, etc.), automobiles, and sensory acquisition systems (Alaba et al. 2017). However, the improved connectivity and accessibility of devices present major security concerns for all the parties connected to the network regardless of whether they are humans or machines. The infiltration launched by the Mirai malware on the Domain Name System (DNS) provider Dyn in 2016 through a botnet-based Distributed Denial of Service (DDoS) attack to compromise IoT devices such as printers, IP cameras, residential gateways and baby monitors represents the fertile ground for cyber threats in the IoT domain (Frustaci et al. 2018). Moreover, the cyber-attack launched at the Ukrainian power grid in 2015 targeting the Supervisory Control and Data Acquisition (SCADA) caused a blackout for several hours and is a prime example of the gravity of resulting devastation possible through

The Wiley 5G REF: Security. Edited by Rahim Tafazolli, Chin-Liang Wang, Periklis Chatzimisios and Madhusanka Liyanage.
© 2021 John Wiley & Sons Ltd. Published 2021 by John Wiley & Sons Ltd.

modern day attacks (Kim and Lee 2017). The main reasons for the security challenges of current information-centric automated systems is their insecure unlimited connectivity with the internet and the non-existent access control mechanisms for providing secure and trustworthy communication. Furthermore, the problem of vulnerabilities in IoT systems arises because of the physical limitations of resource-constrained IoT devices (in terms of computing power, on-board storage and battery-life), lack of consensus/standardization in security protocols for IoT, and widespread use of third-party hardware, firmware, and software. These systems are often not sufficiently secure, especially when deployed in environments that cannot be secured/isolated by other means. The resource constraints on typical IoT devices make it impractical to use very complex and time-consuming encryption/decryption algorithms for secure message communication. This makes IoT systems highly susceptible to various types of attacks (Alaba et al. 2017; Jurcut et al. n.d., 2008, 2018; Deogirikar and Vidhate 2017; Pasca et al. 2008). Furthermore, addressing the security vulnerabilities in the protocols designed for communication is critical to the success of IoT (Liyanage et al. 2017; Jurcut et al. 2012, 2014a,b, 2017; Kumar et al. 2017; Braeken et al. 2019).

This article focuses on security threats, attacks, and authentication in the context of the IoT and the state-of-the-art IoT security. It presents the results of an exhaustive survey of security attacks and access control mechanisms including authentication and authorization issues existing in IoT systems, its enabling technologies and protocols, while addressing all levels of the IoT architecture. We surveyed a wide range of existing works in the area of IoT security that use a number of different techniques. We classify the IoT security attacks and proposed countermeasures based on the current security threats, considering all three layers: perception, network, and application. This study aims to serve as a useful manual of existing security threats and vulnerabilities within the IoT heterogeneous environment and proposes possible solutions for improving the IoT security architecture. State-of-the-art IoT security threats and vulnerabilities have also been investigated, in terms of application deployments such as smart utilities, consumer wearables, intelligent transportation, smart agriculture, Industrial Internet of Things (IIoT) and smart city have been studied. The IoT security, particularly the IoT architecture, such as authentication and authorization, considering all layers.

The remainder of this article is organized as follows. Section titled "Attacks and Countermeasures" provides the IoT classification of attacks and their countermeasures according to the IoT applications and different layers of the IoT infrastructure. Section titled "Authentication and Authorization" addresses the importance of authentication with respect to security in IoT and presents in detail the existing authentication and authorization issues at all layers. Section titled "Other Security Features and Related Issues" introduces other security features and their related issues. Additionally, solutions for remediation of the compromised security, as well as methods for risk mitigation, with prevention and suggestions for improvement discussed in the same section. A discussion on the authentication mechanisms in the IoT domain, considering the most recent methodologies, has been presented in section titled "Discussion". Section titled "Future Research Directions" introduces future research directives such as blockchain, 5G, fog and edge computing, quantum, AI and network slicing. Finally, section titled "Conclusions" concludes this study.

Attacks and Countermeasures

Security is defined as a process to protect a resource against physical damage, unauthorized access, or theft, by maintaining a high confidentiality and integrity of the asset's information and making information about that object available whenever needed. The IoT security is the area of endeavor concerned with safeguarding connected devices and networks in the IoT environment. IoT enables the improvement of several applications in various fields, such as, smart cities, smart homes, healthcare, smarts grids, as well as other industrial applications. However, introducing constrained IoT devices and IoT technologies in such sensitive applications leads to new security challenges.

IoT is relying on connectivity of myriads of devices for its operation. Hence, the possibility of being exposed to a security attack is highly probable. In IT, an attack is an attempt to destroy, expose, alter, disable, steal, or gain unauthorized access to an asset. For example, cryptographic security protocols are a key component in providing security services for communication over networks (Jurcut et al. 2014a). These services include data confidentiality, message integrity, authentication, availability, nonrepudiation, and privacy (Jurcut et al. 2008). The proof of a protocol flaw is commonly known as an "attack" on a protocol and it is generally regarded as a sequence of actions performed by a dishonest principal, by means of any hardware or software tool, in order to subvert the protocol security goals. An IoT attack is not much different from an assault against an IT asset. What is new is the scale and relative simplicity of attacks in the IoT – millions and billions of devices are a potential victim of traditional style cyber-attacks, but on a far greater scale and often with limited or no protection.

The most prevalent devices connected to serving IoT applications for infotainment purposes are smart TVs, webcams, and printers. A vulnerability analysis has been conducted in (Williams et al. 2017) on these devices using Nessus[1] tool to observe that approximately 13% of the devices out of 156 680 were vulnerable which were further classified as critical, high, medium, and low. The vulnerabilities that exist, for example, MiniUPnP, Network Address Translation Port Mapping Protocol (NAT-PMP) detection, unencrypted telnet, Simple Network Management Protocol (SNMP) agents, Secure Shell (SSH) weak algorithms and File Transfer Protocol (FTP) inherited by webcams, smart TVs and printers are further identified based on manufacturer models.

In this section, we present the results of our study on the existing vulnerabilities, exploitable attacks and possible countermeasures in the context of the IoT and the state-of-the-art IoT security. We surveyed a wide range of existing work in the area of IoT security that use different techniques. We classified the IoT security attacks and the proposed countermeasures based on the current security threats, considering all three layers: Perception, Network, and Application. Figure 1 illustrates the typical architecture of IoT and entities which are considered under each layer. Table 1 summarizes the taxonomy of attacks and viable solutions of IoT categorized under each layer. These attacks and their corresponding solutions will be further discussed below.

Perception Layer

The devices belonging to the perception layer are typically deployed in Low-power and Lossy networks (LLNs), where energy, memory, and processing power are constricted compared to the localization of network nodes in conventional internet platforms

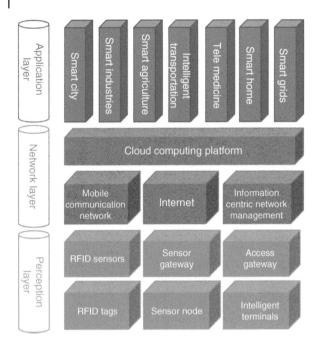

Figure 1 IoT architecture.

(Alaba et al. 2017). Therefore, including secure public key encryption-based authentication schemes would not be feasible because they require high computational power and storage capacity. Hence, developing a lightweight cryptographic protocol would be a challenging task when scalability, context-awareness, and ease of deployment must also be considered (Kim and Lee 2017).

There are several problems and attacks to be considered for the perception layer. We will be addressing these as shown in Table 1 and also by discussing the existing problems and attacks for perception nodes, sensor nodes and sensor gateways.

Perception Nodes

Radio frequency identification (RFID) nodes and tags are typically used as perception nodes. RFID tags could be subjected to Denial of Service (DoS – from radio frequency interference), repudiation, spoofing, and eavesdropping attacks in the communication RF channel (Alaba et al. 2017; Deogirikar and Vidhate 2017; Jing et al. 2014). Moreover, reverse engineering, cloning, viruses (the SQL injection attack in 2006), tracking, killing tag (using a pre-defined kill command to disable a tag), block tag (employing a jammer such as a Faradays' cage) and Side-Channel Attacks (SCA) through power analysis are attacks which could compromise the RFID physical systems (Xiao et al. 2009). These attacks are feasible because of the low resources of RFID devices and comparatively weaker encryption/encoding schemes. Solutions to overcome these vulnerabilities and the corresponding exploitable attacks include access control, data encryption which includes non-linear key algorithms, IPSec protocol utilization, cryptography techniques to protect against SCA (Liyanage et al. 2017; Zhao and Ge 2013), hashed-based access control (Weis et al. 2004), ciphertext re-encryption to hide communication (Kumar

Table 1 Taxonomy of attacks and solutions in IoT layers.

Layer/component	Attacks	Solutions
a. Perception layer		
Perception nodes RFID	Tracking, DoS, repudiation, spoofing, eavesdropping, data newness, accessibility, self-organization, time management, secure localization, tractability, robustness, privacy protection, survivability, and counterfeiting (Jing et al. 2014).	Access control, data encryption which includes non-linear key algorithms, IPSec protocol utilization, cryptography techniques to protect against side channel attack (Liyanage et al. 2017; Zhao and Ge 2013), Hashed-based access control (Weis et al. 2004), Ciphertext re-encryption to hide communication (Kumar et al. 2018a), New lightweight implementation using SHA-3 appointed function Keccak-f (200) and Keccak-f (400) (Kavun and Yalcin 2010)
Sensor nodes	Node subversion, node failure, node authentication, node outage, passive information gathering, false node message corruption, exhaustion, unfairness, sybil, jamming, tampering, and collisions (Zhang et al. 2015; Massis 2016)	Node authentication, Sensor Privacy
Sensor gateways	Misconfiguration, hacking, signal lost, DoS, war dialing, protocol tunneling, man-in-the-middle attack, interruption, interception, and modification fabrication (Liu et al. 2016)	Message Security, Device Onboard Security, Integrations Security (Kumar et al. 2018b)
b. Network layer		
Mobile communication	Tracking, eavesdropping, DoS, bluesnarfing, bluejacking, bluebugging alteration, corruption, and deletion (Alaba et al. 2017; Jurcut et al. 2018; Bekara 2014)	Developing secure access control mechanisms to mitigate the threats by employing biometrics, public-key crypto primitives and time changing session keys.

(*continued overleaf*)

Table 1 (Continued)

Layer/component	Attacks	Solutions
Cloud computing	Identity management, heterogeneity which is inaccessible to an authentic node, data access controls, system complexity, physical security, encryption, infrastructure security and misconfiguration of software (Horrow and Anjali 2012)	*Identity privacy* – Pseudonym (Lin et al. 2008; Lin and Li 2013; Zhou et al. 2015a), group signature (Lin and Li 2013), connection anonymization (Sen 2011; Lu et al. 2010) *Location privacy* – Pseudonym (Lin et al. 2008; Lin and Li 2013; Zhou et al. 2015a), one-way trapdoor permutation (Zhou et al. 2015a,b) *Node compromise attack* – Secret sharing (Zhou et al. 2015b,c; Roman et al. 2011), game theory (Sen 2011), population dynamic model (Zhou et al. 2015b) *Layer removing/adding attack* – Packet transmitting witness (Zhou et al. 2015a,b; Lu et al. 2010), aggregated transmission evidence (Zhou et al. 2015b) *Forward and backward security* – Cryptographic one-way hash chain (Lin et al. 2008; Lin and Li 2013) *Semi-trusted/malicious cloud security* – (Fully) homomorphic encryption (Paillier 1999), zero knowledge proof (Groth and Sahiai 2008)
Internet	Confidentiality, encryption, viruses, cyberbullying, hacking, identity theft, reliability, integrity, and consent (Akhunzada et al. 2016)	Identity Management for confidentiality (Miorandi et al. 2012), Encryption schemes for confidentiality of communication channels (Porambage et al. 2018), Cloud based solutions to establish secure channels based on PKI for data and communication confidentiality (Porambage et al. 2018)
c. Application layer	Data privacy, Tampering Privacy, Access control, disclosure of information (Zhang et al. 2015)	Authentication, key agreement and protection of user privacy across heterogeneous networks (Alaba et al. 2017), Datagram Transport Layer Security (DTLS) for end-to-end security (Garcia-Morchon et al. 2012), Information Flow Control (Roman et al. 2011)

et al. 2018a), new lightweight implementation using SHA-3 appointed function Keccak-f (200) and Keccak-f (400) (Kavun and Yalcin 2010).

Sensor Nodes

Sensor nodes, such as ZigBee, possess additional resources compared to RFID devices with a controller for data processing and interoperability of sensor components, a Radio Frequency (RF) transceiver, a memory, the power source and the sensing element (Alaba et al. 2017). Even though the sensor nodes follow a fairly secure encryption scheme due to the elevated resources, attacks such as node tampering, node jamming, malicious node injection, Sybil, and collisions (Zhang et al. 2015; Massis 2016) could exploit the vulnerabilities due to the nature of transmission technology and remote/distributed localization of them. A malware exploiting a flaw in the radio protocol of ZigBee caused a Save Our Souls (SOS) code illumination in smart Philips light bulbs as a demonstration of weakness in sensor node systems in 2016 (Frustaci et al. 2018). Additionally, GPS sensors are vulnerable to jamming or data-level and signal-level spoofing which results in Time Synchronization Attacks (TSAs) targeted on Phasor Measurement Units (PMUs) of various IoT deployments that rely on GPS for locating or navigation-based services (Khalajmehrabadi et al. 2018). Possible countermeasures for such attacks are node authentication and sensor privacy techniques.

Gateways

Sensory gateways are responsible for checking and recording various properties such as temperature, humidity, pressure, speed, and functions of distributed sensor nodes. User access, network expansion, mobility, and collaboration are provided using sensor gateways.

These channels are also vulnerable to several attacks such as misconfiguration, hacking, signal lost, DoS, war dialing, protocol tunneling, man-in-the-middle attack, interruption, interception, and modification fabrication (Liu et al. 2016). Moreover, perception layer devices could be subjected to SCA such as Differential Power Analysis (DPA), Simple Power Analysis (SPA), timing, and acoustic cryptanalysis (Deogirikar and Vidhate 2017). To ensure security with respect to sensory gateways, message security, device on board security and integrations security are suitable proposed solutions (Kumar et al. 2018b).

Network Layer

Network Layer facilitates the data connectivity to perception layer devices for accomplishing the functionality of various applications in the Application layer. Because this layer is a connectivity provider for other layers, there are probable security flaws which would compromise the operations of the entire IoT architecture.

Mobile Communication

Mobile devices are the main interfaces of human interaction for IoT technology which range from smart phones, PDAs to mini-PCs. The state-of-the-art for mobile devices are extensively resourceful with their location services, biometric sensors, accelerometer/gyroscope, extended memory allocations, etc. The connectivity options range from RF, Low Rate Wireless Personal Area Networks (LR-WPAN/IEEE 802.15.4), Near Field

Communication (NFC), Wireless Fidelity (Wi-Fi) to Bluetooth. However, these devices are vulnerable to DoS, sinkhole, bluesnarfing, bluejacking, blue bugging, alteration, corruption, deletion of data, and traffic analysis attacks (Alaba et al. 2017; Jurcut et al. 2018; Deogirikar and Vidhate 2017; Bekara 2014). In addition, mobile devices are also vulnerable for phenomena such as cloning, spoofing, and various battery draining attacks explained in Fiore et al. (2017). Even the technologies LR-WPAN, Bluetooth, and Wi-Fi are vulnerable to data transit attacks (Frustaci et al. 2018). However, current standards of mobile devices have the means to improve the security through development of secure access control mechanisms to mitigate threats by employing biometrics, public-key crypto primitives and time-changing session keys.

Cloud Computing
A Cloud computing platform is the prime entity in IoT for centralized processing and storage facilitation for IoT applications. Through cloud computing, IoT applications can enable higher computing power with unlimited storage capacity for a low cost, while maintaining versatile accessibility. Reliance on standalone dedicated server-based services is superseded by remote cloud-based server farms with outsourced services. However, outsourcing information to be stored in a remote location could raise security concerns. Privacy preservation is the most inevitable issue with cloud computing among other flaws such as physical security, anonymity, data access control failure, identity management, and direct tampering of the cloud servers (Alaba et al. 2017; Horrow and Anjali 2012). Several security solutions have been proposed in different areas for cloud including: (i) Identity privacy – Pseudonym (Lin et al. 2008; Lin and Li 2013; Zhou et al. 2015a), group signature (Lin and Li 2013), connection anonymization (Sen 2011; Lu et al. 2010); (ii) Location privacy – Pseudonym (Lin et al. 2008; Lin and Li 2013; Zhou et al. 2015a), one-way trapdoor permutation (Zhou et al. 2015a,b); (iii) Node compromise attack – Secret sharing (Zhou et al. 2015b,c; Roman et al. 2011), game theory (Sen 2011), population dynamic model (Zhou et al. 2015b); (iv) Layer removing/adding attack – Packet transmitting witness (Zhou et al. 2015a,b; Lu et al. 2010), aggregated transmission evidence (Zhou et al. 2015b) (v) Forward and backward security – Cryptographic one-way hash chain (Lin et al. 2008; Lin and Li 2013); (vi) Semi-trusted/malicious cloud security – (Fully) homomorphic encryption (Paillier 1999), zero knowledge proof (Groth and Sahiai 2008).

Internet
The term Internet stands for the holistic global networking infrastructure which scopes from private, public, academic, cooperate networks to government networks (Alaba et al. 2017). The connectivity through the Internet is formulated by Transmission Control Protocol/Internet Protocol (TCP/IP) and secured through various protocols such as Secure Socket Layer (SSL)/Transmission Layer Security (TLS), IPSec, and SSH. In IoT, however, Datagram Transport Layer Security (DTLS) is used as the communication protocol (Alaba et al. 2017). Since the Internet is accessible for everyone, the amount and nature of vulnerabilities outweigh the effectiveness of existing secure communication protocols (Jurcut et al. n.d., 2008, 2014a,b, 2017, 2018; Pasca et al. 2008) due to its implosive access capacity. Probable attacks are viruses, worms, hacking, cyber bullying, identity theft, consent, and DDoS (Alaba et al. 2017; Akhunzada et al. 2016). Countermeasures to overcome

these attacks include Identity Management for confidentiality (Miorandi et al. 2012), Encryption schemes for confidentiality of communication channels (Porambage et al. 2018), Cloud-based solutions to establish secure channels based on Public Key Infrastructure (PKI) for data and communication confidentiality (Porambage et al. 2018).

Application Layer

As illustrated in Figure 2, possible applications for IoT are expanded into every industry available in the current era, in addition to myriads of non-industrial applications developed for automation purposes. In general, feasible attacks on the IoT application layer could be represented in two forms. They are software-based and encryption-based attacks. In the software attacks, most attacks are based on malicious software agents, apart from the phishing attacks, where the attacker reveals the authentication credentials of the user by impersonating a trusted authority. Malware, worms, adware, spyware, and Trojans are highly probable occurrences with the heterogeneity of IoT applications and their broader services (Deogirikar and Vidhate 2017). Encryption-based attacks are the approaches taken to exploit the procedural nature of the cryptographic protocols and their mathematical model through extensive analysis. Cryptanalytic attacks, ciphertext only attacks, known plaintext attacks and chosen plaintext attacks exemplify such possible threats (Zhang et al. 2015).

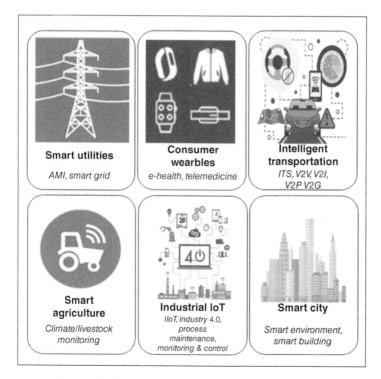

Figure 2 IoT applications.

There are several solutions proposed in the literature for the security of IoT applications such as Authentication, key agreement and protection of user privacy across heterogeneous networks (Alaba et al. 2017), DTLS for end-to-end security (Garcia-Morchon et al. 2012), and Information Flow Control (Roman et al. 2011). The countermeasures for software-based authentication should be taken for mitigating attacks such as phishing attacks, through the verification of the identity of malicious adversaries before proceeding.

Smart Utilities – Smart Grids and Smart Metering

Smart Grids are the future of energy distribution for all industrial and residential sectors. IoT plays a major role in smart grids for establishing the communication and monitoring protocols with consumers of energy. Smart grid is a decentralized energy grid with the ability to coordinate the electricity production in relation to the consumption or consumption patterns of the consumer. It has a monitoring technology Advanced Metering Infrastructure (AMI)/smart metering/net metering, which can measure and update the power consumption parameters to both entities in real time (Kouicem et al. 2018). Additionally, smart grids are incorporating renewable energy sources commissioned in the vicinity of the consumer to cater the bidirectional energy flow for mitigating energy deficiencies (Alaba et al. 2017).

Figure 3 illustrates a Smart Grid Architectural Model (SGAM) proposed by the coordinated group of European Committee for Standardization – European Committee for Electrotechnical Standardization – European Telecommunications Standards Institute (CEN-CENELEC-ETSI), which offers a framework for smart grid use cases

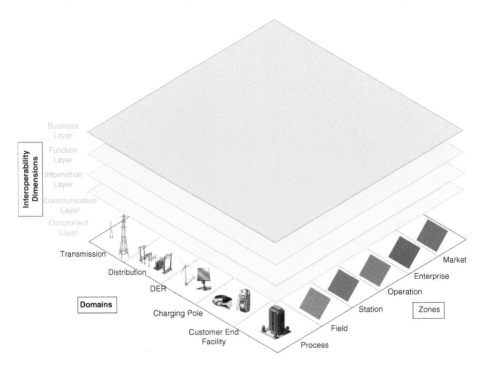

Figure 3 Smart grid architectural model.

(CEN-CENELEC-ETSI Smart Grid Coordination Group 2014). This architecture formulates three dimensions which amalgamate five functional interoperability layers with energy sector domains and zones which account for power system management (Ahmad et al. 2018b). This holistic framework is capable of reinforcing the design stages of the smart energy systems. The IoT technologies could be amalgamated with the SGAM framework to establish the bi-directional communication.

All the monitoring applications are developed with IoT infrastructure, with grid controlling access granted to the grid, controlling officers for pursuing configurations while the consumers can only visualize the consumption details via a mobile device. The information circulated through the AMI may pose a privacy concern for consumers for disseminating information regarding their habits and activities, where the impact could be severe for industries. Due to the heterogeneous nature of communication equipment deployed with IoT, and rapidly increasing population and industries, it would cause scalability issues for security. Smart grids are distributed across the power serving area and are, therefore, exposed to adversaries.

As the energy distribution system is the most critical infrastructure that exists in an urban area, the tendency to convert prevailing wired power-line communication (sending data over existing power cables) based controlling and monitoring channels to the wireless medium, with the introduction of IoT technologies, would expose the entire system to unintended security vulnerabilities. The intruders, using the proper techniques, could perpetuate AMI interfaces stationed at every household or industrial plant. Once access is granted to the hostile operators, potential outcomes can be devastating from disrupting the level of energy flow from a local grid substation to overloading the nuclear reactor of a power station. The availability of the grid could be compromised from IP spoofing, injection, and DoS/DDoS attacks (Kouicem et al. 2018). Thus, access controlling for devices used in AMI and grid-controlling systems should be secured with extra countermeasures.

Consumer Wearable IoT (WIoT) Devices for Healthcare and Telemedicine

IoT-based healthcare systems are the most profitable and funded projects in the entire world. This is mainly due to the higher aggregate of aging people and the fact that health is the most concerning aspect of human life. A sensory system embedded with actuators is provided for individuals to use as a wearable device (i.e. wearable Internet of Things [WIoT] device), illustrated in Figure 4. A WIoT device can be used for tracking and recording vitals such as blood pressure, body temperature, heart rate, blood sugar, etc., (Kouicem et al. 2018). This data can be conveyed and stored in a cloud as a Personal Health Record (PHR) to be accessed by the user and the assigned physicians.

Since the data handled in IoT-based healthcare is personal, privacy is the most demanding security issue. Hence, the access control mechanism for wearable devices as well as for PHRs must be well secured. However, employing strong crypto primitives for enhancing the authentication protocols of PHRs is possible as they are also stored in cloud environments. Hence, the same privacy concerns presented in section titled "Cloud Computing" apply. Moreover, a method for assuring anonymity of patients should be developed in case the PHRs are exposed to external parties, because they are stored in Cloud Service Providers (CSPs). Wearable devices also face the resource scarcity issues for battery power, memory, and processing level (Kouicem et al. 2018). Thus, a lightweight access control protocol should be employed. Similar to all other

Figure 4 WIoT devices.

IoT applications, heterogeneous wearable devices produced by different manufacturers would employ diverse technologies for developing communication protocols. Thus, developing a generic access control policy would be extremely challenging.

Intelligent Transportation

Intelligent Transportation Systems (ITS) are introduced to improve transportation safety and decrease traffic congestions while minimizing environmental pollution. In an ITS system, there are four main components; vehicles, roadside stations, ITS monitoring centers and security systems (Kouicem et al. 2018). All information extracted from vehicular nodes and roadside stations are conveyed to the ITS monitoring center for further processing, while the security subsystem is responsible for maintaining overall security. The entire system could be considered as a vehicular network, while the communications are established between Vehicle-to-Vehicle (V2V), Vehicle-to-Infrastructure (V2I), Vehicle-to-Pedestrian (V2P) and Vehicle-to-Grid (V2G) (Kouicem et al. 2018). These communication links are implemented using technologies like RFID and Dedicated Short Range Communication (DSRC) for launching a large Wireless Sensor Network (WSN) (Alaba et al. 2017). The vehicular nodes and the entire data storing and monitoring infrastructure form a viable IoT deployment.

Figure 5 illustrates an ITS model, which enables communication among vehicular entities traveling through different mediums (airborne, land, and marine) with various technologies such as satellite, mobile, Wireless Local Area Network (WLAN), etc. Such a system would enable services like real-time updated navigation, roadside assistance, automated vehicular diagnostics, accident alerting system and self-driving cars (Rajakaruna et al. 2018). Thus, massive divergence in the applicability of ITS deployment raises the requirement for a ubiquitous wireless connectivity with access points.

As mentioned above, a larger number of entry points to a vehicular network makes it vulnerable to diverse attacks which can be targeted toward many sources

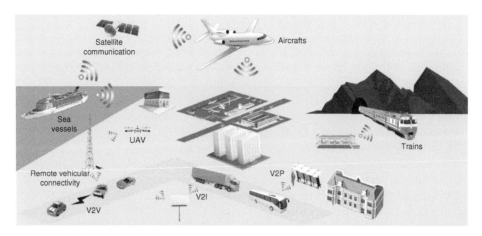

Figure 5 Intelligent transportation System.

(Kouicem et al. 2018). At the same time, the privacy of drivers should be ensured from external observers, though drivers are not participating in any authentication activity. Authentication mechanisms are initiated between V2V interfaces where they can be exploited by an invader impersonating another vehicle or a roadside station. Therefore, a mechanism to verify the identity of the vehicles or roadside stations should be developed as a Trusted Third Party (TTP) with the authentication mechanism.

In some V2V communication systems, an On Board Diagnostic (OBD) unit is utilized to extract information directly from the Engine Control Unit (ECU) (Alaba et al. 2017). The OBD port could be used to manipulate the engine controls of a vehicle and could then be remotely accessed via the systems being developed. Thus, securing the access to the OBD port is vital.

Smart Agriculture

Agriculture is the most crucial industry in the world as it produces food and beverages by planting crops such as corn, rice, wheat, tea, potatoes, oats, etc. With the rapid worldwide population growth accounting for resource depletion, pollution, as well as the scarcity of human labor; agriculture is becoming an demanding industry. Automation is the most probable alternative for improving the effectiveness of the agriculture industry. Thus, IoT could play a vital role in such automation. IoT infrastructure could be deployed to perform climate/atmospheric, crop-status monitoring and livestock tracking. Climatic sensors, water/moisture level sensors and chemical concentration/acidity sensors along with visual sensors could be deployed for crop-status monitoring, while automated water and fertilizer dispersing mechanisms are in place within the bounds of the plantation. Additionally, livestock tracking is another aspect of smart farming implemented through the deployment of Local Positioning Systems (LPSs) on farm animals.

This type of a smart system would provide benefits such as the ability to utilize the fertilizer and water usage while maximizing crop production while mitigating the effects of climatic deficiencies. The fruition science and "Hostabee" are two live cases of smart agriculture solutions used currently by the plantation industry (Rajakaruna et al. 2018).

Because of the diverse nature of sensor devices used in smart agriculture applications, integrating them into a holistic system may raise concerns about the compatibility of technologies among the variety of manufacturers and those protocols in which communication is established. As the plantations or fields are extending to larger areas, the number of IoT-enabled sensory systems to be deployed will be immense. Handling the data flow of such a large number of individual sensors with different data representations dispersed throughout a broad geographical region exerts the requirement for a communication technology with a higher coverage and moderate data rates which could not be satisfied by low-range communication technologies such as Bluetooth or NFC. However, DSRC would be a suitable technology to create a WSN with smart agriculture sensors, as it is compatible with ITSs.

As the IoT devices are disseminated across a larger geographical extent, the probability of any IoT device being compromised is high as they are exposed. Perception level attacks are probable with these devices as they are sensory nodes and would have limited resources for both processing and storing information. The spoofing, impersonation, replay and Man-in-the-Middle (MiM) attacks are probable with this application (Srilakshmi et al. 2018). This required the need for a proper authentication scheme as all perception level attacks could be mitigated with use of such a countermeasure.

Industrial IoT (IIoT)

M2M-based automation systems are quite common for industries such as oil and gas manufacturers. These industries are vast and the machinery employed is massive, expensive, and poses a significant risk to machine operators. Functions such as oil exploration by drilling, refining, and distributing are all conducted using automated machinery controlled through Programmable Logic Controller (PLC) based on SCADA systems. Although, the current M2M infrastructure is ideal for controlling the machinery, remote monitoring and accessibility is limited while a proper data storage and processing mechanism for decision making is not yet available. Thus, the requirement for IoT arises to improve operational efficiency by optimizing control of robots, reducing downtime through predictive and preventive maintenance, increasing productivity and safety through real- time remote monitoring of assets (Rajakaruna et al. 2018). IoT sensor nodes could be deployed at the machinery while monitoring tools could be integrated without affecting the operation of SCADA systems. Hence, SCADA systems could be optimized to enhance productivity.

The Smart Factories term is an adaptation of IIoT, introduced as "Industry 4.0" to represent the Fourth Industrial Revolution (4IR) (Rahimi et al. 2018). This standard signifies a trend of automation and data exchange in manufacturing industries which integrate Cyber-Physical Systems (CPS), IoT, and Cloud Computing based Data analytics (Rajakaruna et al. 2018; Rahimi et al. 2018). The interoperability, information transparency, technical assistance and decentralized decision making are the design principles of Industry 4.0 standard. BOSCH has developed connected hand-held tools which could monitor location, current user, and task-at-hand; analyzed and utilized for improving the efficiency in industrial labor (Rajakaruna et al. 2018). Thus, the deployment of IoT across industry is imminent.

The security of industrial applications is a major concern, as any hostile intrusion could result in a catastrophic occurrence for both machinery and human operators. The SCADA systems are no longer secure (e.g. Considering the recent events (Kim and Lee 2017)) due to their isolated localization and operation. However, main controlling functions are maneuvered within the control station located inside the industrial facility, while limited egress connectivity is maintained via satellite links with VSAT (Very Small Aperture Terminal) or microwave in the case of offshore or any other industrialized plants of such nature.

Due to their offline nature, the probability of any online intrusion is minimal. Though, any malicious entity such as a worm or a virus injected to the internal SCADA network could compromise the entire factory. Once inserted into the system, the intention of the malicious entity would be to disrupt the operations of the facility and its machinery. Thus, limiting the possibility for any malicious insurgence from the internal network and employing effective Intrusion Detection System (IDS) to detect malicious entities, would be the most suitable countermeasure for this application.

Smart Buildings, Environments, and Cities

Smart city is a holistically expanded inclusion of smart buildings and smart environments along with other smart automation systems formed to improve the quality of life for residents in a city. This is, in fact, the most expandable version of any IoT application in terms of cost for infrastructure deployment and geographical extent. In this concept, as shown in Figure 6, sensors are deployed throughout the building, environment, or the city for the purpose of extracting data of varied parameters such as temperature, humidity, atmospheric pressure, air density/air quality, noise level, seismic detection, flood detection and radiation level. CCTV streams and LPSs would be a valuable input for smart building and smart cities to detect intrusions, monitor traffic and emergencies. All other smart systems explained in the previous sections are, in fact, subsystems of a functional smart city.

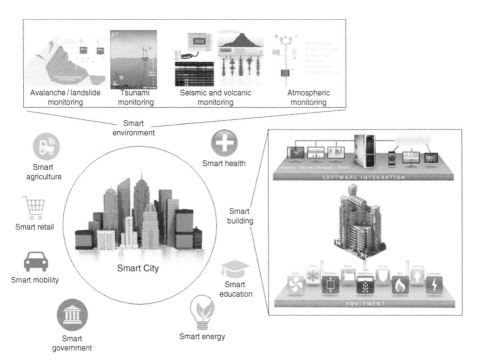

Figure 6 Smart city concept.

Due to various parameters to be gathered from the sensory acquisitions, heterogeneity is immense and the implementation is arduous (Kouicem et al. 2018). At the same time, management of the gathered Big Data content is not scalable. Thus, providing security for all the applications in smart cities would be extremely challenging. Most of the Big Data content extracted from the sensors is forwarded to clouds through M2M authentication. Because of large data transmissions, cryptographic schemes should be lightweight and the authentication mechanism should be dynamic. DoS or DDoS attacks are most probable and could be mitigated with a strong authentication mechanism (Alaba et al. 2017). Individual sensors could be compromised resulting in the initiation of fake emergencies and access control methods should be improved to avoid such inconsistencies at the sensor level.

Desai and Upadhyay (2014) introduces applications of IoT with specific focus on smart homes. The study presented in Desai and Upadhyay (2014) claims that although smart homes are offering comfortable services, security of data and context-oriented privacy are also a major concern of these applications. The security and privacy issues in IoT applications have also been studied in Kumar and Patel (2014).

Authentication and Authorization

Authentication and access control mechanisms hold a great deal of significance in IoT. Without a proper mechanism for access control, entire IoT architecture could be compromised, as IoT devices are highly reliant on the trustworthiness of the other components to which they are connected. Thus, a proper access control mechanism is paramount to mitigate the flaws in the current IoT infrastructure.

Access control mechanisms are comprised of two stages (Figure 7) (Alaba et al. 2017): (i) Authentication and (ii) Authorization.

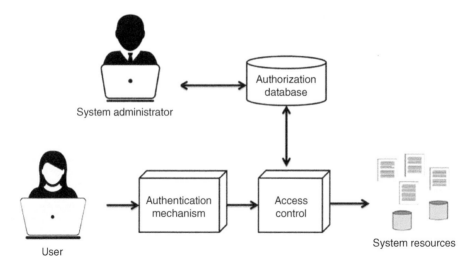

Figure 7 Typical access control System.

Authentication

Authentication is the process of verifying the identity of an entity (Kim and Lee 2017). The entity to be verified could either be human or a machine. Authentication is the first phase of any access control mechanism which can determine the exact identity of the accessing party in order to establish the trust of the system. In most cases, authentication is initiated between a human and a machine in a process to log into the internet banking portal by entering the credentials. However, in this scenario, the access-seeking entity does not have a guarantee regarding the identity of the access granting entity. In order to overcome this concern, mutual-authentication should be established between the entities, by verifying the identity of the access-granting entity with the involvement of a TTP, such as a Certificate Authority (CA) (Kim and Lee 2017). CAs are globally recognized institutions which are responsible for issuing and maintaining secure digital certificates of web entities registered under them. These certificates are imperative for the operation of all modern day authentication protocols such as SSL/TLS, IPSec, and HTTPS.

The process of authentication is merely facilitating credentials of an entity to the access granting system, which are unique to that entity and could only be possessed by them. This mechanism could be enabled with or without a TTP. The credentials used are often categorized as factors. The authentication schemes' accuracy and efficiency depend on the number of factors engaged in the mechanism. The types of factors are listed below.

- Knowledge factor – passwords, keys, PINs, patterns
- Possession factor – Random Number Generators (RNG), ATM card, ID card
- Inherence factor – Biometrics such as fingerprint, palm print, iris, etc.

Recent innovations in embedding biometric sensors to smart handheld devices have enabled the possibility of using multi-factor multi-mode (if more than one bio metric is used for verification) Human-to-Machine (H2M) authentication protocols for IoT devices. However, Machine-to-Machine (M2M) authentication can only be conducted using cryptographic primitives. Moreover, including strong cryptographic primitives (PKI, Hashing, Timestamps, etc.) for the authentication protocols involved is crucial in order to ensure data confidentiality, integrity, and availability, as the credentials being conveyed are highly sensitive and unique for the authenticating entity.

Authorization

Authorization is the process of enforcing limits and granting privileges to the authenticated entities (Stamp 2011). In simple terms, this is determining the capabilities of an entity in the system. In order for an entity to be authorized for performing any action, the identity of that entity should be verified first through authentication. According to Figure 7, an administrator usually configures the authorization database for granting access and rights to system resources. Each resource is assigned with different rights such as read, write, and execute. Depending on the level of authorization (clearance) being set by the administrator, each authenticated entity can perform different actions on resources. A typical access control system has a policy for granting rights. These policies could vary from Discretionary Access Control (DAC), Mandatory Access Control (MAC) or a Multi-Level Security (MLS) model such as Role Based Access Control

(RBAC) (Stamp 2011). In DAC, the administrator specifies the rights, while in MAC there are rules set by the system for assigning rights for subjects. Clearances are granted according to the role of the authenticated entity (Roles: course coordinator, lecturer, or student in a university) in RBAC.

Authentication at IoT Layers

Authentication is the most critical security requirement in IoT for preserving user identity and mitigating the threats mentioned in the previous sections. With each IoT application, more hardware devices are introduced to be integrated to the IoT network. The authentication is the mechanism used to ensure the connectivity of those components to the existing ones. Authentication mechanisms involve cryptographic primitives for transmitting credentials securely. The strength of the scheme is entirely dependent on the crypto primitives being used. However, developing a generic solution would not be feasible, because differing layers attribute different requirements in IoT and the resources available for processing, memory, and energy are diverse. Therefore, we will discuss the authentication requirements for each layer.

Perception Layer

The Perception layer includes all the hardware devices or the Machines to extract data from IoT environments. In most cases, the authentication initiates as M2M connections. Thus, in this layer, authentication could be conducted either as peer authentication or origin authentication (Alaba et al. 2017). In peer authentication, validation occurs between IoT routing peers, preliminary to the routing information exchanging phase, while validating the route information by the connected peer IoT devices with its source is origin authentication. This method enhances the security in M2M communication. Though as mentioned previously, devices in Perception layer are inheriting inadequate resources for generating strong cryptographic primitives.

Perception Nodes These nodes are distributed across the IoT environment. Mostly, they are RFID tags and RFID readers/sensors, where few RFID tags are connected to a RFID reader. The connection establishment between RFID tags and the reader does not involve an authentication mechanism and would be vulnerable if the RFID tags could be cloned. An Identity Based Encryption (IBE) scheme was proposed by Kouicem et al. (2018) for establishing secure communication channels between RFID tags. Due to resource scarcity, an authentication protocol might be implemented using techniques such as Elliptic Curve Cryptography (ECC) based Diffie-Hellman (DH) key generation mechanism (Alaba et al. 2017). The generated keys, once they are transmitted to each end, could be used as the shared symmetric key for information transferring via the medium securely (Stamp 2011). However, MiM attacks are still feasible and could be solved by employing the ephemeral DH method; changing the ECC DH exponents for each connection establishment as a session key.

Sensor Nodes and Gateways Sensor nodes face the similar security flaws as the perception nodes. Thus, deploying a proper authentication scheme could eliminate the possibility of being exposed at a very low level. However, sensors are much more intelligent and resourceful than perception nodes. Hence, M2M authentication could be established as peer authentications and the origin authentication could be established via the sensor gateway. Similarly, to the perception nodes, ECC-based DH key exchange would be ideal for sensor nodes, where the ephemeral exponents are facilitated by the sensor gateway acting as a TTP. Identity validation of the sensor gateway should be conducted prior to any data transfer. Even though using certificates for identity determination is not practical, a similar parameter such as a serial number could be used when registering the sensor node in the IoT environment and all the identities are stored in the sensor gateway for validation. Sensor gateway should also possess a unique identity for mutual authentication to be established between the sensor node and the gateway. Moreover, countermeasures such as integrity violation detection (using Hashed Message Authentication Code – HMAC or Cipher Block Chaining Mandatory Access Control – CBC-MAC) and timestamps should be employed with the authentication protocols involved.

Network Layer

IoT network layer is integrated on top of the existing TCP/IP internet protocols. In this section, we discuss the significance of the authentication for the components of the network layer.

Mobile Communication   Security for mobile communication at the network layer was not a critical necessity until the inception of IoT, as most of the mobile applications were relying on the inbuilt security protocols of the corresponding mobile technology (such as Global System for Mobile Communication – GSM, Wireless Code Division Multiple Access – WCDMA, High Speed Packet Access – HSPA or Long Term Evolution – LTE). With IoT, inbuilt authentication schemes are no longer foolproof, considering the potentiality for integrating technologies embedded in addition to the mobile technologies. Current security level and comprised resources (such as processor, memory, and operating system) in mobile devices are adequate for designing tamper resistance authentication protocols at the network layer (Jurcut et al. 2018). However, the existing key generation algorithms used in TCP/IP protocols for generating large and costly asymmetric keys (Rivest Shamir Adleman (RSA), ElGamal, or Paillier), are still not feasible to be used with mobile devices. Thus, generating unbreachable and lightweight keys would be the most challenging task in mobile communication.

Yao et al. (2015) proposed a lightweight no-pairing Attribute Based Encryption (ABE) scheme based on ECC that is designed for handheld devices. Even though the improved mathematical complexity and linear relationship of the number of attributes with computational overhead are improving the robustness of the proposed ABE scheme, scalability of the scheme would be highly questionable with enormous amount of IoT devices. IBE schemes are also adoptable, if taking the identity parameter as the mobile number or the user Social Security Number (SSN) for developing the authentication scheme integrating with ECC (Kouicem et al. 2018).

Current mobile devices include different biometric sensors for extracting biometrics such as fingerprint, iris, facial, and voice imprints. Biometrics can be used as unique keys that could be used for authentication and can be employed with H2M authentication. As the majority of the mobile devices at operation in an IoT environment are handled by a human user, the authentication design and key generation could be based on biometrics. The security of the biometrics schemes could be enhanced using several biometrics (multi-mode) integrated into multi-factor authentication schemes. These biometrically generated keys could be used as the signatures of each mobile entity, for the verification of their identities and, for conveying a secure session key among the communicating parties with proper encryption schemes. Additionally, authentication credentials should be checked for probable integrity violations in order to avoid MiM attacks.

Cloud Computing   Clouds are the storage facility of IoT architecture and they are quite resourceful in terms of memory and processing. Thus, authentication should employ strong keys that are generated using public-key algorithms such as RSA or ElGamal, which are inviolable cryptographic primitives if the executing authentication mechanism are computationally feasible with the available resources. A symmetric key (advanced encryption standard [AES], triple data encryption standard (TDES), etc.) to be used in data transferring between the IoT devices and the cloud could be generated and shared among the entities engaged in a communication. Existing CAs could be used to validate the identity of the parties involved in communication via mutual authentication schemes for establishing the trust.

However, the main concern in cloud computing is privacy of the user data. A strong authentication scheme does not ensure the misuse of information by the CSP. Thus, approaches such as blockchain and homomorphism should be considered for enhancing the privacy. The authentication schemes would be more secure in these schemes, as blockchain support pseudonymity (the nodes are identified from hashes or public keys – CA not required and simplify the authentication scheme) and the homomorphism facilitates an additional layer of encryption to secure the communication (Kouicem et al. 2018).

Authorization techniques in clouds should be also be considered, as accessing the information in the clouds is vital for the IoT design. Existing access control mechanisms such as RBAC and MAC are no longer scalable or interoperable. Thus, a novel method called Capability-Based Access Control (CapBAC), which uses capability-based authority tokens to grant privileges to entities was proposed by Kouicem et al. (2018).

Internet   Even though authentication in most applications on the internet is pursued by either SSL or IPSec protocols, IoT uses the DTLS as its communication protocol. However, the dependability of CAs for validating authentication parties still exists. Chinese CA WoSign was issuing certificates for false subjects in 2016, leaving an easier access to systems through wrongfully validated certificates for the attackers (Kim and Lee 2017). This happens when the trust of the system is centralized into a single entity. Thus, distributed access control schemes such as OpenPGP (widely used for email encryption) have formidable odds in succeeding in IoT infrastructure. Hokeun et al. in Kim and Lee (2017) introduces a locally centralized and globally distributed network architecture called Auth. Auth is to be deployed in edge devices for providing authorization services for locally registered entities, by storing their credentials and access policies in its database. Since the other instances of Auth are being distributed globally in the network, this maintains the trust relationships among them for granting authorizations for IoT devices acting as a gateway. Providing a solution to the trust issue of CAs is the main concern for the Internet, as the security level in existing protocols is quite adequate.

Application Layer
Heterodyne nature of the IoT predicates the requirement for different approaches of access control mechanisms for different applications. Most of the existing application

layer H2M authentication schemes are two-factor authentication schemes, while the M2M ones are web based such as in SSL. The applicability and effectiveness of existing schemes is evaluated for each IoT application, since a generic solution is infeasible.

Smart Utilities – Smart Grids and Smart Metering When using proper techniques, the intruders could perpetuate AMI interfaces stationed at every household or industrial plant. Once the access is granted to the hostile operators, potential outcomes could be devastating, from disrupting the level of energy flow from a local grid substation to overloading the nuclear reactor of a power station. Thus, access to the smart grids should only be granted to the local grid operator and the monitoring center, avoiding any interfacing through the AMI access points. Local-grid operator authentication mechanism could be employed with a two-factor authentication scheme with a username, password, and RNG. A biometric scheme could be used depending on the availability of biometric extraction devices. As the controlling access is given to the operator, an authorization scheme such as RBAC should be employed, as scalability concern does not exist due to the limited number operators available for a smart grid. An M2M authentication interface is executed between the smart grid and monitoring center for information access. Existing security protocols such as SSLs could be used for authentication.

The access to AMI meter could be given to the residential consumer for the purpose of monitoring statistics. This access could also be based on two-factor authentication or biometrics as access is only given to read the data and not to manipulate it. Smart Grid has the ability to access the AMI meter through M2M authentication and should be secured with strong crypto primitives for preventing any MiM information extraction. Certificates should be issued to all the smart grids by a CA and identities should be validated preferably via a mutual authentication scheme when establishing a grid-to-grid communication channel. A mechanism should be embedded with an authentication protocol to validate the AMI units for detecting possible tampering scenarios.

Consumer Wearable IoT (WIoT) Devices for Healthcare and Telemedicine In a telemedicine system, the parties to grant access are solely the patients and their physicians. Thus, access should be limited. Authentication protocols should be always H2M when accessing the information, while M2M authentication operates when updating sensory information from wearable devices to the server. Access to the patient should be granted in a two-factor authentication scheme if a PC is being used for access. If the patient is using a mobile device to access the server, three-factor authentication scheme could be employed by integrating biometrics. Though, storing all the credentials including biometric templates at the authentication database would not be scalable with expanded healthcare services. Still, authentication should be thorough because accessing PHRs is private and confidential. Cloud servers' access to physicians could be granted from a two-factor authentication scheme. Storing and accessing PHRs at the cloud could be secured using the blockchain concept to counter any obvious privacy concerns with CSPs. An IBE scheme could be adopted to enhance the message transferring in the authentication protocol.

Intelligent Transportation and Logistics Since the vehicles attribute high mobility, the connectivity of an established wireless link across vehicular entities may vary rapidly.

Hence, the availability of a consecutive/fixed inter-link would be uncertain. Thus, dynamic handover mechanisms should be adopted between vehicular nodes for maintaining a consistent connection with each communicating vehicular node. Hence, those handover-based connections might require a lightweight approach for authentication as they are highly dynamic.

Each vehicle should have an Identity-based private key (embedded with its credentials – chassis no., registration no., manufacturer, model, etc.). However, the keys should be generated from an IBE or ABE lightweight mechanism unlike public-key encryption schemes which require costly resources to generate. Authentication protocols are more likely to be M2M mechanisms, where the machines are the vehicles. Therefore, verifying the identity of each vehicular node engaged in communication is paramount to avoid malicious node invasions through a TTP-based identity verification. An ECC-based ephemeral DH scheme could be employed for establishing a shared symmetric session key once the authentication phase is concluded following validation of vehicle identities. All V2V, V2I, V2G, and Vehicle to Cloud (V2C) connections could be implemented in the same manner.

Additional to the approaches discussed earlier, Software Defined Networks (SDN) and Blockchain concepts are highly recommended to ensure the security requirements in the Application layer (Alaba et al. 2017; Kouicem et al. 2018).

Smart Agriculture As mentioned in the previous section regarding attacks, agriculture IoT devices intrinsically require a lightweight authentication protocol as they are vulnerable to external intervention and sparse resources with perception level nodes. With a lesser resourced platform, implementing a mutual-authentication scheme would be questionable. In Srilakshmi et al. (2018), a logic based on the Burrows-Abadi-Needham (BAN) modal logic was proposed and tested using Automated Validation Information Security Protocol Application (AVISPA) for verification, which was validated for MiM and replay attacks. However, a frequently changing session key usage is a vital necessity to prevent perception level attacks. This session key establishment could be employed with a technique such as ephemeral DH or ECC for lesser resource utilization.

IIoT Most IIoT processes are M2M due to their automated platforms. Further, IIoT process operations are continuous as their work cycles might extend to hours. With the amount of controlling data flowing through the communication channels, simultaneous authentication of each sensory node might lessen the efficiency of the entire smart factory. Thus, a methodology for a scheduled authentication scheme, which does not affect industrial performance, should be established. However, the authentication at each sensory node could be evaded, as there could be hundreds of minor sensors connected to massive machines, which would not be feasible for authentication of each node frequently. Only the control information transfer of machines that is subject to authentication, as a single controlling command, could continuously last for hours. These authentication phases could employ heavy cryptographic primitives as there is no scarcity of computational resources.

Smart Buildings, Environments, and Cities Designing a generic authentication protocol for smart cities is not practically feasible. This concept is formed from an entity such as a smart home, inadequate security measures could compromise the privacy of users at any level of use (Williams et al. 2017). However, this application could be visualized from the perspective of the three layers in IoT. Similar methods proposed for access control in the perception layer could be adopted for the sensory system in smart environments. The Network layer accompanies all the internet integrated data connections and routing devices along with severs (clouds), in addition to the mobile devices. Mobile devices could use three-factor authentication schemes incorporated with web-based SSL or DTLS protocols, while cloud servers and routing nodes could be authenticated with cryptographically generated keys. Authentication protocols in smart cities are likely to change with the requirements and applications, as all other applications mentioned under this section are sub-applications of a smart city.

Other Security Features and Related Issues

IoT systems have their own generalized features and requirements regardless of the diversified nature of its applications such as heterogeneity, scalability, Quality of Service (QoS)-aware, cost minimization due to large-scale deployment, self-management including self-configuration, self-adaptation, self-discovery, etc. The last, but not least, general feature/requirement of an IoT system is to provide a secure environment to gain robustness against communication attacks, authentication, authorization, data-transfer confidentiality, data/device integrity, privacy, and to form a trusted secure environment (Borgia 2014). IoT systems are fundamentally different from other transitional WSN systems (Xu et al. 2017) in many ways. (i) The diversity of the types of applications, the capabilities, and attributes of the IoT devices and deployed environments (ii) The holistic design of the system is mostly driven by the applications and it is essential to consider who the users are, what are the purposes and expected outcomes of the applications, etc. An IoT system is required to manage a large variety of devices, technologies, and service environments as the system itself is highly heterogeneous, where the connected IoT devices or equipment can range from simple temperature sensors to high-resolution smart cameras. The communication, computing, and power capability of each device can be unique and unique from others. These resource and interoperability constraints limit the feasibility for a standard security solution.

The Simplified Layer Structure

The traditional Open Systems Interconnection (OSI) has Seven layers: (i) The Physical Layer (Layer 1) is responsible for the transmission and reception of wire level data. (ii) The Data Link Layer (Layer 2) is responsible for link establishment and termination, frame traffic control, sequencing, acknowledgment, error checking, and media access management. (iii) The Network Layer (Layer 3) is implemented for routing of network traffic. (iv) The Transport Layer (Layer 4) is responsible for message segmentation, acknowledgement, traffic control, and session multiplexing. (v) The Session Layer (Layer 5) is responsible for session establishment, maintenance, and termination. (vi) The Presentation Layer (Layer 6) is responsible for character code translation, data conversion, compression, and encryption. (vii) The Application Layer (Layer 7) includes resource sharing, remote file access, remote printer access, network management, and electronic messaging (email). Since IoT systems normally have a huge variety, ranging from the choice of the hardware to the type of applications, the traditional seven network layers are simplified to three layers: perception layer, networking layer and application layer, as shown in Figure 1. The perception layer can be seen as the combination of the traditional physical layer and the MAC layer. It can include 2D bar code labels and readers, RFID tags and reader-writers, camera, GPS, sensors, terminals, and sensor network. It is the foundation for the IoT system (Wu and Lu 2010). The networking layer is responsible for the data transmission and communication inside the system and with the external Internet. It should be aware of the different underlying networks no matter whether it is wired, wireless, or cellular. It can provide support for different communication modes including base station, access point based or Machine-to Machine type based. The application layer provides services to the end users and collects data from different scenarios. IoT has high potential to implement smart and intelligent application for any scenario in nearly every field. This is mainly because IoT can offer both (i) data collection through sensing over natural phenomena, medical parameters, or user habits and (ii) data analysis and predictive modeling for tailored services. Such applications will cover aspects including personal, social, societal, medical, environmental, and logistics, having a profound impact on both the economy and society (Borgia 2014). The perception and network layer together are considered the foundation for the whole IoT system. Together, these two layers provide the backbone and fundamental infrastructure of an IoT system. However, the architecture design and detailed implementation can normally only be confirmed after knowing the application layer design. Where the system will be deployed, what size the field will be and what kind of data will be collected are all issues involved in the applications, but highly affect the decision making on the perception layer and network layer.

The Idea of Middleware

Researchers from academia and industry are exploring solutions to enhance the development of IoT from three main perspectives: scientific theory, engineering design, and user experience (Feki et al. 2013). These activities can enrich the technologies for IoT, but also increase the complexities, when implementing such a system in the real world. For this reason, the concept of IoT middleware has been introduced and many systems are already available (Contiki n.d.; Brillo n.d.; Tinyos n.d.; Openwsn n.d.; Riot n.d.). However, when describing the formal definition for IoT middleware, researchers have different understandings. In some circumstance, IoT middleware is equivalent to IoT Operating System (OS). In general, middleware can simplify and accelerate a development process by integrating heterogeneous computing and communication devices, as well as supporting interoperability within the diverse applications and services (Porambage et al. 2019). Most existing implementations for middleware are designed for WSN and not for a service-oriented IoT system. Though, certain IoT-Specific middleware exists (Perera et al. 2014; Zhou 2012). In reality, middleware is often used to bridge the design gap between the application layer and the lower infrastructure layers. The requirements for middleware service for the IoT can be categorized into functional and non-functional groups. Functional requirements capture the services or functions such as abstractions and resource management (Xu et al. 2014). Non-functional requirements capture QoS support or performance issues such as energy efficiency and security (Ngu et al. 2017).

The Internet of Everything (IoE) aims to connect the objects, buildings, roads and cities and also to make the platform accessible. However, this feature will significantly increase the vulnerabilities of the system and, the inherent complexity of the IoT further complicates the design and deployment of efficient, interoperable, and scalable security mechanisms. It has been clearly stated that all typical security issues (authentication, privacy, nonrepudiation, availability, confidentiality, integrity) exist across all layers and the entire function box to a certain degree. However, when implementing security solutions, different layers of a variety of systems will have specialized priorities (Rajakaruna et al. 2018).

An essential task of the middleware is to provide secure data transmission between the upper and lower layers. For inner system communication, it should guarantee that the data passed to the application layer from the infrastructure is safe and reliable to use – integrity. Integrity in this scenario involves maintaining the consistency, accuracy, and trustworthiness of data over the transmission. Conversely, the middleware should also ensure that the control comments and queries from the applications/end users are verified and it is harmless for the system to take actions – non-repudiation. Non-repudiation features ensure that users cannot deny the authenticity of their signature for their documents and footprints for their activities. In addition, the middleware must protect the data transmission and information exchange between the upper and lower layers from illegal external access by any arbitrary user. The data must not be disclosed to any unauthorized entities – confidentiality.

Cross-Layer Security Problem

It has been frequently argued that although layered architectures have been a great success for wired networks, they are not always the best choice for wireless networks. To address this problem, a concept of cross-layer design is proposed and it is becoming popular. This concept is based on an architecture where different layers can exchange information in order to improve the overall network performance. A substantial amount of work has been carried out on state-of-the-art cross-layer protocols in the literature recently (Srivastava and Motani 2005). Security can be considered as one of the most critical QoS features in IoT systems. Wireless broadcast communication is suffering security risks more than others while multi-hop wireless communication is in a worse situation, as there is no centralized trusted authority to distribute a public key in a multi-hop network because of the nature of its distribution. Current proposed security approaches may be effective in a particular security issue in a specific layer. However, there still exists a strong need for a comprehensive mechanism to prevent security problems in all layers (Zhang and Zhang 2008). Security issues like availability need to be addressed not only at each layer, but a good cross-layer design and communication is encouraged. IoT systems are generally large and complex systems with many interconnections and dependencies, such as in smart cities (Zanella et al. 2014).

If the availability of any of the three layers (perception, network, and application) fails, the availability of the whole system collapses. The lower layer infrastructure must protect itself from malicious behavioral patterns and harmful control from unauthorized users. The application layer should be available for all authorized users continuously without any service overloading-type interruption from unauthorized users.

Privacy

As the new European General Data Protection Regulation (GDPR)[2] has become enforceable on the 25 May 2018, protecting user data and securing user privacy are urgent and predominant issues to be solved for any IoT application. Users' data can neither be captured nor used without their awareness. Privacy has the highest priority for all existing and future application development, including IoT systems (Liyanage et al. 2018b). User identities must not be identifiable nor traceable. Under the new legislation, data processing must involve:

1) Lawful, fair, and transparent processing – emphasizing transparency for data subjects.
2) Purpose limitation – having a lawful and legitimate purpose for processing the information in the first place.
3) Data minimization – making sure data is adequate, relevant and limited, and organizations are sufficiently capturing the minimum amount of data needed to fulfill the specified purpose.
4) Accurate and up-to-date processing – requiring data controllers to make sure information remains accurate, valid, and fit for purpose.
5) Limitation of storage in a form that permits identification – discouraging unnecessary data redundancy and replication.
6) Confidential and secure – protecting the integrity and privacy of data by making sure it is secure (which extends to IT systems, paper records and physical security)
7) Accountability and liability – the demonstrating compliance. As a well-known statement in security, there are security issues at all perception, network, and application layers.

Some other security problems can be addressed effectively and efficiently on a certain layer level, such as implementing privacy components on the application layer. In a healthcare system, patients should be totally aware who is collecting and using their data. They also should have control over the data and who they want to share it with, as well as how and where their data is being used. The applications should provide services and interface to allow users to manage their data. Users must have tools that allow them to retain their anonymity in this super-connected world. The same scenario can be applied to systems such as smart home, smart transportation, etc. IoT applications may collect users' personal information and data from their daily activities. Many people would consider that data or information predicted from the data as private. Exposure of this information could have an unwanted or negative impact on their life. The use of the IoT system should not cause problems of privacy leaking. Any IoT applications which do not meet with these privacy requirements could be prohibited by law. The IoT system must seriously consider the implementation of privacy by the 7 data protection principles, providing user-centric support for security and privacy from its very own foundations (Roman et al. 2013).

Risk Mitigation

Mitigating the risk of an intrusion attempt or attack against an IoT device is not an easy thing to do. Having a higher degree of security protection at every level will discourage the attacker to pursue his/her goals further, by causing a higher amount of effort and time needed versus the benefits. Mitigation needs to start with prevention, by involving every actor in the market, from manufacturers to consumers and lawmakers, and to make them understand the impact of the IoT security threats in a connected world. Another way to mitigate risk is to keep abreast of the times by improving and innovating, from the ground up, and by finding new methods and designs to outgrow the shortcomings of the market.

Discussion

Authentication for IoT is a paramount necessity for securing and ensuring the privacy of users, simply due to the fact that an impregnable access control scheme would be impervious for any attack vector originating outside of the considered trust domain, as explained in the previous sections of this article. Authentication schemes in IoT applications are generally implemented at the software level, where it exposes unintentional hardware and design vulnerabilities (Frustaci et al. 2018). This fact constitutes the requirement of a holistic approach for securing access to the systems via the employment of impregnable authentication schemes. However, developing a generic authentication scheme to counter all possible attack scenarios would be improbable and an arduous attempt due to the heterogeneity of the IoT paradigm. A layered approach that identifies the distinct authentication requirement is desired to formalize a holistic trust domain.

For perception level entities, IBE or ECC would be ideal authentication schemes to generate commendable cryptographic credentials with available resources. The mobile entities, where actual users are interfacing to IoT systems are storing personalized credentials such as photos, medical stats, access to CCTV systems, GPS location (GPS), daily routines, financial statistics, banking credentials, emergency service status and online account statistics, are emphasizing the need for privacy preservation at this level. As proposed in section titled "Perception Nodes", adopting IBE, ABE, ECC, or biometric-based mechanisms should ensure security. Novel mechanisms such as Cap-BAC could be employed to launch a scalable access control scheme for cloud computing platforms for IoT applications. However, the potential for deploying edge computing paradigms in the edge of the network indemnifies the cloud computing services from external direct access, as the access control would be migrated to the edge along with the service platform. The internet technologies of IoT-enabled systems are more secured than the perception level and mobile level entities with the deployed protocols such as DTLS, SSL, and IPSec. Due to the dependency of a CA or TTP for employing such strong and secure protocols, the future of Internet security enhancements would be focused on developing distributed access control schemes to eliminate the single point of failure. Each IoT application composes different devices and systems to accomplish the intended outcome which attributes diverse protocols in hardware and software. Thus, the authentication schemes should be application specific and context aware of resource constraints associated with the diversified deployments. As privacy is the main concern on IoT to be ensured through impregnable access control schemes, the GDPR initiative is a timely solution established to constrict the IoT service providers (both software and hardware) from developing and marketing products with vulnerabilities.

Current researches have focused on developing novel methods for authentication in the IoT domain. We are briefly introducing a few of these recent approaches to demonstrate the state-of-the-art technologies.

In Ning et al. (2015), Ning et al. has proposed an aggregated proof-based hierarchical authentication (APHA) scheme to be deployed on existing Unit IoT and Ubiquitous IoT (U2IoT) architecture. Their scheme employs two cryptographic primitives; homomorphic functions and Chebyshev polynomials. The proposed scheme has been verified formally using Burrows-Abadi-Needham (BAN) logic. However, the scalability of the scheme with the extent of multiple units has not been verified with a physical prototype.

There are various initiatives on Physical Unclonable Functions (PUF) to be used for IoT device authentication. A PUF is an expression of an inherent and unclonable instance-specific unique feature of a physical object which serves as a biometric for non-human entities, such as IoT devices (Aman et al. 2017). Hao et al. are proposing a Physical Layer (PHY) End-to-End (E2E) authentication scheme which generates an IBE-based PHY-ID which acts as a PUF with unclonable PHY features RF Carrier Frequency Offset (CFO) and In-phase/Quadrature-phase Imbalance (IQI) extracted from collaborative nodes in a Device to Device (D2D) IoT deployment (Hao et al. 2018). This mechanism is ideal for perception-level nodes to be impervious to impersonation or malicious node injection attacks, as it is using physical measurements which are unique for each entity and for its location of operation in generating an identity for devices. However, the proposed scheme relies on a TTP called Key Generation Center (KGC). KGC generates the asymmetric key credentials for the nodes in its contact. The reachability of a certain KGC is limited due to the low power D2D connectivity. Thus, multiple KGCs deployed to accomplish the coverage should be managed with a centralized control entity. This enables the attack vectors on decentralized KGC entities. Moreover, the reliance on CFO and IQI features require the nodes to be stationary. This would be an issue considering most IoT devices are mobile and their RF-based characteristics vary in a timely manner. Aman et al. proposed a PUF-based authentication protocol for scenarios when an IoT device is connecting with a server and a D2D connectivity focused on its applicability in vehicular networks. Authentication is based on a Challenge Response Pair (CRP), where the outcome of the CRP is correlated with the physical microscopic structure of the IoT device, which emphasizes its unique PUF attributes with the inherent variability of the fabrication process in Integrated Circuits (ICs). The proposed protocol was analyzed using Mao and Boyd logic, while Finite State Machine (FSM) and reachability analysis techniques have been adopted for formal verification. Even though the performance of the protocol has been analyzed in terms of computational complexity, communication overhead and storage requirement, its scalability with simultaneous multiple IoT device connections to the server have not been addressed. However, this approach would be a feasible solution for V2E applications as the PUF could be successfully integrated with vehicles.

A human-gait pattern based on the biometric extraction scheme WifiU has been proposed in Shahzad and Singh (2017) as a case study that uses Channel State Information (CSI) of the received Wi-Fi signals for determining the gait pattern of the person carrying the transmitter. The gait patterns are becoming a novel biometric mode and this solution is a cost-effective approach which does not employ any floor sensors or human wearables. However, the applicability of WifiU for IoT devices raises concerns over scalability, accuracy of the gait-pattern extraction from CSI, reliability of CSI measurement and Wi-Fi interference. Chauhan et al. in Chauhan et al. (2018) proposes a Recurrent Neural Network (RNN) based on human breath print authentication system for mobile, wearable, and IoT platforms employing a derived breath print as a biometric through acoustic analysis. Even though this approach depicts a viable biometric solution for human interfacing IoT applications, the breath print extraction would be dependent on the health, climatic circumstances and physical stability of the user.

If the proposed authentication schemes are not fully holistically applicable for IoT deployments, optimum solutions at different layers and specific applications could be aggregated to form an impregnable access control system, where the interconnectivity across them should be maintained by decentralized trust domain managers. However, the access control mechanism optimum for each application should be investigated for each case in order to ensure robustness.

Future Research Directions

This section proposes several new research approaches and directions that could have a high impact for the future of the IoT security.

Blockchain

The blockchain is a distributed database of online records. Typically used in financial transactions for the Bitcoin cryptocurrency, the peer-to-peer blockchain technology records transactions without exception, in exchange, to form an online ledger system. Blockchain technologies are immutable, transparent, trustworthy, fast, decentralized, and autonomic, providing solutions that can be public, consortium, or private. Due to the success of Bitcoin, people are now starting to apply blockchain technologies in many other fields, such as financial markets, supply chain, voting, medical treatment and security for IoT (Lin and Liao 2017; Manzoor et al. 2018). There are expectations that blockchain will revolutionize industry and commerce and drive economic change on a global scale (Underwood 2016).

Blockchain technology leads to the creation of secure mesh networks, where IoT devices will interconnect while avoiding threats such as impersonation or device spoofing. As more legitimate nodes register on the blockchain network, devices will identify and authenticate each other without a need for central brokers and certification authorities. The network will scale to support more and more devices without the need for additional resources (Dickson 2016).

Smart contracts open the way to defining a new concept, a decentralized autonomous organization (DAO), sometimes labeled as a Decentralized Autonomous Corporation, an organization that runs through rules maintained on a blockchain. The legal status of this new brand of business organization is rather seen as a general partnership, meaning that its participants could bear unlimited legal liability. Ethereum blockchain, for example, is a public blockchain network optimized for smart contracts that use its cryptocurrency, called Ether (ETH). There is a huge interest in Ethereum, as a blockchain technology for the future. In 2017, Enterprise Ethereum Alliance was formed and already counts close to 250 members, like Samsung, Microsoft, J.P. Morgan, Toyota, ING, Consensys, BP, Accenture and many others. Etherum has become the second highest traded cryptocurrency in 2017, after Bitcoin, with a volume of transactions for over half of million euros in a single 24-hour period.

As with each disruptive concept that turns into an effective offering, the blockchain model is not perfect and has its flaws and shortcomings. Scalability is one of the main issues, considering the tendency toward centralization with a growing blockchain. As the blockchain grows, the nodes in the network require more storage, bandwidth, and computational power to be able to process a block, which currently leads to only a handful of the nodes being able to process a block. Computing power and processing time is another challenge, as the IoT ecosystem is very diverse and not every device will be able to compute the same encryption algorithms at the desired speed. Storage of a continuously increasing ledger database across a broad range of smart devices with small storage capabilities, such as sensors, is yet another hurdle. The lack of skilled people to understand and develop the IoT-blockchain technologies together is also a challenge. The lack of laws and a compliance code to follow by manufacturers and service providers is not helping both the IoT and blockchain to take off as expected.

IOTA solves some problems that the blockchain does not. One of them is centralization of control. As history shows, small miners create big groups to reduce the variation of the reward. This activity leads to concentration of power, computational, and political, in possession of just a handful of pool operators and gives them the ability to apply a broad spectrum of policies, like filtering on or postponing certain transactions.

5G

For the first time in history, LTE has brought the entire mobile industry to a single technology footprint resulting in unprecedented economies of scale. The converged footprint of LTE has made it an attractive technology baseline for several segments that had traditionally operated outside the commercial cellular domain. There is a growing demand for a more versatile M2M platform. The challenge for industrial deployment of IoT is the lack of convergence across the M2M architecture design that has not materialized yet. It is expected that LTE will remain as the baseline technology for wide-area broadband coverage also in the 5G area. The realization of 5G network is affecting many IoT protocols' initial design, especially at perception and network layers (Chiang and Zhang 2016). Mobile operators now aim to create a blend of pre-existing technologies covering 2G, 3G, 4G, WiFi, and others to allow higher coverage and availability, as well as higher network density in terms of cells and devices with the key differentiator being greater connectivity as an enabler for M2M services (Liyanage et al. 2018a). 3GPP standard/5G-based backhaul has become a popular solution for connectivity problems in IoT systems. Munoz et al. (2016) indicates that the next generation of mobile networks (5G), will need not only to develop new radio interfaces or waveforms to cope with the expected traffic growth but also to integrate heterogeneous networks from End-to-End (E2E) with distributed cloud resources to deliver E2E IoT and mobile services. Fantacci et al. (2014) has provided a backhaul solution through mobile networks for smart building applications. The proposed network architecture will improve services for users and will also offer new opportunities for both service providers and network operators. As 5G has become available and is being adopted as the main backhaul infrastructure for IoT system, it will play a huge role in IoT perception and networking layers (Ahmad et al. 2018a). 5G has moved the focus to a user-centric service from a network-centric service unlike 4G and 3G. With massive multiple-input and multiple-output (MIMO) technologies deployed in 5G, network selection and rapid handovers are becoming essential in terms of supporting QoS and Quality-of-user Experience (QoE) aware services (Xu et al. 2016). The handover between different network interfaces should be authenticated and the information exchange during the handover should be protected and private. Currently, SDN is considered as the mainstream for a higher efficiency through its centralized control capability in the 5G communication process (Liyanage et al. 2015). With SDN, the control logic is removed from the underlying infrastructures to a management platform. Software and policies can be implemented on the central SDN controller to provide consistent and efficient management over the whole 5G network. One advanced and beneficial feature offered by SDN is that it can separate the control plane and data source by abstract, the control logic from the underlying switches and routers to the centralized SDN controller (Liyanage et al. 2016). To address the Machine Type Communication (MTC) in IoT systems based on 5G network, several approaches are available (Tehrani et al. 2014; Shariatmadari et al. 2015):

1) A higher level of security for devices is achievable by utilizing new security mechanisms being embedded with Subscriber Identity Module (SIM).
2) It is recommended to implement and employ physical-layer security adopting RF fingerprinting.
3) Using asymmetric security schemes to transfer the burden of required computations to the network domain or gateways with high computing capabilities.

Fog and Edge Computing

Although powerful, the cloud model is not the best choice for environments where internet connectivity is limited or operations are time-critical. In scenarios such as patient care, milliseconds have fatal consequences. As in the vehicle-to-vehicle communications, the prevention of collisions and accidents relies on the low latency of the responses. Cloud computing is not consistently viable for many IoT applications, and so, it is replaced by fog computing. Fog computing, also known as fogging, is a decentralized computing infrastructure in which the data, compute, storage, and applications split in an efficient way between the data source and the cloud.

Fog computing extends cloud computing and services alike, to the edge of the network, by bringing the advantages and the power of the cloud to where the data initially arise. The main goal of fogging is to improve efficiency and also to reduce the quantity of data that moves to the cloud for processing, analysis, and storage. In fogging, data processing takes place in a router, gateway, or data hub on a smart device, which sends it further to sources for processing and return transmission, therefore reducing the bandwidth payload to the cloud.

The back-and-forth communication between IoT devices and the cloud can negatively affect the overall performance and security of the IoT asset. The distributed approach of fogging addresses the problem of the high amount of data coming from smart sensors and IoT devices, which would be costly and time-consuming to send to the cloud each time. Among other benefits, fog computing offers better security by protecting the fog nodes with the same policy, controls, and procedures used in other parts of the IT environment and by using the same physical safety and cyber security solutions (Cisco n.d.). Fog networking complements cloud computing and allows for short-term analytics at the edge while the cloud performs resource-intensive, longer-term analytics. Computation moves even closer to the edge and becomes deeply-rooted in the very same devices that created the data initially, and so, generates even greater possibilities for M2M intelligence and interactions.

The movement of computation from the fog to the actual device opens the path-to-edge computing. That is a distributed architecture in which the processing of client data takes place at the outer edge of the network, in the proximity of the originating source. Mobile computing, low cost of computer components and the absolute quantity of IoT devices drive the move toward edge computing. Time-sensitive data is processed at the point of origin by an intelligent and resource-capable device or sent to a broker server located in close geographical proximity to the client. Less time-sensitive data travels to the cloud for historical analysis, big data analytics, and long-term storage. One of the greatest benefits of edge computing is that it removes network bottlenecks by improving time to action and response time down to milliseconds, while also conserving network resources.

The edge computing concept is not without its flaws though. Edge computing raises a high amount of security, licensing and configuration challenges and concerns. The vulnerability to some attack vectors like malware infections and security exploits increases because of the nature of the distributed architecture. Smart clients can have hidden licensing costs, where the base version of an edge client might initially have a low price, additional functionalities could be licensed separately and will drive the price up. Also, decentralized and poor device management leads to configuration drift by the administrators. They can inadvertently create security holes by not consistently updating the firmware or by failing to change the default password on each edge device (Rouse 2016).

Quantum Security, AI, and Predictive Data Analytics

With the technological advancements of quantum computing, Artificial Intelligence (AI), and cognitive systems, and with the continuous development and mass adoption of IoT ecosystem, the current security practices and methodologies will become a part of the past. Quantum computing, not only can it break through any form of security that is known to humankind, but it can also offer the solution to finding the formula for tight security. IoT will vastly benefit from these technology advancements, especially from the quantum mechanics science on a microchip. Further research is recommended, once the technology matures and evolves, to discover how the security of the future impacts on things around and especially on the IoT ecosystem.

Network Slicing

Network slicing is the concept of slicing a physical network into several logical planes to facilitate the various IoT services to customize their differentiated on-demand services with the same physical network (Ni et al. 2018). The main aim of this paradigm is to reinforce different service requirements such as latency, bandwidth, and reliability of heterogeneous IoT applications to utilize the resources such as storage, computing, and bandwidth of the IoT device platforms (Ravindran et al. 2017). The complexity of the IoT service integration with core network resources could be alleviated using a standardized network slicing mechanism as proposed by the Next Generation Mobile Network (NGMN). A typical network slicing process could be described under three layers, namely service instance layer, network slice instance layer and resource layer which follow the principles automation, isolation, customization, elasticity, programmability, end-to-end, and hierarchical abstraction (Siriwardhana et al. 2019).

The evolvement of the network slicing concept has reached the depths of 5G Information Centric Networking (ICN) model, which consists of five functional planes (FPs), namely; FP1 – service business plane, FP2 – service orchestration and management plane, FP3 – IP/ICN orchestrator plane, FP4 – domain service orchestration and management plane and FP5 – infrastructure plane. FP1 interfaces with external 5G users in providing various service Application Programme Interfaces (APIs) which realize the objective and relevant services to accomplish that objective with inputs such as service type, demand patterns, Service Level Agreements (SLAs)/QoS/QoE requirements. The service requests forwarded by FP1 are communicated to the FP3 as service requirements by FP2. The FP3 interfaces with a domain controller to virtualize compute, storage, and network resources to meet the service requirements conveyed from FP2. FP4 supports the management of IP and ICN services belonging to different technological domains such as 4G/5G Radio Access Network (RAN), Multi-Protocol Label Switching (MPLS) and edge technologies, while FP5 enables the service rules in end-to-end manner.

The entities operating in network slicing infrastructure, such as network slice manager and host platforms are attributing the vulnerabilities exploitable by impersonation attacks, DoS, SCA attacks and the interoperability of different security protocols and policies (Harel and Babbage 2016). An IoT user may access different slices depending on both the requirements and the intended outcomes. Thus, the access granting control for different slices is a critical juncture in the perspective of security. The plausibility for isolating the slices for constricting the deliberate hacking attempts at resources operating at each plane should be focused. Due to the facts that a network slice is a composite of the actual physical infrastructure and the processes should be dynamic, adaptive, and flexible for servicing the intended functions, the assurance of user confidentiality, privacy, integrity, and availability are challenging. However, authentication is the most effective mechanism to be used for enhancing the robustness of the network slices toward attacks. Among the 5G Security-as-a-Service (SaaS) concepts, micro-segmentation, deception of the attacker and AI deployments for monitoring, attack detection and remediation are emerging initiatives for securing network slices (Dotaro 2018).

Conclusions

IoT technology is the most discussed paradigm in the research community these days. Its potential to connect all the devices in the world and to create a large information system that would offer services to improve the quality of human beings exponentially has made the concept much more popular. The integration of various technologies and devices with different architectures are creating interoperability issues with the components in the IoT architecture. These issues and the highly diversified type of services are creating security concerns which disperse into all three layers of IoT architecture: Perception, Network, and Application. Hence, the security measures to be taken should be developed while analyzing the threats and vulnerabilities at each layer.

Mitigating risks associated with security breaches are possible, if security receives consideration from early product planning and design, and if some basic prevention mechanisms are in place. Enactment and standardization will simplify the manufacturing and development processes, give the market an incentive for mass-adoption and also increase the security posture of IoT products and services. Security will have to be inbuilt so that IoT can withstand a chance against the threats that technological advancements will bring.

Endnotes

1 https://www.tenable.com/products/nessus/nessus-professional
2 https://www.eugdpr.org

References

Ahmad, I., Kumar, T., Liyanage, M. et al. (2018a). Overview of 5G security challenges and solutions. *IEEE Communications Standards Magazine* 2 (1): 36–43.

Ahmad, I., Kumar, T., Liyanage, M. et al. (2018b). Towards gadget-free internet services: a roadmap of the naked world. *Telematics and Informatics* 35 (1): 82–92.

Akhunzada, A., Gani, A., Anuar, N.B. et al. (2016). Secure and dependable software defined networks. *Journal of Network and Computer Applications* 61: 199–221.

Alaba, F., Othman, M., Hashem, I., and Alotaibi, F. (2017). Internet of things security: a survey. *Journal of Network and Computer Applications* 88: 10–28.

Aman, M.N., Chua, K.C., and Sikdar, B. (2017). Mutual authentication in IoT systems using physical unclonable functions. *IEEE Internet of Things Journal* 4 (5): 1327–1340.

Bekara, C. (2014). Security issues and challenges for the IoT-based smart grid. *Procedia Computer Science* 34: 532–537.

Borgia, E. (2014). The internet of things vision: key features, applications and open issues. *Computer Communications* 54: 1–31.

Braeken, A., Liyanage, M., and Jurcut, A.D. (2019). Anonymous lightweight proxy based key agreement for IoT (ALPKA). *Wireless Personal Communications* 106 (2): 1–20.

Brillo (n.d.). https://developers.google.com/brillo (accessed 16 July 2019).

CEN-CENELEC-ETSI Smart Grid Coordination Group (2014). SGCG/M490/G Smart Grid Set of Standards Version 3.1, Oct-2014. ftp://ftp.cencenelec.eu/EN/EuropeanStandardization/HotTopics/SmartGrids/SGCG_Standards_Report.pdf (accessed 25 June 2019).

Chauhan, J., Seneviratne, S., Hu, Y. et al. (2018). Breathing-based authentication on resource-constrained IoT devices using recurrent neural networks. *Computer* 51 (5): 60–67.

Chiang, M. and Zhang, T. (2016). Fog and iot: an overview of research opportunities. *IEEE Internet of Things Journal* 3: 854–864.

Cisco (n.d.). Fog computing and the Internet of Things: extend the cloud to where the things are. http://www.cisco.com/c/dam/en_us/solutions/trends/iot/docs/computing-overview.pdf (accessed 25 June 2019).

Contiki (n.d.). http://www.contiki-os.org (accessed 16 July 2019).

Deogirikar, J. and Vidhate, A. (2017). Security attacks in IoT: a survey. *2017 International Conference on I-SMAC (IoT in Social, Mobile, Analytics and Cloud)*, Coimbatore, India (February 2017).

Desai, D. and Upadhyay, H. (2014). Security and privacy consideration for internet of things in smart home environments. *International Journal of Engineering Research and Development* 10 (11): 73–83.

Dickson, B. (2016). How blockchain can change the future of IoT. https://venturebeat.com/2016/11/20/how-blockchain-can-change-the-future-of-iot (accessed 16 July 2019).

Dotaro, E. (2018). 5G Network Slicing and Security. IEEE SDN newsletter. https://sdn.ieee.org/newsletter/january-2018/5g-network-slicing-and-security (accessed 25 June 2019).

Fantacci, R., Pecorella, T., Viti, T., and Carlini, C. (2014). A network architecture solution for efficient iot wsn backhauling: challenges and opportunities. *IEEE Wireless Communications* 21: 113–119.

Feki, M.A., Kawsar, F., Boussard, M., and Trappeniers, L. (2013). The internet of things: the next technological revolution. *Computer* 46: 24–25.

Fiore, U., Castiglione, A., De Santis, A., and Palmieri, F. (2017). Exploiting battery-drain vulnerabilities in mobile smart devices. *IEEE Transactions on Sustainable Computing* 2 (2): 90–99.

Frustaci, M., Pace, P., Aloi, G., and Fortino, G. (2018). Evaluating critical security issues of the IoT world: present and future challenges. *IEEE Internet of Things Journal* 5 (4): 2483–2495.

Garcia-Morchon, O., Hummen, R., Kumar, S. et al. (2012). Security considerations in the ip-based internet of things, draft-garciacore-security-04.

Groth, J. and Sahiai, A. (2008). Efficient noninteractive proof systems for bilinear groups. *Advances in Cryptology ®EUROCRYPT*, Istanbul, Turkey (13–17 April 2008): Springer.

Hao, P., Wang, X., and Shen, W. (2018). A collaborative PHY-aided technique for end-to-end IoT device authentication. *IEEE Access* 6: 42279–42293.

Harel, R. and Babbage, S. (2016). 5G Security Recommendations Package #2: Network Slicing, published by NGMN Alliance, Ver. 01. https://www.ngmn.org/fileadmin/user_upload/160429_NGMN_5G_Security_Network_Slicing_v1_0.pdf (accessed 25 June 2019).

Horrow, S. and Anjali, S. (2012). Identity management framework for cloud based internet of things. *Proceedings of the First International Conference on Security of Internet of Things, SecurIT' 12*, Kollam, India (17–19 August 2012): ACM.

Jing, Q., Vasilakos, A.V., Wan, J. et al. (2014). Security of the internet of things: perspectives and challenges. *Wireless Networks* 20 (8): 2481–2501.

Jurcut, A., Coffey, T., Dojen, R., and Gyorodi, R. (2008). Analysis of a key-establishment security protocol. *Journal of Computer Science and Control Systems* 2008: 42–47.

Jurcut, A.D., Coffey, T., and Dojen, R. (2012). Symmetry in security protocol cryptographic messages – a serious weakness exploitable by parallel session attacks. *7th IEEE International Conference on Availability, Reliability and Security (ARES'12)*, Prague, Czech Republic (August 2012).

Jurcut, A., Coffey, T., and Dojen, R. (2014a). Design requirements to counter parallel session attacks in security protocols. *12th IEEE Annual Conference on Privacy, Security and Trust (PST'14)*, Toronto Canada (July 2014).

Jurcut, A.D., Coffey, T., and Dojen, R. (2014b). Design guidelines for security protocols to prevent replay & parallel session attacks. *Journal of Computers & Security* 45: 255–273.

Jurcut, A.D., Coffey, T., and Dojen, R. (2017). A novel security protocol attack detection logic with unique fault discovery capability for freshness attacks and interleaving session attacks. *IEEE Transactions on Dependable and Secure Computing* doi: 10.1109/TDSC.2017.2725831.

Jurcut, A.D., Liyanage, M., Chen, J. et al. (2018). On the security verification of a short message service protocol. *2018 IEEE Wireless Communications and Networking Conference (WCNC),* Barcelona, Spain (April, 2018).

Jurcut, A.D., Coffey, T., and Dojen, R. On the prevention and detection of replay attacks using a logic-based verification tool. In: *Computer Networks,* vol. 431 (ed. A. Kwiecień, P. Gaj and P. Stera), 128–137. Switzerland: Springer International Publishing.

Kavun, E.B. and Yalcin, T. (2010). A lightweight implementation of keccak hash function for radio-frequency identification applications. *International Workshop on Radio Frequency Identification: Security and Privacy Issues,* New York, USA (23–24 June 2015): Springer.

Khalajmehrabadi, A., Gatsis, N., Akopian, D., and Taha, A. (2018). Real-time rejection and mitigation of time synchronization attacks on the global positioning system. *IEEE Transactions on Industrial Electronics* 65 (8): 6425–6435.

Kim, H. and Lee, E.A. (2017). Authentication and authorization for the internet of things. *IT Professional* 19 (5): 27–33.

Kouicem, D., Bouabdallah, A., and Lakhlef, H. (2018). Internet of things security: a top-down survey. *Journal of Computer Networks* 141: 199–121.

Kumar, J.S. and Patel, D.R. (2014). A survey on internet of things: security and privacy issues. *International Journal of Computer Applications* 90 (11).

Kumar, T., Braeken, A., Liyanage, M., and Ylianttila, M. (2017). Identity privacy preserving biometric based authentication scheme for naked healthcare environment. *2017 IEEE International Conference on Communications (ICC),* Paris, France (21–25 May 2017): IEEE.

Kumar, T., Liyanage, M., Ahmad, I. et al. (2018a). User privacy, identity and trust in 5G. In: *A Comprehensive Guide to 5G Security* (ed. M. Liyanage, I. Ahmad, A.B. Abro, et al.), 267. Wiley.

Kumar, T., Porambage, P., and Ahmad, I. (2018b). Securing gadget-free digital services. *Computer* 51 (11): 66–77.

Lin, X. and Li, X. (2013). Achieving efficient cooperative message authentication in vehicular ad hoc networks. *IEEE Transactions on Vehicular Technology* 62 (7): 3339–3348.

Lin, I.C. and Liao, T.C. (2017). A survey of blockchain security issues and challenges. *International Journal of Network Security* 19: 653–659.

Lin, X., Sun, X., Wang, X. et al. (2008). TSVC: timed efficient and secure vehicular communications with privacy preserving. *IEEE Transactions on Wireless Communications* 7 (12): 4987–4998.

Liu, Y., Cheng, C., Gu, T. et al. (2016). A lightweight authenticated communications scheme for a smart grid. *IEEE Transactions on Smart Grid* 7 (3): 1304–1313.

Liyanage, M., Gurtov, A., and Ylianttila, M. (ed.) (2015). *Software Defined Mobile Networks (SDMN): Beyond LTE Network Architecture.* New York: Wiley.

Liyanage, M., Abro, A.B., Ylianttila, M., and Gurtov, A. (2016). Opportunities and challenges of software-defined mobile networks in network security. *IEEE Security and Privacy* 14 (4): 34–44.

Liyanage, M., Braeken, A., Jurcut, A.D. et al. (2017). Secure communication channel architecture for software defined Mobile networks. *Journal of Computer Networks* 114: 32–50.

Liyanage, M., Ahmad, I., Abro, A.B. et al. (ed.) (2018a). *A Comprehensive Guide to 5G Security*. Wiley.

Liyanage, M., Salo, J., Braeken, A. et al. (2018b). 5G privacy: scenarios and solutions. *2018 IEEE 5G World Forum (5GWF)*, Silicon Valley, USA (9–11 July 2018): IEEE.

Lu, R., Lin, X., Zhu, H. et al. (2010). Pi: a practical incentive protocol for delay tolerant networks. *IEEE Transactions on Wireless Communications* 9 (4): 1483–1492.

Manzoor, A., Liyanage, M., Braeken, A. et al. (2018). Blockchain based proxy re-encryption scheme for secure IoT data sharing. *IEEE International Conference on Blockchain and Cryptocurrency (ICBC 2019)*, Seoul, South Korea (14–17 May 2019): IEEE.

Massis, B. (2016). The internet of things and its impact on the library. *New Library World* 117 (3/4): 289–292.

Miorandi, D., Sicari, S., De Pellegrini, F., and Chlamtac, C. (2012). Internet of things: vision, applications and research challenges. *Ad Hoc Networks* 10 (7): 1497–1516.

Munoz, R., Mangues-Bafalluy, J., Vilalta, R. et al. (2016). The cttc 5g end-to-end experimental platform: integrating heterogeneous wireless/optical networks, distributed cloud, and iot devices. *IEEE Vehicular Technology Magazine* 11: 50–63.

Ngu, A.H., Gutierrez, M., Metsis, V. et al. (2017). Iot middleware: a survey on issues and enabling technologies. *IEEE Internet of Things Journal* 4: 1–20.

Ni, J., Lin, X., and Shen, X.S. (2018). Efficient and secure service-oriented authentication supporting network slicing for 5G-enabled IoT. *IEEE Journal on Selected Areas in Communications* 36 (3): 644–657.

Ning, H., Liu, H., and Yang, Y. (2015). Aggregated-proof based hierarchical authentication scheme for the internet of things. *IEEE Transactions on Parallel and Distributed Systems* 26 (3): 657–667.

Openwsn (n.d.). http://openwsn.atlassian.net (accessed 16 July 2019).

Paillier, P. (1999). Public key cryptosystems based on composite degree residuosity classes. *Eurocrypt '99 Proceedings of the 17th International Conference on Theory and Application of Cryptographic Techniques*, Prague, Czech Republic (2–6 May 1999): ACM.

Pasca, V., Jurcut, A., Dojen, R., and Coffey, T. (2008). Determining a parallel session attack on a key distribution protocol using a model checker. *ACM Proceedings of the 6th International Conference on Advances in Mobile Computing and Multimedia (MoMM '08)*, Linz, Austria (24–26 November). New York, USA: ACM.

Perera, C., Jayaraman, P.P., Zaslavsky, A. et al. (2014). Mosden: an internet of things middleware for resource constrained mobile devices. *47th Hawaii International Conference on System Sciences*. Hawaii, USA (6–9 January 2014): IEEE.

Porambage, P., Okwuibe, J., Liyanage, M. et al. (2018). Survey on multi-access edge computing for internet of things realization. *IEEE Communication Surveys and Tutorials* 20 (4): 2961–2991.

Porambage, P., Manzoor, A., Liyanage, M. et al. (2019). Managing mobile relays for secure E2E connectivity of low-power IoT devices. *IEEE Consumer Communications & Networking Conference*, Las Vegas, USA (11–14 January 2010): IEEE.

Rahimi, H., Zibaeenejad, A., and Safavi, A.A. (2018). A novel IoT architecture based on 5G-IoT and Next Generation Technologies. GlobeCom-IoT. https://arxiv.org/ftp/arxiv/papers/1807/1807.03065.pdf (accessed 25 June 2019).

Rajakaruna, A., Manzoor, A., Porambage, P. et al. (2018). Lightweight dew computing paradigm to manage heterogeneous wireless sensor networks with UAVs. *arXiv preprint arXiv* 1811: 04283.

Ravindran, R., Chakraborti, A., and Amin, S. (2017). 5G-ICN: delivering ICN services over 5G using network slicing. *IEEE Communications Magazine* 55 (5): 101–107.

Riot (n.d.). http://www.riot-os.org (accessed 16 July 2019).

Roman, R., Alcaraz, C., Lopez, J., and Sklavos, N. (2011). Key management systems for sensor networks in the context of the internet of things. *Computers and Electrical Engineering* 37 (2): 147–159.

Roman, R., Zhou, J., and Lopez, J. (2013). On the features and challenges of security and privacy in distributed internet of things. *Computer Networks* 57 (10): 2266–2279.

Rouse, M. (2016). Edge computing. http://searchdatacenter.techtarget.com/definition/edge-computing (accessed 25 June 2019).

Sen, J. (2011). Privacy preservation Technologies in Internet of things. *Proceedings of International Conference on Emerging Trends in Mathematics, Technology, and Management*, (18–20 November 2011).

Shahzad, M. and Singh, M.P. (2017). Continuous authentication and authorization for the internet of things. *IEEE Internet Computing* 21 (2): 86–90.

Shariatmadari, H., Ratasuk, R., and Iraji, S. (2015). Machine-type communications: current status and future perspectives toward 5g systems. *IEEE Communications Magazine* 53: 10–17.

Siriwardhana, Y., Porambage, P., Liyanage, M. et al. (2019). Micro-Operator driven Local 5G Network Architecture for Industrial Internet. *IEEE Wireless Communications and Networking Conference*, Marrakech, Morocco (15–18 April 2019): IEEE.

Srilakshmi, A., Rakkini, J., Sekar, K.R., and Manikandan, R. (2018). A comparative study on internet of things (IoT) and its applications in smart agriculture. *Pharmacognosy Journal* 10 (2): 260–264.

Srivastava, V. and Motani, M. (2005). Cross-layer design: a survey and the road ahead. *IEEE Communications Magazine* 43: 112–119.

Stamp, M. (2011). *Information Security*, 2e. Hoboken, N.J: Wiley.

Tehrani, M.N., Uysal, M., and Yanikomeroglu, H. (2014). Device-to-device communication in 5g cellular networks: challenges, solutions, and future directions. *IEEE Communications Magazine* 52: 86–92.

Tinyos (n.d.). http://www.tinyos.net (accessed 16 July 2019).

Underwood, S. (2016). Blockchain beyond bitcoin. *Communications of the ACM* 59: 15–17.

Weis, S.A., Sarma, S.E., Rivest, R.L., and Engels, D.W. (2004). Security and privacy aspects of low-cost radio frequency identification systems. In: *Security in Pervasive Computing* (ed. D. Hutter, G. Muller, W. Stephan and M. Ullmann), 201–212. Springer.

Williams, R., McMahon, E., Samtani, S. et al. (2017). Identifying vulnerabilities of consumer internet of things (IoT) devices: a scalable approach. *2017 IEEE International Conference on Intelligence and Security Informatics (ISI)*. Beijing, China (22–24 July 2017): IEEE.

Wu, M., Lu, T.-J., Ling, F.-Y. et al. (2010). Research on the architecture of internet of things. *2010 3rd International Conference on Advanced Computer Theory and Engineering (ICACTE)*, Chengdu, China: (20–22 August 2010).

Xiao, Q., Gibbons, T., and Lebrun, H. (2009). RFID technology, security vulnerabilities, and countermeasures. doi: 10.5772/6668. https://www.researchgate.net/publication/

221787702_RFID_Technology_Security_Vulnerabilities_and_Countermeasures (accessed 25 June 2019).

Xu, L., Lillis, D., O'Hare, G.M., and Collier, R.W. (2014). A user configurable metric for clustering in wireless sensor networks. *SENSORNETS*, Lisbon, Portugal (7–9 January 2014): SciTePress.

Xu, L., Xie, J., Xu, X., and Wang, S. (2016). Enterprise lte and wifi interworking system and a proposed network selection solution. *2016 ACM/IEEE Symposium on Architectures for Networking and Communications Systems (ANCS)*. Santa Clara, USA (17–18 March 2016): ACM/IEEE.

Xu, L., Collier, R., and O'Hare, G.M.P. (2017). A survey of clustering techniques in wsns and consideration of the challenges of applying such to 5g iot scenarios. *IEEE Internet of Things Journal* 4: 1229–1249.

Yao, X., Chen, Z., and Tian, Y. (2015). A lightweight attribute-based encryption scheme for the internet of things. *Future Generation Computer Systems* 49: 104–112.

Zanella, A., Bui, N., Castellani, A. et al. (2014). Internet of things for smart cities. *IEEE Internet of Things Journal* 1: 22–32.

Zhang, Q. and Zhang, Y.Q. (2008). Cross-layer design for qos support in multihop wireless networks. *Proceedings of the IEEE* 96: 64–76.

Zhang, Y., Shen, Y., Wang, H. et al. (2015). On secure wireless communications for IoT under eavesdropper collusion. *IEEE Transactions on Automation Science and Engineering* 13 (3): 1281–1293.

Zhao, K. and Ge, L. (2013). A survey on the internet of things security. *2013 9th International Conference on Computational Intelligence and Security (CIS)*, Leshan, China (December 2013).

Zhou, H. (2012). *The Internet of Things in the Cloud: A Middleware Perspective*, 1e. Boca Raton, FL, USA: CRC Press, Inc.

Zhou, J., Dong, X., Cao, Z. et al. (2015a). 4S: a secure and privacy-preserving key management scheme for cloud-assisted wireless body area network in m-healthcare social networks. *Information Sciences* 314: 255–276.

Zhou, J., Dong, X., Cao, Z. et al. (2015b). Secure and privacy preserving protocol for cloud-based vehicular DTNs. *IEEE Transactions on Information Forensics and Security* 10 (6): 1299–1314.

Zhou, J., Cao, Z., Dong, X., and Lin, X. (2015c). TR-MABE: White-box traceable and revocable multi-authority attribute-based encryption and its applications to multi-level privacy-preserving e-healthcare cloud computing systems. *2015 IEEE Conference on Computer Communications (INFOCOM)*. IEEE.

Further Reading

Singh, J., Pasquier, T., Bacon, J. et al. (2015). Twenty security considerations for cloud-supported internet of things. *IEEE Internet of Things Journal* 3 (3): 269–284.

14

Privacy in the 5G World: The GDPR in a Datafied Society

Sebastiao Teatini and Marja Matinmikko-Blue

University of Oulu, Oulu, Finland

Introduction

We may be embarking upon a societal journey that will lead us toward a techno-surveillance dystopian future and the technological revolution that will allow for this scenario to come to fruition is 5G or fifth-generation wireless technology. As we march toward a globalized world where digital data fuels the economy, cities will become hyper-connected and the necessity for data collection will proliferate. As a consequence, regulators around the world will be confronted with the reality that privacy may be on its death bed and 5G will deliver the final blow. With the increase of data transfer speed and reliability comes the risk of privacy deterioration as 5G interlaces with ubiquitous computing creating a sociotechnical all-encompassing global web of personal digital data extraction, storage, and transfer that will directly impact the way society organizes itself.

The focus of this article is on identifying and understanding the possible threats to privacy protection that may be encountered in the deployment of 5G. The datafication of society is likely to increase in the upcoming 5G world and legal safeguards are necessary to ensure our privacy rights are protected. The purpose of this article is to analyze some of the unique characteristics of 5G and how the EU General Data Privacy Regulation (GDPR) stands as the last legal guardian against unwanted personal data violations.

Sweeping aside the hyperbole, it can be stated with some degree of confidence that we are in the midst of a technological revolution in the field of wireless communication that will transform the way society organizes itself. The coming hyper-fast wireless communication infrastructure, 5G, will enable future business ecosystems to flourish in the fields of augmented reality, virtual reality and artificial intelligence, autonomous driving, health care, among others (Habib et al. 2019; Frias and Martinez 2018). Owing to its low latency, reliability, high capacity, energy efficiency, and speedy data transmissions, 5G will directly impact the way we interact, communicate, work, and participate in society (Jacobson 2019; Lemstra 2018). For this breakthrough to take place a new generation of communication infrastructure is being designed, created, and implemented in many urban areas in Europe, North East Asia, North America, and other regions of the world. In addition, more intangible elements, such as software, protocols, and standards are being put in place that combined with form the 5G technical ecosystem. Not only the

The Wiley 5G REF: Security. Edited by Rahim Tafazolli, Chin-Liang Wang, Periklis Chatzimisios and Madhusanka Liyanage.

volume, velocity, and variety of digital data will increase in the 5G world but also the value. However, in the fast-approaching datafied world, privacy is rapidly becoming a cherished commodity rather than an inherited human right (Cave 2018).

The main concern for regulators and developers of this infrastructure is how to better protect individual data privacy in the coming ubiquitous computing connected world (Liyanage et al. 2018). With increased connectivity comes increased fear that our personal data may fall in the hands of unscrupulous or uninvited actors, such as hackers, unwanted advertisers, or overzealous security agencies. Complete data security is impossible and the protection of personal data remains especially challenging in a world where our personal data is being extracted, organized, classified, manipulated, and commercialized (Zuboff 2019). Over the past decade, we have observed an overflow of digital devices and sensors in our cities, work, homes, and increasingly in our bodies. Our entire surrounding functions in a constant data production mode. The thirst for the optimization of data extraction, analysis, and manipulation is justified by the logic of surveillance capitalism, where all personal digital data that can be monetized will be (Silverman 2017).

Today, more than half of the world population is connected to the Internet, producing searches, posting videos, playing games, paying bills, sending messages, making calls, and the number of users who consume and produce digital data is likely to increase. The scope, speed, and depth of data collection will dramatically accelerate as 5G will allow for billions of sensors (Internet of Things, IoTs) to be connected to the Internet. From our home appliances to automobiles, factories, machines, clothes, power grids, and smart cities, "everything" will be connected to the Internet and constantly producing digital data (Cave 2018).

The structure of this article is as follows: in the first part, we discuss what distinguishes 5G from its predecessors and some regulatory challenges for the coming technology. In the next section, we analyze in detail the appropriateness of the current data privacy legal infrastructure (GDPR) in Europe and how some individual nation states may be coping with its implementation. We then identify some potential obstacles to ensuring data privacy in the 5G world. A synthesis of our analysis will be presented in the conclusion.

The 5G World. What Sets It Apart from Previous Communication Technologies?

Most mobile data users are not yet familiar with 5G because the technology will not be commercially available on a large scale until 2020. According to the EU 5G Action Plan, which delineated the strategies and goals for the building and deployment of the 5G infrastructure in member states, a roadmap was established to facilitate the implementation of 5G by via spectrum allocation by national governments and to facilitate the coordination between industry and the public sector via the EU Public–Private Partnership (5G-PPP). Until 5G is available to consumers, for the most part, users will rely on 4G (long-term evolution, LTE) and its evolutions, or even 3G, to access the internet, send messages and place phone calls from their gadgets. So what distinguishes 5G from its predecessors?

5G is expected to change the traditional mobile communication business ecosystem by connecting billions of devices and ultimately digitizing the entire society which can have a big impact on productivity and competitiveness. Policy makers have globally recognized the importance of widespread deployment and timely take-up of very high capacity networks for realizing the full economic and social benefits of the digital transformation. The development of 5G networks will be based on dense deployments of small cell networks in specific high demand locations.

Traditionally, a small number of mobile network operators (MNO) has dominated the mobile connectivity market with high infrastructure investments and long-term spectrum licenses granted by the national regulators (Al-Debei et al. 2013). The mobile communication sector has recently gone through a transition where the MNO market dominance has been shaken with the appearance of over the top (OTT) services that have substituted MNOs' voice and text services (Weber and Scuka 2016). These internet services can operate independently of the infrastructure, leaving the MNOs to act as bit pipes providing mobile broadband connectivity to deliver the services. Regulations, on the other hand, have not evolved along with the complex technical and market developments resulting in different regulatory requirements for the mobile connectivity domain and the internet domain. National regulators are currently in the process of adapting their regulatory mechanisms for the adoption of next-generation networks and implementing the European Electronic Communications Code (EECC 2018; Briglauer et al. 2017).

5G networks aim at providing new high-quality wireless services to meet stringent and case-specific needs of various vertical sectors beyond traditional mobile broadband offerings. 5G networks are expected to disrupt the traditional mobile communication market by lowering the entry barrier to new entrants by sharing of required resources (Agiwal et al. 2016; Samdanis et al. 2016). This development has the potential to open the mobile market for a large number of local 5G networks (Matinmikko et al. 2018).

A key regulation aspect shaping the 5G market is spectrum. Globally, new spectrum is made available for the deployment of 5G networks through World Radiocommunication Conferences (WRCs) and existing bands for cellular networks are being transformed to allow the deployment of 5G networks. New 5G spectrum awards by the national regulators are showing a divergence in the approaches taken in different countries either strengthening the existing MNO market dominance or allowing new entry for different stakeholders to locally deployed 5G networks (Matinmikko-Blue et al. 2019).

Data Privacy and the GDPR

In the coming 5G world, privacy concerns are likely to increase as the speed and quantity of personal data being transmitted over the network will increase significantly (Lemstra 2018; Cave 2018). The legal system, specifically the laws and regulations related to privacy rights and privacy protection, is the most effective safeguard individuals have to protect them from unscrupulous actors that may attempt to extract, manipulate, commercialize, classify, or misuse personal data without the users' explicit consent.

Just as authorities were slow to react to the growth of the internet in the 1990s and social media in 2010s, 5G will again pose a challenge to regulators trying to get a grasp of the direction this technology will take society from its inception (Wu 2011). Data is the fuel of the digital economy. Data privacy laws address the way in which data is collected, stored, classified, and disseminated (Bygrave 2014). In Europe, the current law that covers personal data protection was approved by Parliament in April 2016 and after a two-year grace period, to allow national governments and regulators to get ready, was finally put into practice on 25 May 2018. The General Data Protection Regulation (GDPR) was enacted in order to enable EU citizens and institutions to get better control of their personal data.

This legislation, which attempts to address the increasing concerns of EU residents about the use of the data, may also have some adverse effects. Owing to its broad reach, many business operators are struggling to adjust to the new legal playing field (Reuters 2018). One of the focus of the law is to ensure transparency and accountability in order to minimize risks of individuals' data from being misused. Organizations, whether European or from abroad, operating in the EU are now required to abide by this legislation and this presents a challenge as some global institutions, such as NGOs and multinational corporations, will have to comply with several regulatory bodies (Bygrave).

There are some important cultural differences that impact the way privacy laws are interpreted in the US and Europe. In the US, privacy protection concerns are mostly addressed by directives, whereas in Europe, the strong concern for personal self-determination and privacy calls are addressed by laws and regulations (Table 1). As we demonstrate in the next section, even inside the EU national governments have chosen different approaches in adopting the GDPR as regulators have intentionally left room for member states to interpret the law.

The 5G roll-out is likely to impact privacy in many realms. In real-time interactions, anything that can be connected to the Internet will be. From transportation, power grid management, home appliances, and even wearables (Jacobson 2019) the coming digital hyper-connected ecosystem implies that everything that can be connected, will be. Thanks in great part to 5G, in the next five years, there will be hundreds of billions of sensors connected and an enormous amount of digital data will be produced. In 2016 alone, there was as much digital data produced as the entire history of the universe going back to the big bang (Table 2).

The datafication of society is operating at full throttle. Every day, 5 billion searches are made, 65 billion messages are sent on WhatsApp, 4 terabytes of data are created from connected automobiles, 294 billion emails are sent and 4 petabytes of data are produced on Facebook. It is estimated that by 2025, 463 exabytes of digital data will be created daily around the hyper-connected world, and the 5G infrastructure will be an integral part of the coming digital society (VisualCapitalist 2019).

The GDPR is becoming the global standard for how privacy and privacy protection laws should be molded. In essence, it sets the rules by which companies treat our personal data. Legislators who wrote this legislation cherished the idea of consent, which means companies often have to ask users for permission to use their data. The law also stipulates that third party data sharing will be more restricted since they will have to offer a reasonable explanation for why and how long they need the data and EU residents now have the right to request their personal data from companies.

When it comes to the protection of data of EU residents, the law (GDPR) guarantees that not only organizations established in the EU, but also abroad, such as Amazon, Facebook, and Weibo, are required to comply with the terms of the law. The scope of the law is broad to the extent that it covers not only digital data but also paper or other form, so long as the type of data can be used, directly or indirectly, to identify an individual.

Table 1 Summary of Data Protection Regulation in Europe.

Data Protection Regulation in Europe	
Finland	New privacy regulation approved by the Finnish government in 2018 goes a step further than the GDPR protecting. The new legislation also increases the power of regulators to administer steep fines on individuals and institutions that breach the law. Based on the new law, children's date and age of consent states that public and private institutions, including individuals, will no longer be able to retrieve data of minors younger than 13 years old
Germany	Germany has historically had some of the most comprehensive data protection laws in Europe. The German Federal Data Protection Act (Bundesdatenschutzgesetz) was adopted in 1970. In the following decade, Constitutional Court drew a distinction between the right to information self-determination from the right to respect for personality. In 2001, the parliament amended the Federal Data Protection Act by creating a provision, which incorporated the recommendations of EU Directive 94/46/EC. Since 2009, Germany has had some of the strictest data protection laws in Europe. However, as the GDPR supersedes national law, German regulators are required to apply GDPR standards when necessary
France	The French Data Protection Bill was introduced by the Ministry of Justice in December of 2017. The new proposed legislation revises the previous 1978 French Date Protection Act. The new bill attempts to balance the increased need for access to personal data with the necessity to protect the privacy of some critical data, such as medical records, criminal records, data of underage citizens, genetic data, etc. In 2017, France passed a data protection law which called for the lowering of the age of consent from 16 to 15 years old. In addition, in 2018, France adopted a law that imposes hefty fines, up to 125 000 euros, on operators that fail to provide adequate data protection to users
Italy	Rather than passing new a legislation, a decree was signed in 2018 that requires data operators to comply with the GDPR by introducing new code of conducts and guidelines. The decree maintained GARANTE as the national data protection agency in charge of guaranteeing compliance with the new EU legislation. The decree also stipulates that the age of consent was reduced to 14 years old and data controllers are required to design simple, clear, concise, and objective consent forms for children
Spain	The privacy and data protection law was enacted in December of 2018. The Protection of Personal Data and the Guarantee of Digital Rights targets five specific issues: political parties and personal data processing, digital rights at work, object of the law, data subject rights, and data protection officers. The Spanish legislation goes a step beyond the EU law by offering increased personal data protection. The expansion of data rights is stipulated in the law by addressing the right of parents to access, modify, suppress, and oppose on behalf of their children
Portugal	In June of 2018, Portugal passed the Execution Law of the General Data Protection Regulation. A regulation approved by the Portuguese Data Protection National Commission lists the types of activities to be covered by the Data Protection Impact Assessment (DPIA). The purpose is to mitigate the threats posed by unnecessary exposing of personal data during the implementation of projects, systems, protocols, strategies, and policies. The types of data included are health data electronic devices, large scale profiling data, locators and trackers of individual subjects by organizations, biometric data for identification and genetic data

(continued overleaf)

Table 1 (Continued)

Data Protection Regulation in Europe	
The Netherlands	In the case of the Netherlands, the GDPR replaced the Dutch Data Protection Act. The Dutch Data Protection Authority (DPA) proactively instituted rules and compliance obligations to GDPR ahead of its EU counterparts. The new rules determine that failure to comply may result in the incurrence of fines up to 1 million Euros, depending on the type and severity of the infraction. Dutch authorities also streamlined the process for individuals or companies to report data breach or misuse of data by contacting the DPA website and reporting the violation
Poland	The Polish Data Protection Act (PDPA) was passed in order to facilitate the implementation of the EU's GDPR. However, the PDPA lacks enforcement mechanisms as authorities are not allowed to institute fines when an infraction has been detected. In addition, Poland is in the process of adjusting other laws, such as telecommunications, commerce, and copyright in order to comply with GDPR. This work falls under the jurisdiction of the Ministry of Digitization. One of the concerns brought forth is the low fines instituted for public agencies, which is capped at 25 000 Euros
Denmark	In 2000, the Danish government enacted the Danish Act on Process of Personal Information in which it stated that personal data should be collected only for specific, legal and explicit reasons. It also stated that it should be accurate and not be excessive. The Danish Data Protection Act was passed in 2018 and it adopts and amends the GDPR by including in the regulation sections that were specifically designed to be interpreted by nation states

Table 2 Daily Digital Data Production in 2020 (estimation).

500 million tweets
294 billion emails
4 billion email users around the globe
350 million photos
100 million hours of video
95 million photos and videos shared on Instagram
28 petabyte of data extracted from wearables
5 billion searches
200 billion connected devices (IoTs)

Within the jurisdiction of the law an enterprise (private of public), an organization or individuals can collect private data for these purposes

- To execute a contract, such as a purchase of service or an employment agreement
- To fulfill a legal obligation, as in the case of the place of employment providing personal data to welfare agencies
- To protect vital interests, for example, to collect personal information to protect one's life
- To perform bureaucratic duties, such as schools and hospitals

- To carry out legitimate interests, as in the case of a bank using personal data of clients to benefit them, such as offering lower rates

Any other form of personal data extraction and use without the consent of the individual is strictly forbidden. When an institution requires the consent of citizens to access their personal data, this consent has to be explicitly given. According to the law, for example, it would not be enough to simply click on an icon to indicate interest to be excluded from receiving emails. A clear and explicit agreement has to be made in order to authorize the other party to access and use the data. Before the decision is made, individuals have the right to receive detailed information of who is requesting the data and for what purpose. Individuals also have the right to know how long the information will be used and stored and if the data will be shared with third parties. The law also states that this information should be clear and easy to understand.

In the case when a specific consent has been given to an organization to access the person's data, an individual can at any moment contact the institution and revoke the consent, at which point the organization will no longer be legally allowed to extract or use this data for any purpose. In a few rare specific cases, such as in a critical scientific study, or if the data extraction and manipulation is performed by public officials for the purpose of national security, the public interest may prevail over individual rights.

It has taken the European Commission almost five years to get the law in the books. There have been a few cases that served to illustrate the importance of having a law that would address the pertinent issue of data protection. The case of Google and Facebook in privacy by design (Rubinstein and Good 2012) fell under the EU laws that were written and adopted more than 20 years ago when both companies did not even exist and the technologies that today power the Internet were in its infancy and most people did not use online banking, reserved trips online, or talked to their friends on the Internet.

In the current law, the idea of fair information practice was to be incorporated in the design and build-in of software, hardware, and services, in order to guarantee that privacy would be addressed from the inception of new technological systems. The most efficient way to ensure the implementation of this legal specification was to establish specific and concrete measurable. In addition to the development of software interfaces that complied with privacy regulation. In his research, Rubinstein found that the main obstacle to implementing privacy by design was not the technical barriers or cumbersome regulatory infrastructure, but the competing interests between business and individuals. He concluded that "business interests overshadow privacy concerns."

The current GDPR is exactly what it states, a regulation and not a directive. There are some significant differences between them. The directive is a piece of legislation that has been brought out by the EU, but it needs to be implemented into member states legislation before it effectively becomes law. A regulation is also approved on the EU level but automatically becomes law without the necessary national ratification process. The law, however, was written in a way that intentionally leaves some room for national interpretation on specific items, either due to cultural awareness or legal practice. With this in mind, it is important to remember that the GDPR is legally binding and should be observed and respected by all member states.

The type data that concerns the law is personal data, otherwise, it falls outside the scope of the law. What would be the legal interpretation of what is personal and what is not? This law has attempted to settle this debate by offering its own understanding of the concept. Personal data, under the new regulation, can be considered to be personal any information that may identify individuals, such as audio, address, video, texts, but also online identifies, such as IP addresses.

The new EU law also introduces and discusses the idea of pseudonymous data, which is data in which attempts have been made to identify the subject. Profiling is also covered under the law. So if genuine efforts have been made to pseudonymize personal data, the law looks favorably at these efforts in court cases. Effectively, GDPR will apply to all types of data collected, whether it is directly identifiable or quantitative and technical data from or about any EU resident.

One of the new aspects of GDPR is the scope and reach of this law that extends to other countries and territories under the concept of extra-territoriality. For that matter, institutions, whether private or public, with or without a presence in the EU, will fall under this legislation, so long as these organizations offer goods or services to EU residents or monitor the behavior of EU residents.

The GDPR legislation supersedes all nation states privacy laws. However, EU regulators were conscious about the necessity for nation states to interpret some part of the law in accordance with national culture and practice so for that purpose EU countries have adopted their own strategy in implementing GDPR at home.

The Risks and Threats to Privacy in 5G

There is still a significant amount of confusion about personal data protection in 5G. Are enhanced security features available in 5G going to assure the protection of my data? Engineers recognize that there are no infallible or invulnerable networks and the maintenance, improvements, and protection of the technical infrastructure is an ongoing process. One of the main features of 5G is the ability users will have to download and upload vast quantities of data at hyper speeds with very low latency. However, the underlying internet infrastructure will remain the same. The same mechanisms that data companies, hackers, and intelligence agencies exploit today to access our personal data will continue to exist. In addition, there will be a slow transition from 3G, 4G, or 4.5G to the new 5G protocols and engineers designed the infrastructure with the purpose to accommodate the previous protocols (Cave 2018; Jacobson 2019).

Vulnerabilities and risks to the networks will continue to exist and will likely be exploited as nations transition to 5G. Owing to the structure of millimeter waves, one of the requirements of 5G is that antennas must be located much closer to one another which will allow for a precise geographic location of individuals and devices. This could pose a significant risk to the privacy and security of users, especially those who are more vulnerable.

As 5G and IoTs interlace, billions of devices will be producing and transmitting data without any interference from users. From automated cars, to cellphones, to wearables, many personal devices are now being designed by default to be constantly connected to the internet. As ubiquitous computing becomes the norm, some of these personal devices will have different levels of security and any design flaw, such as hardcoded and embedded credentials or unpatched vulnerabilities can be exploited (Miller 2019).

Researchers from the University of Oulu in Finland have identified several privacy issues related to 5G (Liyanage et al. 2018). Based on their analysis, we shed light and discuss some of the more significant matters that could represent a threat to privacy in 5G. Moreover, we discuss the threat presented by overzealous security agencies which pose a danger to civil liberty and privacy by operating on the outskirts of the legal system in order to extricate intelligence from digital data.

Data confidentiality: It is related to the protection of data and its access without authorization. For confidentiality to be absolute only those who were granted access should be able to reach the data. In 5G, many organizations in vertical industries will be involved in the process and manipulation of user's data, and it is critical that privacy agreements are established between all parties involved, including network operators.

Data ownership: In 5G and cloud computing, the idea of who owns the data is as important as where is the data. The EU data economy is on its infancy but it is vital that legal parameters are established to determine who owns the digital data. Clearly defined legal arrangements should be undertaken between operators and users.

Shared environment and trust: Not unlike other networks, the 5G infrastructure will be shared between several operators and users. In some cases, even competitors will be sharing network resources. In this scenario where multiple actors with different intents and purpose operate, unauthorized exploitation, and breach of personal data, such as distributed denial of service (DDoS) may occur. In this scenario, where several actors operate with distinctive goals, there is a risk that the level of engagement in network security and data privacy will be uneven or cumbersome.

Loss of visibility: Communication service providers (CSPs) determine the level of security to their systems but may be reluctant to share the security strategy and measures with the mobile operator. As a consequence, privacy concerns may falter as operators lose full visibility of the network.

The globalization of data flow: As we live in a globalized digital world, many networks and computer systems are interconnected around the planet. In 5G, data will move between countries at a hyper speed and the necessity to protect this data will be imperative. Operators and regulators have to clearly define what constitutes private data, how it will be transferred, stored, and organized. What may a controversial privacy concern in a country, such as sexual orientation or political affiliation, may be a completely normal issue in another. This is a challenge the GDPR has attempted to address but its efforts may fall short as some nations have distinctive legislation that does not address privacy concerns in similar ways as GDPR.

Hacking: It usually occurs when an unauthorized intrusion to a computer or network takes place. 5G will be susceptible to hacking as cloud computing, IP, and web-based attacks will exploit the vulnerabilities of the technology. Technical networks are never completely bullet-proof to prevent attacks and hackers will use their software and hardware expertise to extract data for nefarious purpose that could compromise individuals' privacy. Hacking may be carried out by individuals, criminal organizations, or government agencies under the banner of national security.

Security agencies: Although this could be included in hacking, we decided to distinguish this threat from the others presented by the researchers based on its geopolitical significance. Over the past few years, some former security agents turned whistleblowers, such as former NSA contractor Edward Snowden and William Binney, have shed light into the unscrupulous and overzealous methods utilized by the NSA, CIA, and other intelligence agencies around the world on how and to what extent they are willing to bend the law in order to achieve their goals. US government programs such as

PRISM, where private enterprises, among others Microsoft, Facebook, Google, Apple, and Yahoo, as well as national governments, such as Israel, New Zealand, Australia, UK, Canada, and others cooperated with the US intelligence community to collect data of individuals without their consent (Connor 2019). At the moment, the politics of 5G is on overdrive as the US and China attempt to position their technological advancements as a catalyst to global hegemonic power (Watts 2019).

Concluding Thoughts

EU Government officials and technology developers are eagerly promoting the benefits of the 5G with negligible regards for the public's concern about privacy. However, 5G is not a panacea, and it comes with a price tag society may not be fully informed about or ready to accept. Issues related to the continued and accelerated corrosion of privacy are likely to become 5G's biggest challenges to a successful implementation and deployment. The GDPR in Europe is the current safeguard most EU consumers have to ensure their personal data is not being unintentionally used by uninvited parties. As it is often the case, laws tend to address recognizable issues. In the case of 5G and the unknown vertical industries it may foster, the GDPR will likely be inadequate to address yet undetected potential threats to privacy that will arise in the near future as we move steam ahead into a surveillance society driven by the logic of monetization of personal data (Zuboff, Silverman).

The coming 5G infrastructure will connect billions of data-producing devices and sensors worldwide and this datafication of society may significantly alter the way we live. In addition to lighting speeds to communication technologies, countries are investing heavily in the deployment of 5G and the deployment of the infrastructure. The US, China, the EU, Japan, South Korea, and others are engaged in a fierce global competition to be the first to take full advantage of this technology. In order to accomplish these lofty goals, countries must ensure the development and maintenance of this technology will not present significant threats to the public and society in general.

5G will be more than just a disruptive technology (Cave 2018; Jacobson 2019).

5G networks will facilitate the development of advanced robotics, Artificial Intelligence, Big Data, IoTs, and the possibilities for new verticals in the sectors of automobile, energy, industry, and health care demonstrate the reach and significance 5G will have for society. Consequently, the amount of digital data produced will dramatically increase in the near future.

China is pressing ahead in the development of this technology and the US and others are concerned with the speed, structure, and pace of China's recent technological advancements. The issue of 5G technology plays into the scenario of global geopolitics and countries' national security interests. One way where this has transpired in the global scene has been in the US and its allies pursue of banning Chinese products (Villas Boas 2019).

The argument presented in this article is that the coming 5G technology and the infrastructure required for its deployment and operation, should be considered as an integral and indispensable part of a functioning society and as such, measures to protect its integrity and its users' privacy must be a priority not only for national governments but also the private sector and civil society.

Related Article

5G Mobile Networks Security Landscape and Major Risks

References

Agiwal, M., Roy, A., and Saxena, N. (2016). Next generation 5G wireless networks: a comprehensive survey. *IEEE Communications Surveys & Tutorials* 18: 1617–1655.

Al-Debei, M.M., Al-Lozi, E., and Fitzgerald, G. (2013). Engineering innovative mobile data services: developing a model for value network analysis and design. *Business Process Management Journal* 19: 336–363.

Briglauer, W., Cambini, C., Fetzer, T., and Hüschelrath, K. (2017). The European electronic communications code: a critical appraisal with a focus on incentivizing investment in next generation broadband networks. *Telecommunications Policy* 41: 948–961.

Bygrave, L. (2014). *Data Privacy Laws: An International Perspective*. UK: Oxford University Press.

Cave, M. (2018). How disruptive is 5G? *Telecommunications Policy* 42: 653–658.

Connor, B.T. (2019). *Government and Corporate Surveillance: Moral Discourse on Privacy in the Civil Sphere*. Information, Communications & Society. https://www.tandfonline.com/doi/full/10.1080/1369118X.2019.1629693 (accessed 5 July 2019).

Desjardins, J. (2019). How much data is generated each day. Visual Capitalist. https://www.visualcapitalist.com/how-much-data-is-generated-each-day (accessed July 2019).

European Electronic Communication Code (EECC) (2018). Directive (EU) 2018/1972 of the European Parliament and the Council of the European Union.

Frias, Z. and Martinez, J.P. (2018). 5G networks: will technology and policy collide? *Telecommunications Policy* 42: 612–621.

Habib, M.A., Ahmad, A., Jabbar, S., et al. (2019). Security and privacy based access control model for Internet of connected vehicles. *Future Generation Computer Systems* 97. 687–696.

Jacobson, A. (2019). The 5G future will be powered by AI. *Network Computing*. https://www.networkcomputing.com/wireless-infrastructure/5g-future-will-be-powered-ai (accessed 26 June 2019).

Lemstra, W. (2018). Leadership with 5G in Europe: two contrasting images of the future, with policy and regulatory implications. *Telecommunications Policy* 42: 587–611.

Liyanage, M., Salo, J., Braeken, A. et al. (2018). *5G Privacy: Scenarios and Solutions*. IEEE. doi: 10.1109/5GWF.2018.8516981.

Matinmikko, M., Latva-aho, M., Ahokangas, A., and Seppänen, V. (2018). On regulations for 5G: micro licensing for locally operated networks. *Telecommunications Policy* 42: 622–635.

Matinmikko-Blue, M., Yrjölä, S., Seppänen, V. et al. (2019). Analysis of spectrum valuation elements for local 5G networks: case study of 3.5 GHz band. *IEEE Transactions on Cognitive Communications and Networking* 5: 741–753.

Miller, M. (2019). Hardcoded and embedded credentials are an IT security hazard. *Beyond Trust*. 26 February. https://www.beyondtrust.com/blog/entry/hardcoded-and-embedded-credentials-are-an-it-security-hazard-heres-what-you-need-to-know (accessed 19 July 2019).

Reuters (2018). Top 5 concerns with DGPR compliance. https://legal.thomsonreuters.com/en/insights/articles/top-five-concerns-gdpr-compliance (accessed 28 June 2019).

Rubinstein, I.W., Good, N. (2012). Privacy by design: a counterfactual analysis of Google and Facebook privacy incidents. New York University Public and Legal Theory Working Paper 347.

Samdanis, K., Costa-Perez, X., and Sciancalepore, V. (2016). From network sharing to multi-tenancy: the 5G network slice broker. *IEEE Communications Magazine* 54: 32–39.

Silverman, J. (2017). Privacy under surveillance capitalism. *Social Research* 84 (1): 147–164.

Villas Boas, A. (2019). Wuawei has been blacklisted by the US Government. *Business Insider*. 20 March. https://www.businessinsider.com/huawei-us-ban-similar-to-zte-us-ban-20195?r=US&IR=T (accessed 17 June 2019).

Watts, G. (2019). US is losing the 5G war to China. *Asia Times*. https://www.asiatimes.com/2019/07/article/us-is-losing-the-5g-war-tochina (accessed 1 July 2019).

Weber, A. and Scuka, D. (2016). Operators at crossroads: market protection or innovation? *Telecommunications Policy* 40: 368–377.

Wu, T. (2011). *The Master Switch: The Rise and Fall of Information Empires*. NY: Random House.

Zuboff, S. (2019). *The Age of Surveillance Capitalism: The Fight for a Human Nature in the New Frontier of Power*. NY, USA: Hachette Book Group.

Further Reading

Ahmad, I, Kumar, T., Liyanage, M., et al. (2017). 5G security: analysis of threats and solutions. IEEE Conference on Standards for Communications and Networking (CSCN), Helsinki, 2017, pp. 193–199.

Bräutigam, T. and Miettinen, S. (eds.) (2016). *Data Protection, Privacy and European Regulation in the Digital Age*. Faculty of Law University of Helsinki.

Council of Europe (2008). On the identification and designation of European critical infrastructures and the assessment of the need to improve their protection. *Official Journal of the European Union*. 23 December. Council Directive 2008/114/EC, 8 December. https://eurlex.europa.eu/LexUriServ/LexUriServ.do?uri=OJ:L:2008:345:0075:0082:EN:PDF (accessed on 25 June 2019).

European Commission (2019). Security of 5G Networks. Commission Recommendation, Strasbourg, 26 March 2019. https://ec.europa.eu/digital-single-market/en/news/cybersecurity-5g-networks.

FitzGerald, D. (2019). 5G race could leave personal privacy in the dust. *The Wall Street Journal* (11 November). https://www.wsj.com/articles/5g-race-could-leave-personal-privacy-in-the-dust-11573527600

Gandy, O.H. Jr. and Nemorin, S. (2019). Toward a political economy of nudge: smart city variations. *Information Communication and Society* 22 (14): 2112–2126.

Hiltunen, K., Matinmikko-Blue, M., and Latva-aho, M. (2018). Impact of interference between neighbouring 5G micro operators. *Wireless Personal Communications* 100 (1): 127–144. doi: 10.1007/s11277-018-5617-5.

Khajuria, S. and Skouby, K.E. (2017). Privacy and economics in a 5G environment. *Wireless Personal Communications* 95: 145. doi: 10.1007/s11277-017-4421-y.

Klitou, D. (2014). *Privacy-Invading Technologies and Privacy by Design: Safeguarding Privacy, Liberty and Security in the 21st century*. The Hague: T.M.C. Asser Press.

Liyanage, M., Salo, J., Braeken, A. (2018). 5G Privacy: Scenarios and Solutions. doi: 10.1109/5GWF.2018.8516981.

Liyanage, M., Ahmad, I., Abro, A.B. et al. (2018). *Comprehensive Guide to 5G Security*. Hoboken, N.J.: John Wiley & Sons.

Marr, B. (2018). How much data do we create everyday? The mind-blowing stats everyone should read. *Forbes*. 18 May. https://www.forbes.com/sites/bernardmarr/2018/05/21/how-much-data-do-we-create-every-day-the-mind-blowing-stats-everyone-should-read/#5fb6cf1860ba (accessed 17 July 2019).

Matinmikko, M., Latva-aho, M., Ahokangas, P. et al. (2017). Micro operators to boost local service delivery in 5G. *Wireless Personal Communications* 95 (1): 69–82.

Sgora, A. (2018). 5G spectrum and regulatory policy in Europe: an overview. 2018 Global Information Infrastructure and Networking Symposium (GIIS), Thessaloniki, Greece, 2018, pp. 1–5.

Smith, O.A. (2018). Finland's beefed-up Data Protection Act to take effect on January 1st. *Helsinki Times*. 28 December. http://www.helsinkitimes.fi/finland/finland-news/politics/16070-finland-sbeefed-up-data-protection-act-to-take-effect-on-january-1st.html (accessed 15 June 2019).

Tikkinen-Piri, C., Rohunen, A., and Markkula, J. (2018). EU General Data Protection Regulation: Changes and implications for personal data collecting companies. *Computer Law & Security Review* 34 (1): 134–153.

Van Schalkwyk, F. and Verhulst, G. (2017). *The Social Dynamics of Open Data*. Oxford: African Minds.

Zhang, S., Wang, Y., and Zhou, W. (2019). Towards secure 5G networks: a survey. *Computer Networks* 162: 106871.

15

Structural Safety Assessment of 5G Network Infrastructures

Rui Travanca[1], Tiago de J. Souza[2], and João André[3]

[1] *DIGAMA Engineering Consultants, Lisbon, Portugal*
[2] *American Tower of Brazil, São Paulo, Brazil*
[3] *National Laboratory for Civil Engineering, Lisbon, Portugal*

Introduction

The advent of digital technologies has dramatically changed every aspect of our lives. Computer-based communications became a key factor in our modern society, allowing what is called an inclusive digital society and deeply changing individuals' daily social and economic lives. The 5G is only the next big step.

Despite the amount of opportunities created, it is important to underline that our common future also has new challenges. As an example, the strong dependency of any other socioeconomic system upon this particular sector (Esposito et al. 2018; Gomes et al. 2016; Mauthe et al. 2016; Rak et al. 2016; Travanca and André 2018). For instance, nowadays critical infrastructures used in the energy, transportation, or even water-supply sectors are already being monitored, supported, and controlled by telecommunication systems.

Both natural and man-made disasters represent serious threats to communication structures (Figure 1), which can result in the unavailability of services and/or low-quality services. Considering that, during a disaster, citizens need to be able to communicate and receive updated information, or to inform rescue teams about their location, it is clear that the resilience of telecommunication systems is crucial, and communication structures are critical assets needed for achieving it (Esposito et al. 2018; Gomes et al. 2016; Mauthe et al. 2016; Rak et al. 2016; Travanca and André 2018).

The failures involving communication structures are among the most common types of incidents observed in the telecommunication sector, and often leading to both economic and social consequences; whether partial or complete, these structural failures can lead to severe damage to critical assets, such as the interruption of vital socioeconomic activities, or causing the delay of critical emergency and relief efforts. In both cases, the consequences are unpredictable and difficult to estimate (Esposito et al. 2018; Gomes et al. 2016; Mauthe et al. 2016; Rak et al. 2016; Travanca and André 2018). Despite the increasing reliability and resiliency of communication networks, the level of risk associated with failures remains severe (Esposito et al. 2018; Gomes et al. 2016; Mauthe et al. 2016; Rak et al. 2016; Travanca and André 2018).

The Wiley 5G REF: Security. Edited by Rahim Tafazolli, Chin-Liang Wang, Periklis Chatzimisios and Madhusanka Liyanage.
© 2021 John Wiley & Sons Ltd. Published 2021 by John Wiley & Sons Ltd.

(a)   (b)   (c)

Figure 1 Examples of communication structures. (a) Monopole 50-metre-high. (b) Lattice tower 50-metre-high. (c) Guyed mast 100-metre-high. Source: Courtesy of DIGAMA Engineering Consultants.

A clear example of the above are the wildfires that occurred in June 2017, in Portugal, one of the worst natural events in the recent Portuguese history. These wildfires severely affected telecommunication services, causing a significant obstacle for an efficient and effective planning, command, and execution of fire-fighting and rescue operations. It is generally accepted that these communication issues increased the overall number of human life losses (Esposito et al. 2018). Another notable example is the Hurricane Katrina on 29 August 2005. This hurricane struck the Gulf Coast of United States, causing catastrophic damage to the telecommunication and power infrastructures. Devastating effects on both infrastructures delayed rescue efforts, blocking early responses, and made impossible in the worst-hit areas to call for aid (Miller 2006; The White House 2006).

Furthermore, direct threats to the integrity of these structures are drastically increasing. Climate change is a reality, not a particular event which can be addressed on a case-by-case basis. Climate change, natural catastrophes, and the failure of critical infrastructures are ranked at the top of the 2019 Global Risks (WEF 2019) database prepared by the World Economic Forum, and for which less progress has been made in the past years. The latest UK National Risk Register of Civil Emergencies (UK Cabinet Office 2017) also considers the failure of critical infrastructures to have a high risk level and, therefore, a priority risk.

Extreme weather and climate changes result in communication structures being exposed to recurrent, different, and more extreme hazards. In other words, a situation once considered exceptional but acceptable, may become unexceptional and therefore unacceptable (European Commission 2013a), resulting in the collapse of the structure, and causing the disruption of critical services. In this expected scenario, climate change will greatly increase future losses.

As a robust physical infrastructure is an essential need to economic functioning and growth, the worldwide economy is exposed to increased risks. Therefore, there is an urgent need to limit the consequences of failures and accelerate service response capabilities, through both engineering solutions and managing consumer expectations (The Royal Academy of Engineering 2011). Irrespective of the success of mitigation efforts, the impact of climate change will increase in the coming decades because of the delayed impacts of past and current greenhouse gas emissions. Thus, while the efforts must continue toward mitigate its effects, adaptation measures to deal with economic, environmental, and social costs must also be taken.

The growing importance of the telecommunication sector and the increasing external threats to the infrastructures happen simultaneously with other aggravating factors, such as (i) the attainment of the maturity of existing communication structures; and (ii) the aggressive conditions of today's markets, with the sector no longer managed by public entities, but instead being privatized and fragmented. In addition, many structures used in this sector have been used as an additional income resource to telecom companies, by way of leasing their exploitation to third parties. Furthermore, communication structures suffer from the lack of regulation and general accountability, and clear understanding of systems' interdependencies.

Although the main responsibility for setting goals for the protection of critical infrastructure rests primarily with governments, the application of the necessary measures to reduce the vulnerability of privately owned assets depends mainly on the private-sector action, but currently there are not adequate incentives to fund it. Indeed, by adopting measures to improve resiliency, companies may face a backlash from the markets for taking such actions if they raise costs and reduce profits in the short term (Auerswald et al. 2005).

The present reality requires a paradigm shift in relation to the analysis, design, construction, and maintenance of communication structures. Standing on the edge of the 5G technological revolution, because of its scale, scope, and complexity, its impact on our society surely involves outstanding opportunities and benefits but, unquestionably, also large and dangerous threats. It is not yet possible to know how it will develop, but one thing seems clear: this process must be an integrated and comprehensive initiative, involving all stakeholders, both from the public and private sectors, to academia, and civil society.

Discussion

Telecommunication Sector Background

In 2017, the 54 telecom companies listed on the Forbes Global 2000 (Touryalai and Stoller 2018) generated nearly US$1.5 trillion in revenue, and is expected that this market continues to gain space and value, since the number of mobile connections worldwide is predicted to reach nine billion by 2020, about twice the amount of 2009.

The recent years created the opportunity for a new market to emerge: telecom tower companies, which are infrastructure companies that purchase assets (i.e. infrastructures) from a third company and lease them back. This market has been growing since 2009, creating the opportunity to telecommunication companies to raise capital which can then be reinvested in other markets or technologies. This new market created a nearly US$300 billion infrastructure asset class, which represents almost 70% of the world's investible communication structures.

While there is certainly a great diversity in business models for these infrastructure companies, they can be categorized into two main groups: (i) completely independent companies, e.g. privately owned builders that started retaining and acquiring assets in the 1990s; and (ii) operator-led infrastructure companies, in which 51% or more of the equity is retained by the parent mobile network operators. A third group worth mentioning is power-as-a-service infrastructure companies, which provide a full service including both communication structures and power supply.

In general, there are large uncertainty levels and complexities involved in developing strategic plans for the future in the telecom sector. Mainly due to market pressures, investments in communication structures are mostly short-term intended, typically focusing on replacing and renewing as needed, rather than modernizing key structures. Expenditure only takes place as a response to a crisis, rather than proactively planning and managing infrastructures, by considering resilience and climate change effects, and the main focus is in operating near the maximum capacity of the structure, which is seen as being an optimal and efficient management decision.

The above, however, causes the system to be less resilient against any anticipated or unknown event that may occur within its lifespan; optimization for one set of conditions creates vulnerabilities to changes in those same conditions. In fact, service disruptions caused by the physical destruction tend to be more severe than those caused by disconnection or congestion, mainly due to the time and funding needed to repair or even replace such structures (Figures 2 and 3).

Research and innovation toward communication structures, such as experiments in real monopoles (Figure 4), require time and cannot keep up with the fast pace of the telecommunication sector, hence are given low priority. Network operators often expect that the above issues can be addressed when old assets are replaced, using improved or, more predictable, adapted technology.

Wireless networks have a higher degree of variability in their vulnerability to physical destruction of nodes and to loss of service; as an example, broadcasting facilities are typically centralized at the metropolitan scale, making them extremely vulnerable,

(a)   (b)   (c)

Figure 2 Construction of a 40-m-high lattice tower. (a) Lattice tower foundation. (b) Lower parts of the tower. (c) Upper parts of the tower. Source: Courtesy of DIGAMA Engineering Consultants.

(a)   (b)   (c)

Figure 3 Construction of a 30-m-high monopole. (a) Monopole foundation. (b) Lower part of the monopole. (c) Upper part of the monopole. Source: Courtesy of DIGAMA Engineering Consultants.

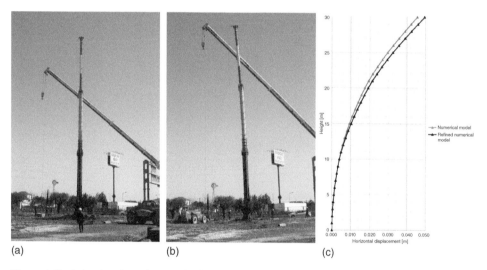

(a)   (b)   (c)

Figure 4 Static load testing of a 30-m-high monopole. (a) Unloaded structure. (b) Loaded structure. (c) Displacements for 15 kN. Source: Courtesy of DIGAMA Engineering Consultants.

as the 11 September 2001 terrorist attack on the World Trade Center clearly showed (Townsend and Moss 2005; National Commission on Terrorist Attacks 2004).

Widely seen as the most sophisticated urban infrastructure, telecommunication networks are damaged in nearly every major urban disaster. However, it is not the size of the disaster that is the only decisive factor, but also how the destruction location and degree coincides with both old and new facilities used for communications.

Consequently, it should be of no surprise that, historically, telecommunication systems have been highly vulnerable to disasters. Earthquakes and other extreme events can destroy structures, cables, flood underground equipment, and disrupt wireless networks (Esposito et al. 2018; Gomes et al. 2016; Mauthe et al. 2016; Rak et al. 2016; Travanca and André 2018).

Communication Structures Background

Due to cost-effective and functional reasons relating to their very nature, communication structures are lightweight and flexible, especially tall monopoles, and lattice towers with a narrow footprint. The foundations are typically isolated from any other structure and may be shallow (footings or semi-deep caissons) or deep (driven, bored, or auger cast piles). The headframe is generally a lightweight steel structure located at the top of the structure, and it is used for mounting multiple antennas or other equipment, allowing access for maintenance purposes (Figure 5).

With the implementation of new technologies, like upgrades to the 4G/LTE, and the necessary dialogue between the structural and the radio engineer (Travanca et al. 2013), the distribution of ancillaries has been corrected in many cases, resulting in safer and more efficient structures, as presented in Figure 6.

Using the nation of Portugal as a case study, covering various typologies of communication structures, in a total of 385, it was possible to conclude that 20% (78) of the observed structures have anomalies (Travanca et al. 2013). This problem was

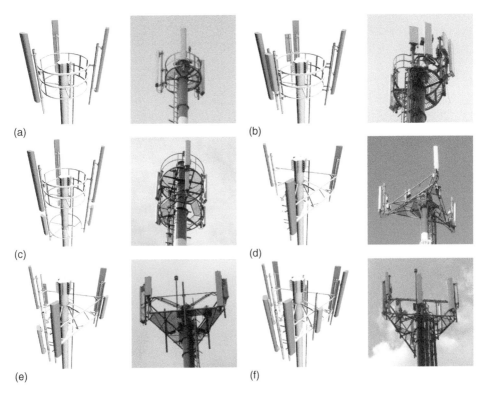

(a)

(b)

(c)

(d)

(e)

(f)

Figure 5 Examples of antenna configurations used on monopoles. (a) Scenario A-01. (b) Scenario A-02. (c) Scenario A-03. (d) Scenario A-04. (e) Scenario A-05. (f) Scenario A-06. Source: Courtesy of DIGAMA Engineering Consultants.

(a)   (b)

Figure 6 Ancillaries arrangements of an upgrade to 4G/LTE. (a) Before the 4G/LTE upgrade. (b) After the 4G/LTE upgrade. Source: Courtesy of DIGAMA Engineering Consultants.

clearly identified in independent studies developed at the University of Aveiro and at the Portuguese National Laboratory for Civil Engineering, which have concluded that the current standards are not prepared for the specific analysis of these structures (Travanca and André 2018; Travanca et al. 2013).

Remarkably, 20 years ago, Solari (Solari and Pagnini 1999) clearly described this issue when referring to monopoles: "Specific standards exist based on empirical calculation criteria definitely emanating from actual physical phenomena. The use of old codes, developed by considering other kinds of structures, is unjustified in this sector. The Eurocode on wind actions and effects on structures as well as other standards of the new generation cannot be applied to these constructions."

The continuous development of higher strength materials used in the construction sector and, as a consequence, the changes in stiffness, mass, and damping of structures should lead to new analysis and design methods for increasing slender structures (Travanca and André 2018; Travanca et al. 2013; Solari and Pagnini 1999; Hawkins 2010; Smith 2007, 2009; Støttrup-Andersen 2009).

In order to adequately and efficiently design and manage communication structures, it is mandatory to obtain a correct understanding of their structural behavior. Therefore, there is an urgent need for the application of advanced research methods, such as wind tunnel testing; monitoring of existing structures, numerical modeling and calibration, in situ testing, to develop a comprehensive risk framework, and review the current standards.

Safety Assessment of Communication Structures

For a better understanding of the present and future of the telecommunication sector, it is important to look to the past. Up to the 1970s specific design procedures were not available for the analysis and design of communication structures. Since 1959, there were American, Canadian, and German standards in existence (EIA 1959; CSA 1965;

DIN 1969). However, wind loading was based on general loading codes of practice and, specific requirements for these type of structures were not properly considered (Travanca et al. 2013; Smith 2007).

Regrettably, there has been a long history of failures of communication structures, far larger than when compared with other structures of equal economic and social importance (Travanca et al. 2013; Solari and Pagnini 1999; Hawkins 2010; Smith 2007, 2009; Støttrup-Andersen 2009). This was identified by the industry in the early 1960s, when a rapid growth of these structures was observed to accommodate the large expansion verified in radio and television broadcasting (Smith 2007). Consequently, a Working Group (namely the WG4) of the International Association for Shell and Spatial Structures (IASS) was set up to study the behavior of these structures, and to produce recommendations for its analysis and design. These recommendations (IASS Working Group 4 1981) were published in 1981, and they were the basis for the development of many national and international standards in this subject over the past 35 years. At that time, these recommendations contained innovative ideas, such as the structural reliability safety format, the adoption of dynamic response procedures, and the consideration of uncertainties associated to the wind action in the design (Travanca et al. 2013; Smith 2007).

In the last decade there has been a growing interest in the field of structural health monitoring (SHM), resulting in the development of new techniques and equipment, such as optical sensors based on fiber Bragg gratings (FBGs). Thus, the recent improvement of optical sensors to study the dynamic behavior of structures is a valuable tool for the evaluation of structural integrity, for the study of the dynamic response of communication structures (Figure 7), and for calibration and validation of design methods (Antunes et al. 2011a,b, 2012a,b,c, 2013).

Several simplified design methods are proposed in the specialized literature, but presently with the advent and development of computer technology, the finite

(a)   (b)

Figure 7 Structural health monitoring of a 45-m-high monopole. (a) General view. (b) Detail of sensor at the top. Source: Courtesy of DIGAMA Engineering Consultants.

element method (FEM) has surpassed any other method in the analysis and design of communication structures.

Regarding predominant actions for the analysis and design of this type of structures, these are the wind action, accumulation of ice, earthquakes, or the combination of them.

Wind is a dynamic action, and therefore communication structures with fundamental frequency below 1 Hz are vulnerable to the turbulent component of the wind, because they can exhibit a resonant response under low wind speeds. Consequently, dynamic analysis is essential for the determination of the response of such structures (Travanca and André 2018; Travanca et al. 2013).

The alternating detachment of vortices from opposite sides creates a force that is variable and perpendicular to the direction of the air flow; such phenomenon mainly affects monopoles, which nowadays represents the majority of the existing communication structures used worldwide in the telecommunication sector. As a consequence, cross-vibrations in the structure may occur if the frequency caused by vortex shedding is similar to the natural frequency of the structure. Since this condition occurs for low and frequent wind speeds, the number of stress-loading cycles may be enough to cause a fatigue failure to happen before the design life of the structure has been reached. Although the abovementioned phenomenon is widely known, and has caused several failures (Travanca and André 2018; Travanca et al. 2013; Solari and Pagnini 1999; Repetto and Solari 2010), the existing procedures for its evaluation are still simplified and empirical.

In the 1980s, wind tunnel testing was carried out using small-scale models, considering different combinations of nonstructural elements such as ladders, feeders, platforms, and/or antennas (Smith 1985). In these studies, divergences were detected between the test results and measurements made on real lattice towers, namely, a higher wind resistance than that measured in real structures. The causes for the above divergences are fundamentally related with the roughness of the various structural elements, the turbulent component of the wind, and parameters that influence the transition between subcritical and supercritical flow zones, but also due to the difficulties in defining the wind action on ancillaries (Georgakis et al. 2009).

An example of such difficulties is the shield effect. Despite being a well-known phenomenon, there is no well-defined method to quantify it, which would lead, in general, to a reduction of the current design values of wind pressure, especially in areas with headframes and multiple antennas, which are currently determined by a simple sum of the wind action on each element (Travanca and André 2018; Travanca et al. 2013; Filipe et al. 2014; Carril et al. 2003).

Ice loading has the following three effects in these structures: (i) increases the mass on structure, so decreases its natural frequency and changes its dynamic response; (ii) thickens the structural members, increasing the wind load; and (iii) the foundation will take more load due to both previous points.

Earthquakes are rarely presented as a cause for damage and/or collapse of communication structures (McClure 1999). However, on 26 December 2004, an earthquake with magnitude 9.2 struck the southern Asian region causing a tsunami. Both events caused the complete destruction of communication systems, seriously damaging local communication structures and equipment (Gomes et al. 2016; Travanca and André 2018; Bendimerad 2013). Likewise, on 12 May 2008, a strong earthquake occurred in Sichuan, China, causing serious damage to the communication systems (Gomes et al. 2016;

(a)   (b)   (c)

Figure 8 Nondestructive tests used for soil/foundations. (a) Detail of the hole being drilled. (b) Parallel seismic test. (c) Magnetometer test. Source: Courtesy of American Tower of Brazil.

Travanca and André 2018; Bendimerad 2013). Therefore, it is important to study whether in recent years there has been an increase of seismic risk, mainly as a result of increased exposure and vulnerability.

Semi-probabilistic approaches based on global and partial safety factors are currently used in the design of communication structures. However, it requires a very careful consideration of the complex nature of the soil–structure interaction (SSI). In some cases, namely involving pile foundations, it is more suitable to base the design in the explicit calculation of the failure probability (Ang and Tang 1984).

The analysis and design of foundations used for communication structures follows the conventional engineering practices but considering additional and specific requirements. The effect of uplift and overturning are frequently critical, especially for self-supporting towers or monopoles, and contrasting to downward loading, as normally encountered in the analysis and design of conventional structures. Diverse and necessary input/output information are required for bearing capacity and settlement analyses, such as stratigraphic profile, foundation type and dimensions, and constructive methods. Both the spatial variability of soil mass properties and the variability of load applied to the superstructure are the determinant factors that control the choice of the foundation to be used (Chaudhari and Chakrabarti 2012; Davisson and Robinson 1965; Broms 1964).

It is not uncommon, whenever designing or evaluating the bearing capacity of communication structures that the foundation engineer is confronted with lack of information, untrustworthy design, or the absence of as-built documentation (Gandolfo et al. 2015, 2017). The need to perform in-situ characterization of soil/foundations is frequent (Figure 8), with special emphasis for the use of geophysical methods, in particular for deep foundations (Souza et al. 2016, 2017, 2018).

Regardless of the topic in discussion, whether the superstructure or the foundations, one issue is common and needs to be addressed. Current standards have been calibrated to ensure performance expectations by focusing on a single level of assets and based on historical practice.

However, there are a number of examples of new issues and new problems which demonstrate that past experience is often not an appropriate guide because of structural or economic aspects. In addition, large uncertainties may arise in the process of extrapolating past experiences to current problems given that there are differences in methodologies used over time. An obvious example where such problems are emerging will be in the planning, analysis, design, construction, and management of the communication structures, considering resilience and climate change.

Over the last few years, performance-based philosophies have been gradually introduced into design standards, yet still not enough effort has been made with respect to philosophies based on resilience. Structural design standards are typically aimed at ensuring the adequate structural performance of the physical infrastructure within the range of hazards it will be exposed to, rather than the integrity of the service.

As a consequence, in certain limiting situations, being desirable that structures sustain the occurrence of an extreme event, it is not necessary that those structures remain functional. However, in the case of communication structures, service interruptions due to physical damage can create severe delays in the response, relief, and recovery efforts. The longer the time of services disruption is, from few hours to several weeks, the harder it is to recover, thereby amplifying losses.

When minimum levels of communication services are defined, they are based on a maximum number of services interruptions per year, or based on an availability criterion, where the maximum service outage hours per year are fixed. Yet, the concrete application of this approach does not find any support in a comprehensive and holistic framework, lacks in the contribution of interdisciplinary disciplines, in particular, structural engineering, and does not take into account the resilience of systems, the effects of climate change or any other unexpected event.

The telecommunication and energy systems are highly interconnected, either due to the physical proximity of infrastructures or because of their operational interdependence; not only is the telecommunication industry wholly dependent on power supply, but also the energy industry strongly depends on telecommunications to manage their extensive network and grid distribution.

It has long been recognized that understanding these interconnections will help to identify points of critical vulnerability. In this context, even if a single system exhibits a high resilience, it is known that when interdependencies are taken into account the system of systems resilience level can be much lower.

Extreme events highlight that system interdependency can increase the vulnerability to cascade failure of individual systems; for example flooded energy power stations leading to power cuts which thereby could seriously affect the telecommunication networks (Buldyrev et al. 2010; Min et al. 2007; Pederson et al. 2006). However, there is still insufficient knowledge to comprehend how different infrastructure systems influence and are influenced by each other, especially with a wider and interconnected society.

Investment in climate change adaptation and in resilience was proven to be far smaller than the rebuilding costs post-disaster. It has been estimated that an investment of US$1 in resilience saves US$7 in reconstruction costs (UN 2014). A study from the World Bank (Hughes et al. 2010) highlights this statement: "Our primary conclusion is that the cost of adapting to climate change, given the baseline level of infrastructure provision, is no more than 1–2% of the total cost of providing that infrastructure. While there are differences across regions and sectors, the pattern is clear and unambiguous – the cost of adaptation is small in relation to other factors that may influence the future costs of infrastructure." Another study (European Commission 2013b), focused on EU, estimates that the climate change adaptation and resilience-related measures on the energy infrastructure will cost about some €500 million per year until 2020, with a best-guess estimate of the associated benefits from such investments of €870 million per year.

Yet, research shows that the great majority of organizations managing critical infrastructure networks do not include climate change mitigation options, neglecting adaptation strategies and resilience assessments in their strategic plans. Insurers begin to consider this reality as a serious vulnerability (CEA 2009), and insurance premiums will surely rise for those who fail to demonstrate that they developed appropriate infrastructure resilience strategies, including climate change adaptation measures.

Additionally, the economic risk from natural or man-made causes perceived by network operators often represents a small percentage of the capital at risk. This might often be the case, but does not take fully into consideration the follow-up consequences, for example, the loss of life, very high economic losses when compared with potential investment costs, and impacts on GDP due to disruption of flows of goods and/or service, reputational damage, contractual penalties, and the potential for litigation.

There are sound reasons to believe that the communication engineering will continue to innovate and advance technologically in the future; 5G is just the next step. Consequently, communication structures will continue to be subjected to difficult challenges as they need to adapt to service requirements in an increasingly competitive market, in particular with the growing demand and importance of this service. Therefore, it is essential that stakeholders can turn the page to inefficient past practices and commit themselves to comprehensive and continuous planning and management policies of critical infrastructure assets, with the primary goal of reducing uncertainties, risks, and magnitude of adverse consequences; hence, increasing safety, resilience, and sustainability.

Major Concerns to 5G when Compared with Pre-5G Networks

The United Nations International Strategy for Disaster Reduction (UNISDR) defines disaster risk as a function of hazard, exposure, vulnerability, and capacity (UN 2016). Hazard is the process, phenomenon, or human activity that may cause loss of life, injury, or other health impacts, property damage, socioeconomic disruption, and/or environmental degradation. Exposure is another component of disaster risk and defines the situation of people, infrastructures, housing, production capacities, and other tangible assets located in hazard-prone areas; exposure can be described as the human or economic value of the elements exposed to the hazard and it comprises the direct costs and losses of the physical destruction in case of extreme events, and also the indirect costs caused due to service interruption, socioeconomic costs, reputational damage, etc. Vulnerability is defined as the condition determined by physical, social, economic, and environmental factors or processes which increases the susceptibility of an individual, a community, assets, or systems to the impact of hazards. The determination of these costs, losses, and/or consequences, for communication structures, has already been discussed throughout this article. Capacity is the combination of all strengths, attributes, and/or resources available within an organization, community, or society to manage and reduce disasters risks and strengthen resilience, and it will be discussed in the last section of this article.

After these brief terminology presentation it is possible to better understanding of the major issues worth mentioning in terms of the impacts of 5G networks to physical infrastructure that can be identified, at the present time, when compared to pre-5G networks, and that are essentially three, namely (i) increased exposure due to the growing importance of communications systems; (ii) different and additional number of network equipment required to be mounted on existing structures, which may lead to increased vulnerabilities; and (iii) overall increase of the number of the sites required for 5G networks that also may lead to increased vulnerabilities.

The increased exposure is a direct consequence of a growing dependency of our society on digital communications, in general, and on 5G networks, in particular; the determination of these costs and consequences has already been discussed throughout this article.

Also, it is expected that changes in the equipment mounted in communication structures, due to 5G networks, may lead to an increase of the wind loading, e.g. bigger antennas, the need for more equipment installed, co-existence with pre-5G networks equipment, etc.; these equipment's changes when merged with the issues mentioned in previous sections, i.e. structures already operating near the maximum capacity, and the existing difficulties in an accurate and adequate quantification of nonstructural elements such as the combination of antennas, feeders, ladders, or headframes, poses serious difficulties for an accurate analysis and design of existing structures, and/or suitable safety assessment of the future 5G communication structures.

Finally, the expected overall increase in the number of sites needed for 5G networks. In fact, this requirement was already observed a few years ago, when upgrading for 4G/LTE networks. Despite their unit costs are limited, the increasingly large numbers of sites required represent a relevant economic problem. Additionally, as abovementioned, market pressures for economical and faster solutions could result in the use of structures that are not suitable for the intended purpose. Therefore, special care should be exercised when engineering communication structures to host the additional connection points needed for 5G networks, and not use existing structures that were not initially designed for carrying this type of equipment, such as lighting columns or advertising poles, and which are obviously unfitted for this purpose (Figure 9).

Conclusions and Recommendations

The US Department of Homeland Security defines critical infrastructure as: "systems and assets, whether physical or virtual, so vital to the United States that the incapacity or destruction of such systems and assets would have a debilitating impact on security, national economic security, national public health or safety, or any combination of those matters" (USA Congress 2001). Likewise, the European Council defines critical infrastructure as: "an asset, system or part thereof located in Member States which is essential for the maintenance of vital societal functions, health, safety, security, economic or social well-being of people, and the disruption or destruction of which would have a significant impact in a Member State as a result of the failure to maintain those functions" (European Council 2008; Mattioli and Levy-Bencheton 2014).

(a)   (b)

Figure 9 Lighting column 9-m-high used for 4G/LTE. (a) General view. (b) Detail at the top. Source: Courtesy of DIGAMA Engineering Consultants.

Therefore, critical infrastructure comprises networks, systems, sites, facilities, and/or businesses that deliver crucial goods and services to citizens, and support the economy, environment and, more comprehensively, the safety and social well-being.

Communication structures are critical infrastructures and one of today's most pressing challenges is to allocate limited resources in order to reduce as low as practicable the risks posed by natural or man-made hazards to these physical infrastructures. Failure to do so will certainly have enormous consequences, as the value at risk is massive. Not due to the economic value of these specific structures but because present and future society welfare and sustainable development strongly depend on the capability of these structures to continue to provide reliable public services and adapt to our civilization evolution.

In the recent years, many steps have been taken toward understanding climate change and its effects, and developing sectoral plans targeting resilience. Examples are the UK and EU risk registers, national and sectoral plans for resilience and adaptation strategies, research projects at USA with particular emphasis to the investigations carried out at the United States National Infrastructure Simulation and Analysis Center (NISAC), and the recent initiatives at EU level, namely IRRIIS and CIPRnet research projects. However, all these projects lack the contribution from structural engineering.

Design standards often incorporate clauses taking into account the resilience of a specific system, although they rarely address issues concerning system interdependencies. In addition, design standards also include exception clauses for extreme weather events and/or unexpected operating conditions, which often lead to the overlooking of resilience by the organizations that manage these infrastructures. Moreover, penalties payable for eventual losses do not necessarily reflect the tangible costs and/or inconveniences to the customer (Cabinet Office 2011). These limitations will be greatly magnified in the future due to the effects of climate change.

The contribution of multiple engineering fields to a system-of-systems approach toward increasing the safety and resilience of the infrastructures used in the telecommunication sector is a strategic approach that needs follow-up. These infrastructures are of upmost importance to the sustainable development, vulnerability, and resilience of our current societies, which are highly dependent on the availability of this service. The following recommendations aim to promote excellence and innovation from the conceptual design to the operation phase of physical infrastructures in these sector, mainstreaming the topic of physical infrastructure resilience, climate change adaptation, and taking advantage of the opportunities associated, namely to develop new design guidelines, innovative structural monitoring techniques, in-situ tests, simulation and analysis tools, comprehensive planning, and management methodologies:

1) *Managing Risks Rather than Managing Disasters.* In order to be better prepared to crises related to hazards, it is crucial to make proposals to adapt existing structures and develop new infrastructures standards, applying a number of technological resources and design procedures with the main objective of control risks, protect critical infrastructure, and save human lives in the case of an extreme event.

2) *Sharing Information and Knowledge between Relevant Stakeholders.* Nowadays, there is not enough information available and adequate communication between stakeholders of the telecommunication and other critical infrastructure sectors. This is a serious issue and it is critical that this scenario is reversed. It is important that research centers, policy makers, and leading utility companies share experiences, information, and join efforts to develop a comprehensive approach that take into account the safety issue from the conceptual design of any critical asset to its operation, including the interdependencies of interconnected infrastructure assets.

3) *Development of a Clear Procedure for the Identification of Critical Physical Structures.* In order to proactively target the needs and requirements of modern societies, in particular public bodies with responsibilities regarding utility grids, such as the telecommunication sector, it is essential to identify the critical physical assets from among the plethora of existing and future physical infrastructure assets. The establishment of a consistent, clear, robust, and adaptive classification procedure is an immediate necessity. This also includes the characterization of the key communication structures and the topology of existing networks.

4) *Identify and Characterize the Hazards and Assess the Exposure of Communication Structures to Such Events.* For example, there are currently knowledge gaps concerning the effects of the wind action on communication structures, for instance related to vortex shedding, which is critical to fatigue resistance. Therefore, there is an urgent need of use advanced structural monitoring and experimental testing to increase our knowledge about wind-related hazards and assert about their potential relevance, also assessing the exposure levels of key communication structures to such hazardous events, clearly identifying hotspots, with proper consideration of climate change effects, a topic to which very little research has been performed.

5) *Reduce the Current Uncertainty Levels by Improving the Understanding Concerning Resilience of Telecommunications Systems.* It is important to develop system-of-systems simulation software tools, to be able to estimate the consequences of failures occurring within the system and in interconnected systems, but also to identify what are the critical enablers (characteristics, capabilities, and

resources) of the systems that reduce risk and provide robustness and resilience. This information can then be used to achieve more effective standards and design methodologies for critical infrastructures located in areas vulnerable to man-made or natural impacts.

6) *Take Advantage of the Most Recent SHM Techniques and Advanced Numerical Simulation Methods, In Situ testing, and Predictive Models for the Accurate Management of Communication Structures.* Management of these structures can be achieved by adapting existing standards (CEN 2004, 2005a, b, c, 2006, 2007; BSI 1986a, b, 1995, 1999; TIA 2017), developing new ones, and updating design methodologies for infrastructures located in vulnerable areas, namely by promoting the use of probabilistic design and assessment methods. Advanced SHM techniques, namely using the newly developed FBG-based accelerometers, need to be used to gather data, presently virtually nonexistent, related to the structural behavior of communication structures. These data could then be processed by innovative methods and transformed into usable and relevant information that nowadays is still neglected in structural design procedures, i.e. SSI, degradation of materials over time, or dynamic structural behavior. The information obtained from the SHM and from specific experimental tests could be used to validate numerical models of communication structures and thus achieving a more accurate forecasting of the structural behavior. In particular, concerning the influence of the foundation to the structural behavior, laboratory tests enable the use of more sophisticated nonlinear behavior models of the soil, such as Modified Cam-Clay (MCC) and/or Hardening Soil Model (HSM). Concerning deep foundation assessments, new testing techniques such as the parallel seismic (PS) and magnetometer tests could be used with the same objective.

7) *Establish Adequate Levels of Resilience for Key Communication Structures.* This is critical to enable the system to continue to operate without widespread loss or disruption of essential services. However, the need to protect critical infrastructure and save human lives in the case of a major event must be balanced with the amount of available resources, e.g. technical and/or financial.

8) *Develop Preventive and Proactive Adaptation Strategies to Enhance the Safety and Resilience of Communication Structures.* There is a need to study the impacts of preventive strategies to enhance resilience and post-event recovery actions, in particular, regarding existing structures. Contingency strategies need to be developed and analyzed to limit the risk of cascading and disproportionate failures of interconnected infrastructure assets, to enhance security of citizens and assets in vulnerable areas and reduce the socioeconomic impact of natural catastrophes. SHM should be part of the strategies inventory, mainly used to obtain relevant information about the behavior of the structures, which can be used to assess the structural performance against target operational levels and to early detect damages, making it possible to reduce the economic impact of structural anomalies, and to extend the lifetime of the structures.

9) *Develop Better Risk-Informed Decision-Making Taking into Account Resilience and Climate Change Adaptation.* A comprehensive risk management approach needs to be developed, supported by the previous objectives, and also on a risk-informed decision-making methodology that should be used to compare different alternative solutions and identify priority investments to enhance the safety and resilience of telecommunication systems.

Related Articles

Physical-Layer Security for 5G and Beyond
5G Security – Complex Challenges
Security Monitoring and Management in 5G
Privacy in the 5G World: The GDPR in a Datafied Society

References

Ang, A. and Tang, W. (1984). *Probability Concepts in Engineering Planning and Design. Volume II: Decision, Risk, and Reliability*, 574. New York: John Wiley & Sons.

Antunes, P., Travanca, R., Varum, H. et al. (2011a). Dynamic characterization of a radio communication tower with a FBG based accelerometer. Proceedings of the Conference on Lasers and Electro-Optics (CLEO) Pacific Rim, Sydney, Australia, 28 August–1 September, 1403–1405.

Antunes, P., Travanca, R., Varum, H. et al. (2011b). Dynamic monitoring of a mobile telecommunications tower exposed to natural loading with a FBG biaxial accelerometer. Proceedings of the 16th Opto-Electronics and Communications Conference (OECC 2011), Kaohsiung, 4–8 July, 604–605.

Antunes, P., Rodrigues, H., Travanca, R. et al. (2012a). Structural health monitoring of different geometry structures with optical fiber sensors. *Photonic Sensors* 2 (4): 357–365.

Antunes, P., Travanca, R., Rodrigues, H. et al. (2012b). Dynamic structural health monitoring of slender structures using optical sensors. *Sensors* 12: 6629–6644.

Antunes, P., Travanca, R., Varum, H., and André, P. (2012c). Dynamic monitoring and numerical modelling of communication towers with a FBG based accelerometer. *Journal of Constructional Steel Research* 74: 58–62.

Antunes, P., Leitão, C., Rodrigues, H. et al. (2013). Optical fiber bragg grating based accelerometers and applications. In: *Accelerometers: Principles, Structure and Applications*, 27–56. Nova Science Publishers Inc.

Auerswald, P., Branscomb, L., La Porte, T., and Michel-Kerjan, E. (2005). The challenge of protecting critical infrastructure. *Issues in Science and Technology* 22: 77–83.

Bendimerad, F. (2013). *Earthquake Risk Considerations of Mobile Communication Systems*, 55. Diliman: Earthquakes and Megacities Initiative.

Broms, B. (1964). Lateral resistance of piles in cohesive soils. *Journal of the Soil Mechanics and Foundations Division* 90 (2): 27–64.

BSI (1986a). *BS 8100-1. Lattice Towers and Masts – Part 1: Code of Practice for Loading*, 72. British Standards Institution.

BSI (1986b). *BS 8100-2. Lattice towers and Masts – Part 2: Guide to the Background and Use of Part 1: Code of Practice for Loading*, 180. British Standards Institution.

BSI (1995). *BS 8100-4. Lattice Towers and Masts – Part 4: Code of Practice for Loading of Guyed Masts*, 88. British Standards Institution.

BSI (1999). *BS 8100-3. Lattice Towers and Masts – Part 3: Code of Practice for Strength Assessment of Members of Lattice Towers and Masts*, 32. British Standards Institution.

Buldyrev, S., Parshani, R., Paul, G. et al. (2010). Catastrophic cascade of failures in interdependent networks. *Nature* 464: 1025–1028.

Cabinet Office (2011). *Keeping the Country Running: Natural Hazards and Infrastructure*, 98. London: Cabinet Office.

Carril, C., Isyumov, N., and Brasil, R. (2003). Experimental study of the wind forces on rectangular latticed communication towers with antennas. *Journal of Wind Engineering and Industrial Aerodynamics* 91: 1007–1022.

CEA (2009). *Tackling Climate Change. The Vital Contribution of Insurers*, 63. Brussels: Comité Européen des Assurances.

CEN (2004). *EN 1997-1, Eurocode 7: Geotechnical Design, Part 1: General Rules*, 168. European Committee for Standardisation.

CEN (2005a). *EN 1991-1-4, Eurocode 1: Actions on Structures, Part 1-4: General Actions – Wind Actions*, 146. European Committee for Standardisation.

CEN (2005b). *EN 1998-6, Eurocode 8: Design of Structures for Earthquake Resistance, Part 6: Towers, Masts and Chimneys*, 47. European Committee for Standardisation.

CEN (2005c). *EN 1993-1-9, Eurocode 3: Design of Steel Structures, Part 1-9: Fatigue*, 34. European Committee for Standardisation.

CEN (2006). *EN 1993-3-1, Eurocode 3: Design of Steel Structures, Part 3-1: Towers, Masts and Chimneys – Towers and Masts*, 79. European Committee for Standardisation.

CEN (2007). *EN 1997-2, Eurocode 7: Geotechnical Design, Part 2: Ground Investigation and Testing*, 196. European Committee for Standardisation.

Chaudhari, S. and Chakrabarti, M. (2012). Modeling of concrete for nonlinear analysis using finite element code ABAQUS. *International Journal of Computer Applications* 44: 14–18.

CSA (1965). *Standard S37: Antennas, Towers, and Antenna-Supporting Structure*. Canadian Standards Association.

Davisson, M. and Robinson, K. (1965). Bending and buckling of partially embedded piles. Proceedings of the Sixth International Conference on Soil Mechanics and Foundation Engineering, Montreal, 8–15 September, 243–246.

DIN (1969). *DIN 4131: Steel Radio Towers and Masts*. Deutsches Institut für Normung.

EIA (1959). *EIA RS-222: Steel Antenna Towers and Antenna Supporting Structures*. Electronics Industries Association.

Esposito, C., Gouglidis, A., Hutchison, D. et al. (2018). On the disaster resiliency within the context of 5G networks: The RECODIS experience. European Conference on Networks and Communications, Lubjana, 18–21 June, 4.

European Commission (2013a). *An EU Strategy on Adaptation to Climate Change. Adapting Infrastructure to Climate Change*, 37. Brussels: European Commission.

European Commission (2013b). *An EU Strategy on Adaptation to Climate Change. Impact Assessment – Part 2*. Brussels: European Commission.

European Council (2008). Council directive 2008/114/EC of 8 December 2008. On the identification and designation of European critical infrastructures and the assessment of the need to improve their protection, Brussels. *Official Journal of the European Union* 51: 8.

Filipe, J., Travanca, R., Pipa, M. et al. (2014). Monopoles for telecommunications – The influence of the equipment for wind action definition (in Portuguese). Proceedings of the 5th Portuguese Conference on Structural Engineering, Lisbon, 26–28 November, 16.

Gandolfo, O., Souza, T., Aoki, P. et al. (2015). Evaluation of unknown foundation depth using parallel seismic (PS) test – a case study. Proceedings of the 14th International

Congress of the Brazilian Geophysical Society & EXPOGEF, Rio de Janeiro, 3–6 August, 5.

Gandolfo, O., Souza, T., Hemsi, P. et al. (2017). The parallel seismic method for foundation depth evaluation: A case study in Arthur Alvim, São Paulo, Brazil. Proceedings of the 15th International Congress of the Brazilian Geophysical Society & EXPOGEF, Rio De Janeiro, 31 July–3 August, 5.

Georgakis, C., Støttrup-Andersen, U., Johnsen, M. et al. (2009). Drag coefficients of lattice masts from full-scale wind-tunnel tests. Proceedings of the 5th European and African Conference on Wind Engineering (EACWE5), Florence, 19–23 July, 8.

Gomes, T., Tapolcai, J., Esposito, C. et al. (2016). A survey of strategies for communication networks to protect against large-scale natural disasters. Proceedings of the 8th International Workshop on Resilient Networks Design and Modeling (RNDM), Halmstad, 13–15 September, 11–22.

Hawkins, D. (2010). Discussion of current issues related to steel telecommunications monopole structures. Proceedings of the ASCE/SEI Structures Congress, Florida, May 12–15, 2417–2438.

Hughes, G., Chinowsky, P., and Strzepek, K. (2010). *The Costs of Adapting to Climate Change for Infrastructure*, 51. Washington, DC: The World Bank.

IASS Working Group 4 (1981). *Recommendations for the Design and Analysis of Guyed Masts*. Madrid: International Association of Shell and Spatial Structures.

Mattioli, R. and Levy-Bencheton, C. (2014). *Methodologies for the Identification of Critical Information Infrastructure Assets*, 35. Heraklion: European Union Agency for Network and Information Security.

Mauthe, A., Hutchison, D., Ceetinkaya, E. et al. (2016). Disaster-resilient communication networks: Principles and best practices. Special Session at the 8th International Workshop on Resilient Networks Design and Modeling (RNDM), Halmstad, 13–15 September, 1–10.

McClure, G. (1999). Earthquake-resistant design of towers. Meeting of the IASS Working Group 4: Towers and Masts, Krakow.

Miller, R. (2006). Hurricane Katrina: Communications & infrastructure impacts. In: *Threats at Our Threshold: Homeland Defence and Homeland Security* (ed. B.B. Tussing), 191–203. Pennsylvania: U.S. Army War College.

Min, H., Beyeler, W., Brown, T. et al. (2007). Toward modeling and simulation of critical national infrastructure interdependencies. *IIE Transactions* 39: 57–71.

National Commission on Terrorist Attacks (2004). *The 9/11 Commission Report: Final Report of the National Commission on Terrorist Attacks Upon the United States*, 516. New York: W. W. Norton & Company.

Pederson, P., Dudenhoeffer, D., Hartley, S., and Permann, M. (2006). *Critical Infrastructure Interdependency Modeling: A Survey of U.S. and International Research*, 116. Idaho: Department of Energy.

Rak, J., Hutchison, D., Calle, E. et al. (2016). RECODIS: Resilient communication services protecting end-user applications from disaster-based failures. Proceedings of the 18th International Conference on Transparent Optical Networks, Trento, 10–14 July, 4.

Repetto, M. and Solari, G. (2010). Wind-induced fatigue collapse of real slender structures. *Engineering Structures* 32: 3888–3898.

Smith, B. (1985). *Comparison of the Draft Code of Practice for Lattice Towers with Wind Load Measurements on a Complete Model Lattice Tower*. Building Research Establishment.

Smith, B. (2007). *Communication Structures*, 338. London: Thomas Telford.

Smith, B. (2009). 50 years in the design of towers and masts. From IASS re-commendations to current procedures. Proceedings of the International Association for Shell and Spatial Structures (IASS) Symposium, Valencia, 28 September–2 October, 139–152.

Solari, G. and Pagnini, L. (1999). Gust buffeting and aeroelastic behaviour of poles and monotubular towers. *Journal of Fluids and Structures* 13: 877–905.

Souza, T., Hemsi, P., Gandolfo, O., and Ribeiro, A. (2016). Determination of the unknown length of a pile partially embedded in rock using induction logging testing in Brazil. *Electronic Journal of Geotechnical Engineering* 21: 5283–5292.

Souza, T., Hemsi, P., Gandolfo, O. et al. (2017). Evaluations of the depth of a root-pile and a caisson foundations using the parallel-seismic test. Proceedings of the GeoOttawa 2017 Conference, Ontario, Canada, 1–4 October.

Souza, T., Hemsi, P., Gandolfo, O. et al. (2018). Theoretical simulation and experimental study of the parallel seismic test for a caisson. Proceedings of the 19th International Conference on Soil Mechanics and Geotechnical Engineering, Salvador, 28 August–1 September.

Støttrup-Andersen, U. (2009). Masts and towers. Proceedings of the International Association for Shell and Spatial Structures (IASS) Symposium, Valencia, 28 September–2 October, 127–138.

The Royal Academy of Engineering (2011). *Infrastructure, Engineering and Climate Change Adaptation – Ensuring Services in an Uncertain Future*, 107. London: The Royal Academy of Engineering.

The White House (2006). *The Federal Response to Hurricane Katrina: Lessons Learned*, 217. Washington, DC: The White House.

TIA (2017). *TIA-222-H: Structural Standard for Antenna Supporting Structures, Antennas and Small Wind Turbine Support Structures*, 313. Telecommunications Industry Standard.

Touryalai, H. and Stoller, K. (2018). *Global 2000: The World's Largest Public Companies*. Forbes.

Townsend, A. and Moss, M. (2005). *Telecommunications Infrastructure in Disasters: Preparing Cities for Crisis Communications*, 45. New York: The Center for Catastrophe Preparedness & Response.

Travanca, R. and André, J. (2018). Safety of 5G network physical infrastructures. In: *A Comprehensive Guide to 5G Security* (eds. M. Liyanage, I. Ahmad, A.B. Abro, et al.), 165–193. New Jersey: John Wiley & Sons.

Travanca, R., Varum, H., and Vila Real, P. (2013). The past 20 years of telecommunication structures in Portugal. *Engineering Structures* 48: 472–485.

UK Cabinet Office (2017). *National Risk Register of Civil Emergencies*, 2017e, 71. London: Cabinet Office.

UN (2014). Infrastructure and Disaster. A Contribution by the United Nations to the Consultation Leading to the Third UN World Conference on Disaster Risk Reduction. New York: United Nations.

UN (2016). *Report of the Open-Ended Intergovernmental Expert Working Group on Indicators and Terminology Relating to Disaster Risk Reduction*, 41. New York: United Nations.

USA Congress. (2001). Uniting and strengthening America by providing appropriate tools required to intercept and obstruct terrorism. USA Patriotic Act, 131.

WEF (2019). *Global Risks. Report 2019*, 14e, 107. Geneva: World Economic Forum.

Further Reading

Holmes, J. (2015). *Wind Loading of Structures*, 412 pp. Florida: CRC Press.

Liyanage, M., Ahmad, I., Abro, A. et al. (2018). *A Comprehensive Guide to 5G Security*, 474 pp. New Jersey: Wiley.

Simiu, E., and Scanlan, R. (1996). *Wind Effects on Structures: Fundamentals and Applications to Design*, 668 pp New York: Wiley.

Smith, B. (2007). *Communication Structures*, 338 pp. London: Thomas Telford.

Varum, H. and André, P. (2013). *Accelerometers: Principles, Structure and Applications*, 295 pp. New York: Nova Science Publishers Inc.

Index

The Wiley 5G REF: Security. Edited by Rahim Tafazolli, Chin-Liang Wang, Periklis Chatzimisios and Madhusanka Liyanage.
© 2021 John Wiley & Sons Ltd. Published 2021 by John Wiley & Sons Ltd.